Techniques in Speech Acoustics

Text, Speech and Language Technology

VOLUME 8

The titles published in this series are listed at the end of this volume.

Techniques in Speech Acoustics

by

Jonathan Harrington
Speech Hearing and Language Research Centre,
Macquarie University,
Sydney, Australia

and

Steve Cassidy
Speech Hearing and Language Research Centre,
Macquarie University,
Sydney, Australia

KLUWER ACADEMIC PUBLISHERS
DORDRECHT / BOSTON / LONDON

A C.I.P. Catalogue record for this book is available from the Library of Congress.

ISBN 0-7923-5731-0

Published by Kluwer Academic Publishers,
P.O. Box 17, 3300 AA Dordrecht, The Netherlands.

Sold and distributed in North, Central and South America
by Kluwer Academic Publishers,
101 Philip Drive, Norwell, MA 02061, U.S.A.

In all other countries, sold and distributed
by Kluwer Academic Publishers,
P.O. Box 322, 3300 AH Dordrecht, The Netherlands.

Printed on acid-free paper

Printed in the Netherlands

For Alexander and Catriona
Owen and Nichola

Contents

PREFACE

This book is the development of a series of lectures to undergraduate and postgraduate students at Macquarie University on basic principles in acoustic phonetics and speech signal processing. The first part of the book (Chapters 1 to 4) is intended to provide students with the ability to interpret acoustic records of speech signals in their various forms. These chapters include a review of elementary wave motion and frequency analysis as applied to speech (Chapter 2), a summary of the relationship between speech production and its acoustic consequences (Chapter 3), and finally a review of the principal cues to speech sounds and prosodic units (Chapter 4). The material from these first four chapters (and the related exercises on the accompanying CD-ROM) has formed the basis of a one-semester undergraduate course in acoustic phonetics to students primarily of linguistics, but also of other disciplines including computer science and psychology.

The second part of the book provides an introduction to speech signal processing, which is intended for similar groups of students. It is therefore different from more detailed introductory texts in this area (e.g., Rabiner & Schafer, 1978; O'Shaughnessy, 1987) which assume both a background in engineering/signal processing and a more sophisticated mathematical knowledge (e.g., Parsons, 1987). Part of the motivation for writing this section of the book is to make many of the techniques and algorithms that are discussed in the engineering literature on speech analysis (for example, in the *IEEE Acoustics Speech and Signal Processing* journal) more accessible to both students and researchers of phonetics and speech science whose training is not usually in a scientific discipline. We have therefore tried to keep equations to a minimum and to assume, as far as possible, no more than a very basic understanding of algebra and trigonometry. In this part of the book, we cover fundamental aspects of time and frequency domain processing of speech signals (Chapters 5 and 6). Chapters 7 and 8 are concerned with digital techniques for combining (in digital formant synthesis) and separating (in linear predictive coding) the contributions of the source and filter to the acoustic speech signal. The final chapter and related exercises deal with techniques for the probabilistic classification of acoustic speech data that forms the basis of more advanced work in automatic speech recognition.

We have always felt that students of speech benefit from being able to carry out their own experiments to test some of the theories that they learn about in books such as these. However, preparing experiments that are also appropriate to testing a particular aspect of speech acoustics can be very time-consuming to construct and may not be feasible either because of the difficulties of collecting (and digitising) large quantities of speech data or because of the problem of student access to laboratory machines that are usually dedicated to research projects. In the last three to four years, we have adapted a research tool that has been developed in our laboratory (the *mu+* system Harrington, Cassidy, Fletcher, & McVeigh, 1993) for building and analysing speech corpora to the needs of teaching on our undergraduate and postgraduate courses. In extending this system, which is included together with a number of speech corpora on a CD-ROM in this book, we have sought to develop an interface that would obscure minimally the aim of the exercises, which is to solve problems in speech acoustics rather than in computer programming. The revised software (known as the EMU speech analysis system) runs on both UNIX and Windows 95 platforms. Since the EMU system can be used independently of the exercises in this book for segmenting and labelling utterances, as well as analysing them in the time and frequency domain (including spectrographic analysis), the exercises can be easily modified to include other sets of speech corpora beyond those provided on this CD-ROM. The EMU system also includes a set of extensions to the Xlisp-Stat statistical system to facilitate the analysis of speech data; the exercises on the CD-ROM use Xlisp-Stat extensively to allow students to analyse real data from the various speech corpora included on the disc. The Xlisp-Stat system (but not currently Emu) also runs on Macintosh systems.

We usually teach courses on speech acoustics after students have gained some grounding in phonetics and phonology. Accordingly, we assume some background knowledge of phonetic principles of sound classification and elementary phonological theory, which are given excellent coverage in many other books (see, for example, *recent general phonetics texts* at the end of Chapter 1) and which are not dealt with in detail in this book.

Vowel and Consonant Transcriptions

The columns for the vowels apply to many talkers of American, Australian and Southern British English (Adapted from Ladefoged, 1993).

Vowels

American English	Australian English	Southern British English	Key word
i	i	i	*heed*
ɪ	ɪ	ɪ	*hid*
ɪɹ	ɪə	ɪə	*here*
u	u	u	*who'd*
ʊ	ʊ	ʊ	*hood*
eɪ	eɪ	eɪ	*hay*
ɛ	ɛ	ɛ	*head*
eɹ	ɛə	ɛə	*there*
ə	ə	ə	*the*
ɝ	ɜ	ɜ	*heard*
oʊ	əʊ	əʊ	*hoed*
ɑ	ɒ	ɒ	*hod*
ɔ	ɔ	ɔ	*hawed*
ɔɪ	ɔɪ	ɔɪ	*boy*
ʌ	ʌ	ʌ	*bud*
æ	æ	æ	*had*
aɪ	aɪ	aɪ	*hide*
ɑɹ	a	ɑ	*hard*
aʊ	aʊ	aʊ	*how*

Consonants

Symbol	Key word	Symbol	Key word	Symbol	Key word
p	*pie*	θ	*think*	ʤ	*judge*
t	*tie*	ð	*these*	m	*my*
k	*cat*	s	*so*	n	*no*
b	*bat*	z	*zoo*	ŋ	*sing*
d	*do*	ʃ	*shoe*	w	*we*
g	*go*	ʒ	*measure*	ɹ	*run*
f	*fan*	h	*he*	j	*you*
v	*van*	ʧ	*chew*	l	*leaf*

Contents of the CD-ROM

The CD-ROM which accompanies this book contains software, speech databases and a set of exercises that allow you to explore the techniques described in the text.

The CD-ROM contains four subdirectories (folders):

archive contains an archive of possibly useful software, in particular, a mirror of the Xlisp-Stat software distribution which contains versions of Xlisp-Stat for Unix and Macintosh platforms. The archive also contains versions of Netscape Navigator for Windows and Macintosh systems.

dbase contains seven separate speech databases containing over 400Mb of labelled speech data for use in the exercises and in your own research. Details of the databases are given on the CD-ROM web pages.

html contains the web pages giving installation instructions and other information as well as the worksheets and documentation for the software included on the CD-ROM. All of these pages are accessed through the top level file `techniq.htm`.

software contains software to be installed on your computer. This consists of two main packages: the Emu speech database system, which allows labelling and searching of speech databases; and Xlisp-Stat, which you will use for various demonstrations and for statistical analysis of speech data.

See the file `readme.txt` on the CD-ROM for details of how to install and run the software.

System Requirements

Two software packages are provided on the CDROM: Xlisp-Stat and the Emu speech database system. Xlisp-Stat will run on both Windows (3.1, 95, 98 or NT), Macintosh and Unix systems while the Emu system currently only runs on Windows (95, 98 or NT) and Unix. You will be able to do many of the exercises with only Xlisp-Stat, but some labelling and speech data analysis exercises require Emu.

For the Emu software we recommend a Pentium or better processor running Windows 95/98 or Windows NT with at least 16MB of RAM. The software requires 15Mb of free disk space for installation; you should have around an extra 5Mb free for temporary files while you are working on the worksheets. Since the speech databases and web pages remain on the CD-ROM you will need to keep the CD-ROM in your computer while you work on the exercises. A sound card is required for audio playback.

Versions of Xlisp-Stat are included on the CDROM for various models of Macintosh system. Installation requires about 5Mb of free disk space on your computer.

Emu also runs on Unix (Solaris and Linux) computers. To obtain these versions, and information about new releases, please see our web site `http://www.shlrc.mq.edu.au/techniques`.

1 The Scope of Speech Acoustics

1.1 What is speech acoustics?

The function of speech communication is to convey information. At one level, the information is exchanged using words that a talker must structure and arrange in such a way that a listener can extract the intended information. At another level, information is conveyed about the talker's attitude that may be communicated non linguistically (as in, for example, gestures or facial expressions) but also by means of prosodic variation. In spoken communication, we also learn something about the talker's regional accent and possibly also socioeconomic status. If we cannot see the person (as when talking over the telephone), we can often judge if the person is a male or female and estimate the person's age from the information conveyed by voice quality. All these strands of spoken communication are transmitted in parallel, which, as listeners, we decode effortlessly.

It will be helpful to begin by saying a little more about the sequence of events that takes place in decoding the speech signal. When we produce intelligible speech, the vocal organs have to move in a particular way. This movement creates a pattern of disturbances to the air molecules, which is transmitted outward from the vocal tract in all directions, eventually reaching the ears of the listener.

As listeners, we have become very adept at recognising characteristic patterns of sound. We recognise the difference between bird song and the noise from traffic because these create their own distinctive patterns of vibrations, which are transduced as corresponding movements in the eardrum and as an electrical stimulation in the auditory nerve. Similarly, we hear the differences between the sounds of speech because these also have their own distinctive acoustic patterns.

In speech acoustics, our concern is to study the relationship between these acoustic patterns and the speech sounds associated with them. This aim is analogous to that of articulatory phonetics, which is a field with which readers of this book may already be familiar. In articulatory phonetics, a speech sound is described by the movements and configuration of the vocal organs that characterise it and that differentiate it from other sounds: when we describe a sound like [m], we highlight features such as vocal fold vibration, a lowered velum, and complete closure at the lips because these are its distinctive articulatory characteristics. In speech acoustics, we will start from the assumption that the articulatory patterning that is unique to [m] produces a correspondingly unique acoustic pattern that differentiates it acoustically from other sounds. Describing how these acoustic patterns vary with sounds of different phonetic quality is one of the principal goals of speech acoustics.

In studying the acoustics of speech, we are immediately confronted with the problem of *variability*. Variability can be considered from two perspectives. First, it implies that there is a good deal of information in the acoustic signal that is not only irrelevant to the sound's phonetic identity but that also masks the distinctive acoustic features of the sound that we wish to define. The second and related sense of variability is that no two acoustic records of the production of the same phonetic sound are ever quite the same.

We begin by considering some of the sources of this variability because they have had such an important influence on the way in which research in speech acoustics has been carried out over the last fifty years.

1.2 The variable nature of speech

We start with the highly artificial situation of the same talker producing two repetitions of a single monosyllabic word under identically quiet recording-studio conditions. If we print out acoustic records of the repetitions of these words and compare them, we would notice that there are minor but noticeable differences in their acoustic patterns. It is unlikely that the talker will have produced the vowel to be of exactly the same length. Oral stops may be released in slightly different ways, with variable amounts of frication; vowels before nasal consonants could well be nasalised to different degrees; and even if the talker is a trained phonetician, it is very probable that the successive repetitions will exhibit some minor differences in loudness and pitch.

The fact that this sort of variability occurs, even in this highly constrained setting, should not surprise us. If there were no variability, the brain would be constrained by the unmanageable task of having to instruct the more than 100 different muscles that are active every second in speech to replicate their performance, while taking appropriate account of forces such as inertia and gravity. Still another reason that invariability in the acoustic signal is implausible is because of human listeners' inability to hear all the details that we measure acoustically. For example, when two events are separated by an interval of less than a few milliseconds, we hear them as continuous, even though they are physically discontinuous. Similarly, amplitude and frequency changes that fall below a certain threshold are not perceptible. Since we presumably speak in order to be heard, there would little point in instructing the vocal organs to perform tasks that cannot be perceived.

Once we go beyond the very artificial setting of the same talker producing the same word, we soon realise that the speech signal is affected by many other sources of variability. Different talkers have different-sized and -shaped vocal organs that imprint their own characteristic features on the acoustic speech signal.

This is most apparent in the clearly audible differences between male, female, and child speech. Yet another source of variability is background noise that might include the ringing of a telephone, distant conversation, or the rustling of papers that would all be registered if a conversation were recorded in an office setting. Although the resulting recording may seem to be of high quality when we listen to it, an analysis of the acoustic signal would soon show how such effects can contaminate the linguistic-phonetic content of the speech signal. Even the noise from an air-conditioning unit in a recording studio can severely degrade the quality of a recorded speech signal.

Speech sounds are further influenced by the context in which they occur. Many studies of the production of speech show that talkers produce sounds so as to overlap with each other. This effect, known as *coarticulation* or *co-production*, introduces a further source of variability into the speech signal. Coarticulation has been described as a necessary part of speech production: if sounds were produced in a nonoverlapping way, then the rate of speech would be slowed considerably. Additionally, speech would be more laborious for the talker because the vocal organs would have to move through a greater spatial distance at the boundaries between speech sounds. Because of coarticulation, the sounds are transmitted in parallel and therefore at a faster rate to the listener. At the same time, since sounds are adjusted to their contexts, the vocal organs have less work to do because the articulatory distance in moving from one speech sound to the next is reduced.

Coarticulation is therefore a necessary and natural part of producing speech, but it can have quite dramatic effects on the acoustic speech signal. A good example of this is the variability of /k/ and /g/ in English and many other languages depending on the quality of the following vowel. Before front vowels (such as *keep*, *geese*), the production is advanced to a postpalatal position; before central vowels (e.g., *curd*, *gird*), /k/ and /g/ are realised as velar stops; before back vowels (e.g., *caught*, *gawp*) the place of articulation is retracted to postvelar or preuvular position. From the point of view of speech *production*, we can nevertheless distinguish /k/ from the other voiceless stop classes in English by saying that /k/ is produced with the tongue dorsum, /t/ with the tip of the tongue, and /p/ with both lips. However, from the point of view of the *acoustics* of speech, we find that the various allophones of /k/ tend not to form one homogeneous class but to pattern as two groups (advanced and retracted) that have more in common acoustically with alveolars and bilabials respectively than with each other.

A further complication is that the nature and degree of coarticulation is itself affected by a multitude of factors. It is known that coarticulatory patterning varies across languages, accents within a language, and talkers. Coarticulation is also influenced by the prosody of an utterance. There is evidence to show that at faster rates of speech, the degree of coarticulation between speech sounds increases. Similarly, the size of coarticulatory overlap tends to be greater in unstressed syllables (particularly if these are function words) than in stressed syllables.

Another source of variability that must be mentioned is dialect (accent) variation that is carried mostly by vowel quality differences. This brings us back once again to the effect the talker has on the speech signal. Talkers are a source of variability not only because the vocal organs imprint their own signature on the acoustic speech signal but also because different accents (for example, General American, Southern British English) are characterised by sounds of different phonetic quality that are manifested as corresponding acoustic differences. Independently of both the anatomical and accent attributes of a talker, the style of speaking may introduce a further source of variability: for example, a talker may produce speech with excessive lip-rounding, nasalisation, or larynx lowering, all of which would further confound the relationship between linguistic phonetic units and the acoustic speech signal.

1.3 Experimental designs in speech acoustics

The inherent variability of speech has had a marked influence on the way in which research in speech acoustics has been carried out in the last fifty years or so. In the large majority of studies in speech acoustics, attempts are made to control as far as possible for many of the effects of variability. Most studies of speech acoustics are based on a very idealised form of speech in which the type and number of talkers are carefully controlled for and in which the range of stimuli that are to be produced is very limited. A typical experiment might include a handful of talkers of the same sex, same accent, and possibly also drawn from a similar age group, producing the target sounds that are to be investigated in a highly constrained context (for example, /CVd/ words, where C might vary over one consonant class and V is a number of vowels). In almost all cases, the experiments would be conducted in a sound-treated recording-studio setting.

Another factor that is influential in the design of experiments is that the acoustics of some sounds are easier to study than others. Voiced and voiceless stops are frequently studied not just because they are an integral part of various influential phonetic theories (see below), but also because they have very well-defined acoustic characteristics (for example, a period of silence followed by a release) that can be easily measured. This means researchers can compare results more easily because there is a fair degree of consensus about what is being measured (for example, voice onset time as defined by the duration from the end of the closure to the periodic onset of the following vowel).

It is a feature of most scientific disciplines that the invention of equipment and the discovery of new experimental procedures have a marked influence on the way in which research issues are addressed. The invention of the sound spectrograph in the 1940s (Koenig, Dunn, & Lacy, 1946) is a good example of this in speech acoustics. The spectrograph is a device that records in a display called a spectrogram how the intensity and frequency of the acoustic signal change in time. Since the invention of the spectrograph, spectrograms have become pervasive in speech acoustics. The reason for this is that a spectrogram provides a representation of the salient features of speech in a single display. It can be used to measure the duration of speech sounds, to measure the changes in pitch, and also to trace how the resonances of the vocal tract change in time.

The development of the electronic speech synthesiser has had a major impact on the design of experiments in speech acoustics. One of the earliest devices of this kind was the vocoder demonstrated by Dudley at the New York World's Fair (Dudley, Risez, & Watkins, 1939); some of earliest electronic synthesisers to be used in the laboratory were the machines of Lawrence (1953) and the pattern playback system of the Haskins Laboratories from which speech could be synthesised from hand-drawn spectrograms. One of the reasons that synthesisers are so popular in studies of speech acoustics is that they allow the effectiveness of individual acoustic cues to be tested. This is usually done by repeating a synthetic stimulus in which all the parameters are held constant except for one. Any changes in the way that listeners label the stimuli phonetically can then only be due to the parameter which is changed. Experiments such as these, in which listeners are played a variety of synthetic stimuli that they have to label (as, for example, /b/, /d/, or /g/), have become prolific since the 1950s, and they remain one of our most important sources of knowledge about acoustic cues to speech sounds.

The advent of high-speed digital computers in the 1960s produced yet another range of experimental techniques for the analysis of acoustic speech data. Computers became increasingly popular in speech acoustics throughout the 1970s not just because they allow speech data to be conveniently stored, accessed, and examined in microscopic detail but also because some very useful computational techniques were developed for decomposing speech into a source signal (corresponding either to periodic vocal fold vibration characteristic of voiced sounds or to a turbulent airstream characteristic of voiceless sounds) and a filter signal that models the shape of the supralaryngeal vocal tract. The significance of this decomposition is that the filter signal is more closely correlated with judgements of phonetic quality — so that by removing the contribution from the vibrating vocal folds or a turbulent airstream, we can model more efficiently the mapping between linguistic phonetic units and the speech signal.

Nowadays, computers are ubiquitous in the speech research field. Personal computers have the speed and capacity to carry out most speech analysis procedures. The interrelationship between computer and speech analysis is also underpinned by the emergence of the speech technology field, which is concerned with developing human-machine speech communication systems and which now represents a multimillion-dollar market of the information technology industry.

1.4 Some key areas of research in speech acoustics

Research into the acoustics of speech follows the same pattern as in many other scientific disciplines in which the range of empirical analyses is usually constrained by a set of theoretical models that is to be tested. Certain theories become influential, and a multitude of experiments that seek to test further and refine the theory follow. Because speech is multidisciplinary, being carried out by researchers from fields such as audiology, computer science, engineering, linguistics, speech pathology, and psychology, studies in speech acoustics have been used to address many different kinds of theoretical enquiries. We give some indication of the research orientations that have made use of speech acoustics in the following sections.

1.4.1 Acoustic phonetics

These studies are concerned with quantifying the acoustic differences between naturally produced speech sounds. These investigations are used to provide evidence for phonetic differences across languages, in prosodic analysis, in the development of speech synthesis systems, and many other kinds of studies.

Some landmark studies include Peterson and Barney's (1952) analysis of vowel data, various studies by Lehiste and Peterson (e.g., Lehiste & Peterson, 1961a) on the duration and spectral characteristics of speech sounds, and the analysis by Öhman (1966) of formant transitions in consonants.

1.4.2 Models of speech perception

These studies are more directly concerned with how listeners decode the acoustic speech signal into phonetic or linguistic units. Often a model of speech perception is suggested and then tested based on listeners' judgements of speech stimuli.

A range of different techniques can be brought to bear in perception experiments. Some studies make use of edited or spliced acoustic speech stimuli in a labelling or reaction time experiment, while others use speech synthesis techniques to test whether an acoustic cue, or set of cues, alters listeners' perceptions.

The synthesis and labelling experiments carried out at the Haskins Laboratories (e.g., Liberman, Delattre, & Cooper, 1952) have been a landmark in this field. They used the pattern-playback synthesis system (see Chapter 7) to show that listeners' perceptions of place of articulation in stop consonants was cued by the transitions between the consonant and a following vowel.

1.4.3 Articulatory-acoustic relationships

Very significant advances have been made in modelling the relationship between the shape of the vocal tract and the acoustic speech signal in the last forty years. Part of the motivation for this research is that if we wish to understand how linguistic phonetic units are related to the acoustic signal, we must first develop a model of the intermediate stage of the articulatory-acoustic relationships in speech communication. Research in this area is also of importance for developing an articulatory speech synthesis system in which the control parameters of the synthesiser are directly related to the movements of the vocal organs.

Fant's (1960) acoustic theory of speech production is recognised as the leading work in this field, while Dunn (1950) and Stevens and House (1955) also made significant advances before then. The importance of this research is that it provides an explanation for why the acoustic speech wave is shaped as it is: for example, the mapping from an articulatory model to the acoustic speech signal can be used to account for the downward sloping spectrum that is characteristic of voiced sounds or to explain why sibilant fricatives such as [s] and [ʃ] are characterised by predominantly high-frequency energy.

1.4.4 Auditory analysis of speech

Studies in this area are once again directly concerned with listeners' perceptions, but in this case the focus is on modelling the transformation from an acoustic signal to its corresponding mechanical representation in the inner ear and electrical representation in the auditory nerve. We know that there must be differences between an acoustic and auditory representation of sounds as shown by the fact that a physical doubling of acoustic frequency does not correspond to a perceived doubling of pitch (similarly, acoustic amplitude and judgements of loudness are not linearly, but quasi-logarithmically, related).

Earlier, we had mentioned that one source of variability in the speech signal is due to speaker differences: a male and female production of the same phonetic vowel have quite different acoustic characteristics. The fact that listeners identify sounds as phonetically equivalent despite these acoustic differences is thought by some to be due to the psychoacoustic (auditory) transformations that take place as the speech signal is mechanically and then electrically transduced by the ear.

1.4.5 Speech pathology

An acoustic analysis of speech data is an intrinsic part of classifying speech and voice disorders and it is often used to supplement a more detailed articulatory analysis. One of the reasons that many types of analysis in speech pathology rely on acoustic data is that, in contrast to most articulatory analyses, they are noninvasive and therefore more acceptable to clients. This is a particularly important consideration in analyzing data from children with speech handicaps.

1.4.6 Speech technology

Techniques in acoustics and speech processing are a fundamental part of the growing field of human-machine communication systems. These include systems for generating synthetic speech from text (speech output systems) that have numerous applications in the telecommunications and speech handicap fields and also automatic speech recognition systems (speech input systems), which are concerned with converting the speech signal into text. Speech coding is another branch of speech technology industry that is of importance to the telecommunications industry, in which the aim is transmit the digital speech signal efficiently over a communication channel.

Another growing area of the speech technology industry is automatic talker recognition. This has applications in both forensic speech analysis, which is being increasingly used in a legal context, and in the development of machines that restrict access to a group of talkers by analyzing their voice characteristics.

Further reading

For recent general phonetics texts, see Clark and Yallop (1995), Ladefoged (1993), and Laver (1994). Various papers on *variability and coarticulation* are given in Hardcastle and Marchal (1990) and Perkell and Klatt (1986). See Ladefoged (1967), Lehiste (1967), Kent and Read (1992), and Olive, Greenwood, and Coleman (1993) for an introduction to *acoustic phonetics*. For *instrumentation in speech research*, see Borden, Harris, and Raphael (1994) and Lass (1996). Recent texts on *psychoacoustics and hearing* include Greenberg (1988), Moore (1989), and Schouten (1992). A summary of *trends in speech technology* is given in Ainsworth (1988), Bailly, Benoît, and Sawallis (1992), Rabiner and Juang (1993), and van Heuven and Pols (1993).

2
The Physics of Speech

2.1 Speech waveforms

In order for sounds to be heard, they must be propagated through a medium. The medium is in most cases air, but other substances, such as water, a metal pipe, or a brick wall, can conduct sounds as well. When there is no medium at all, as in a vacuum, sound cannot be conducted.

The speed with which sounds are transmitted depends on the properties of the medium. Under most conditions, the speed of sound in air is approximately 340 metres per second. This means that if we see an explosion at a distance of a kilometre, there will be about a three second delay between seeing the explosion and hearing it. In the gas mixture of helium and oxygen that is typically used for deep sea diving, sound travels at a much faster speed, and this gives divers' voices the attributes of a high-pitched and distorted quality under those conditions.

We must now be more precise about what we mean when we say that sound travels. In order to hear any sound at all, an object must have caused a disturbance to the air molecules. The object might be a book falling on the floor or the ringing of a telephone; in the production of most speech sounds, the disturbance is caused by pushing air out of the lungs. The disturbance caused by an object is measurable as local rises (compressions) and falls (rarefactions) in air pressure. Furthermore, because air is an elastic medium, this pattern of local air pressure changes is propagated outward in all directions from the source that caused the disturbance. We hear the sound when the air pressure changes reach our ears. It is important to realise that it is the air pressure variations that are propagated, not the actual air molecules themselves at the source, which move only a very small distance.

We hear sounds as different because they create different patterns of air pressure variation. The different patterns that are produced by a musical note and noise are especially relevant to our understanding of the acoustics of speech sounds. When a guitar string is plucked, a similar pattern of air pressure variations is repeated at regular intervals. These types of disturbances are often called *periodic*, and we hear them as having *pitch*. On the other hand, the sound that is made by screwing a sheet of paper into a ball does not have pitch in quite the same sense because there is no discernible regularity to the air pressure variations that are produced by such an event.

Any sound must produce a change to the air pressure, which is then prop-

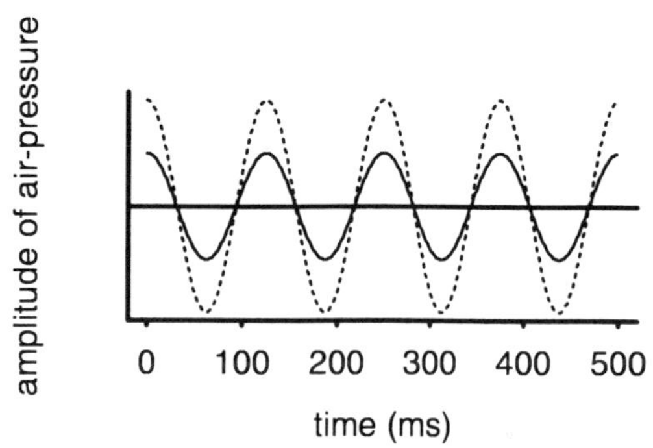

Figure 2.1: The waveform corresponding to the first few cycles of movement of a tuning fork.

agated to our ears. If we want to study the nature of speech sounds, it seems clear that we must begin with a record of the air pressure change itself. Figure 2.1 shows an example of such a record corresponding to the sound that is heard when a tuning fork is struck. The vertical axis of the graph represents the size, or amplitude, of air pressure variation. Points that lie above the horizontal line represent an increase in air pressure, or a compression, relative to atmospheric pressure; points below the line represent a decrease in air pressure or a rarefaction. In general, louder sounds are correlated with greater displacements from the horizontal line: the dotted line represents the same tuning fork, which is struck more vigorously causing bigger air pressure variations and which is perceived to be louder. The unit of air pressure for the vertical axis is usually the *Pascal* (which is equal to 1 Newton per metre squared), but since we will be more interested for the present in looking at the different *patterns* of air pressure variation that characterise different speech sounds, the units can usually be disregarded. The horizontal axis is *time*, which, when analysing speech, is often measured in milliseconds (1 millisecond = 1/1000 second). The type of graph shown in Figure 2.1 is called a *speech pressure waveform* or *waveform*.

Figure 2.2 is a waveform for the word "stamp" that has been divided into separate phonetic units. We can see that the pattern of air pressure variations is different for the various sounds: during [t] and [p], in which there is complete vocal tract closure, the speech waveform tends towards a straight line, indicating there is near silence (that is minimal sound output).

A noticeable feature that is shared by the vowel and following [m] is the repetition of a very similar pattern in the waveform; these sounds are examples of a periodic waveform, which we mentioned earlier. We can see the periodic nature of [æ] more clearly if we expand out a section near the middle of the vowel: the resulting waveform (Figure 2.3) shows four similar patterns in succession. In speech analysis, each individual pattern is known as a *pitch period* and a

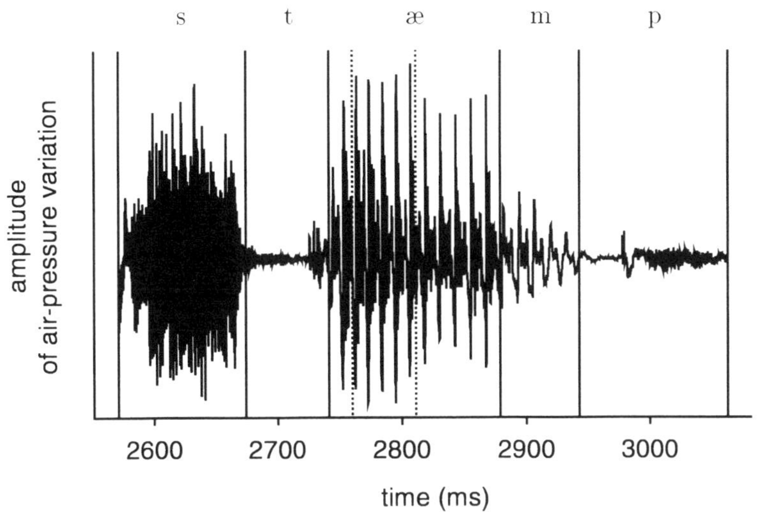

Figure 2.2: Waveform for the word "stamp" divided into phonetic units.

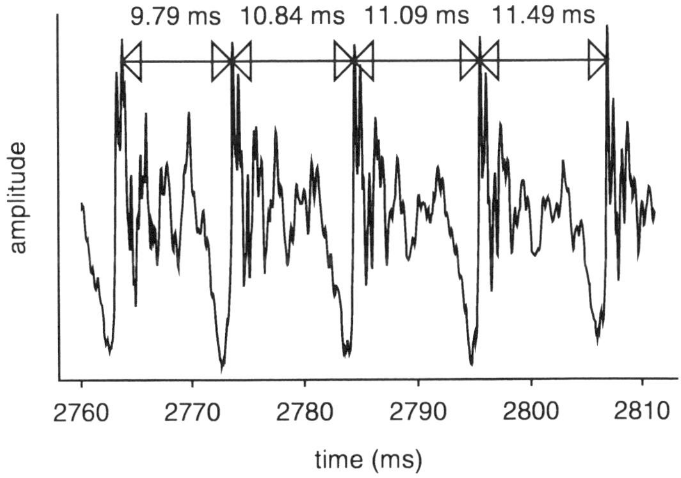

Figure 2.3: A section of the waveform in Figure 2.2.

waveform that consists of a number of pitch periods is said to be *periodic*. In fact, speech waveforms are never *absolutely* periodic: the pitch periods are not all exactly of the same length, and they vary slightly in shape (see Figure 2.3). Nevertheless, when discussing speech waveforms, we shall refer to these as periodic while recognising that there will always be some deviations from absolute periodicity.

A periodic speech waveform is the acoustic correlate of vocal fold vibration that characterises most phonetically voiced sounds and that gives rise to the auditory quality of *pitch*. In Figure 2.2, both [æ] and [m] are voiced sounds, and therefore the waveform is periodic throughout almost all of their temporal extent. Since voiceless sounds are produced without vocal fold vibration, the corresponding waveform is nonrepetitive and there is no pitch.

In periodic waveforms, each pitch period corresponds to a single cycle of vocal fold vibration. Consequently, the time that each pitch period takes, or the duration of each pitch period, can be used to give an estimate of the rate of vocal fold vibration or, in acoustic terms, of the *fundamental frequency*. In Figure 2.3, for example, the duration of the first pitch period is 9.79 ms (0.00979 seconds). The fundamental frequency (abbreviated as f_0) in Hertz (Hz), or cycles per second, can be calculated from the reciprocal of the pitch period duration in seconds. For the first pitch period, we have

$$\begin{aligned} f_0 &= 1/0.00979 \quad \text{Hz} \\ &= 102 \quad \text{Hz (rounded to the nearest integer).} \end{aligned}$$

Therefore, the vocal folds are estimated to be vibrating at 102 times per second over the temporal extent of the first pitch period. Since there is an increase in the duration of the following three pitch periods, the fundamental frequency and therefore the rate of vocal fold vibration must be decreasing. From a listener's point of view, changes in the rate of vocal fold vibration are perceived as changes in pitch. When the vocal fold vibration/fundamental frequency increases, we hear a rising pitch; when it falls, we hear a falling pitch. By determining the pitch period duration, we have shown that the waveform in Figure 2.3 corresponds auditorily to a falling pitch.

2.2 Frequency analysis

Although speech waveforms can provide the speech scientist with much valuable information — in particular, about the voicing status of a speech sound and its fundamental frequency — for many other kinds of analyses we shall need to consider the *frequency* content of a speech signal. Frequency in this context can be thought of as another dimension, or space, in which speech sounds can be compared. So far, we have considered two dimensions for representing speech sounds: *amplitude* (of air pressure variation) and *time*. With the additional dimension of frequency, we can analyse sounds in spaces of *amplitude* × *frequency* or *frequency* × *time* or indeed in a three-dimensional space of *amplitude* × *frequency* × *time*. In order to understand the nature of

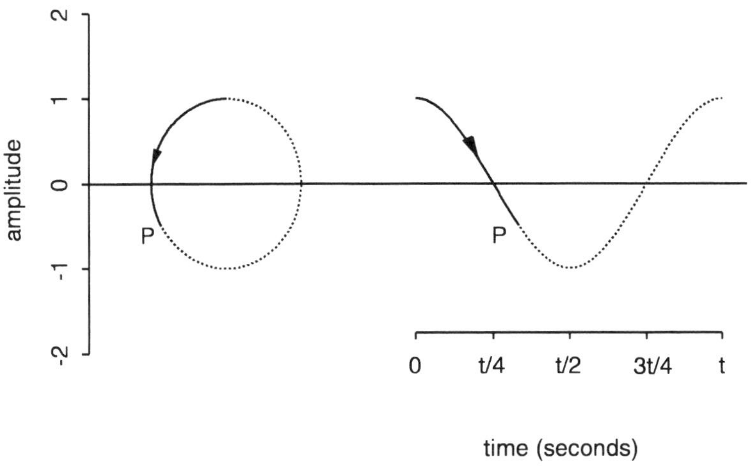

Figure 2.4: The relationship between circular motion and a sinusoid. The height of P above the horizontal line is always equal for both the circle and sinusoid.

frequency, it will first be necessary to discuss a class of waveforms known as *sinusoids* and then to consider how speech waveforms can be decomposed into a set of sinusoids using a Fourier transform.

2.2.1 Sinusoids

A sinusoidal waveform can be produced by plotting the height of a point against time as the point moves around the circumference of a circle at a constant speed. This type of waveform is very close to a record of the air pressure variations produced by a vibrating tuning fork (see Figure 2.1). An example of another sinusoidal waveform is shown in Figure 2.4. In this figure, P moves around the circumference at a constant speed, completing one revolution in t seconds (the circle might represent a turntable, and P a small object attached to the edge of the turntable). If we consider the movement in only one of the dimensions of the circle, labelled *amplitude*, then P moves between ± 1 units because the radius of the circle is itself $+1$ unit: at the beginning of the cycle, P has an amplitude of $+1$ unit; after a quarter of a cycle, the amplitude is zero, after half a cycle -1. The amplitude of P then changes from -1 through zero to $+1$ after which one cycle is completed. The resulting plot of the amplitude of P as a function of time is a sinusoid known as a *cosine wave*, which also has an amplitude that varies from $+1$ through zero to -1, then back through zero to $+1$ corresponding to one revolution of P around the circumference. Each time that P completes a revolution, the waveform is repeated, and it is in this sense that the waveform is periodic.

We can vary the way in which P moves around the circle in three ways. These are summarised below.

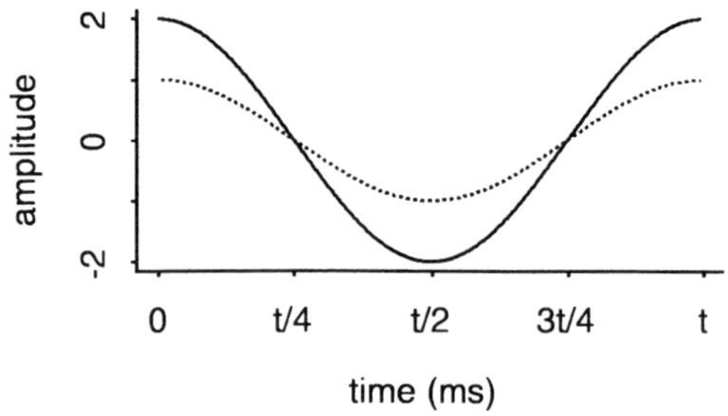

Figure 2.5: The solid-line sinusoid has double the amplitude of the dotted-line one and is derived from a circle of twice the radius.

Amplitude

The maximum amplitude of the cosine wave is equal to the radius of the circle. In Figure 2.4, the radius of the circle is +1 unit, and the maximum amplitude of the sinusoid therefore varies between ±1 units. Increasing the radius of the circle results in a sinusoid of correspondingly greater amplitude (Figure 2.5).

Phase

In Figure 2.4, P arbitrarily starts to move at the top of the circle. It could also begin its movement further on, or earlier, than this point. Varying the point at which P starts corresponds to a change in the *phase*. In Figure 2.6, P starts a quarter of a cycle earlier than in Figure 2.4, the result of which is a *sine wave* (a sine wave is therefore equivalent to a cosine wave that has been shifted by a quarter of a cycle). Phase can vary between plus or minus half a cycle: when the phase is at its maximum or minimum value of ± half a cycle, P starts at the opposite end of the circle.

We have so far defined phase as varying between plus or minus half a cycle relative to the reference point of zero phase at the top of the circle. In fact, phase is usually defined in terms of the angle (in radians) that the point P makes at the centre of the circle relative to the point that is defined to have a phase of zero (Figure 2.7). Since one complete revolution around the circle corresponds to an angle of 2π radians, plus or minus half a cycle corresponds to an angle of $\pm\pi$ radians.

Phase will henceforth be defined as a fraction of π, corresponding to the phase angle in radians relative to the top of the circle (at which the phase is defined to be zero; as in Figure 2.7).

Frequency

Frequency is defined as the number of revolutions made by P around the circle per second. In Figure 2.4, P completes one revolution every t seconds. It therefore completes $1/t$ revolutions per second: its frequency is $1/t$ cycles per

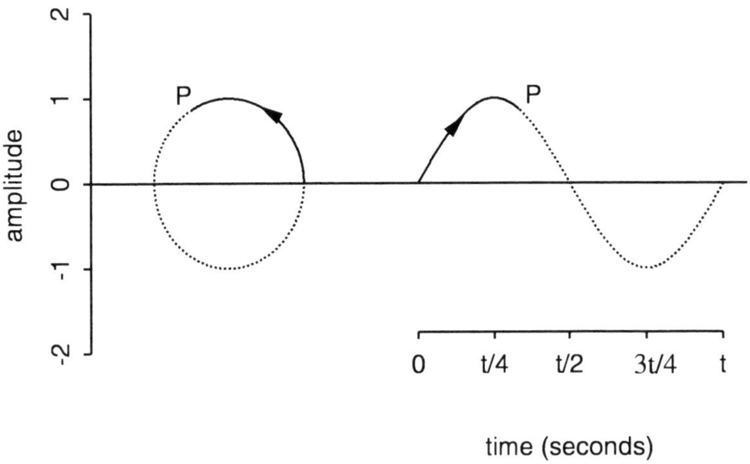

Figure 2.6: Varying the starting point of rotation changes the *phase* of the resulting sinusoid.

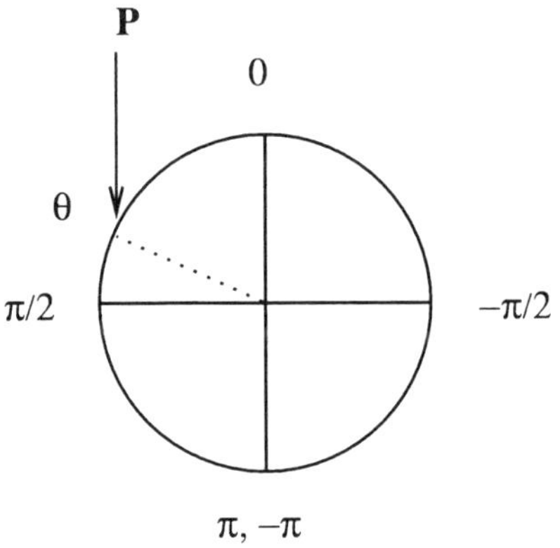

Figure 2.7: The phase-angle can vary between 0 radians (*top of the circle*) and $\pm\pi$ radians (*bottom of the circle*). In this case, P begins its movement at a phase-angle of θ radians.

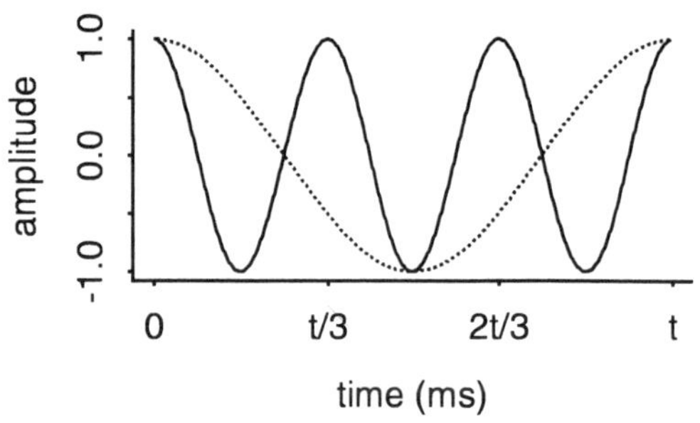

Figure 2.8: The solid line is a sinusoid that has treble the frequency of the dotted line sinusoid.

second or $1/t$ Hz. The time taken for a single cycle (single revolution around the circle) is known as the *period* of the waveform: the period for the sinusoid in Figure 2.4 is therefore t seconds. From this we can see that the frequency of the sinusoid in Hz is the reciprocal of the period in seconds (period $= t$ seconds, frequency $= 1/t$ Hz).

The effect of trebling the frequency is shown in Figure 2.8. In this case, P completes three revolutions in the same amount of time (t seconds) and has a frequency of $3/t$ Hz (and a period of $t/3$ seconds).

A set of sinusoids is said to be *harmonically related* if they occur at multiples of the lowest frequency sinusoid. For example, sinusoids of frequency 150 Hz, 300 Hz, 450 Hz are harmonically related because they occur at multiples of 150 Hz. The lowest frequency sinusoid is the first harmonic, but it is more commonly known as the *fundamental*. The sinusoid at twice the frequency of the fundamental (300 Hz in this case) is the second harmonic; the sinusoid at n times the frequency of the fundamental is the nth harmonic.

2.2.2 Amplitude spectrum and phase spectrum

So far we have only considered one type of display, the speech-pressure waveform of which the axes are amplitude of air pressure and time. In the case of sinusoids, it is also possible to present the data in the form of an *amplitude spectrum* and a *phase spectrum*. An amplitude spectrum is a display of amplitude (on the vertical axis) against frequency, while a phase spectrum is a display of the phase against frequency.

Figure 2.9 shows the speech pressure waveform, the amplitude spectrum, and the phase spectrum of a sinusoid with an amplitude of four units, a frequency of 10 Hz, and a phase of $\pi/4$ radians. Notice that the height of the frequency component at 10 Hz in the amplitude spectrum is equal to the sinusoid's maximum (positive or negative) displacement.

In speech analysis, we shall be almost exclusively concerned with amplitude spectra, since phase contributes relatively little to the perception of speech. Henceforth, when using the term *spectrum* on its own, an amplitude spectrum

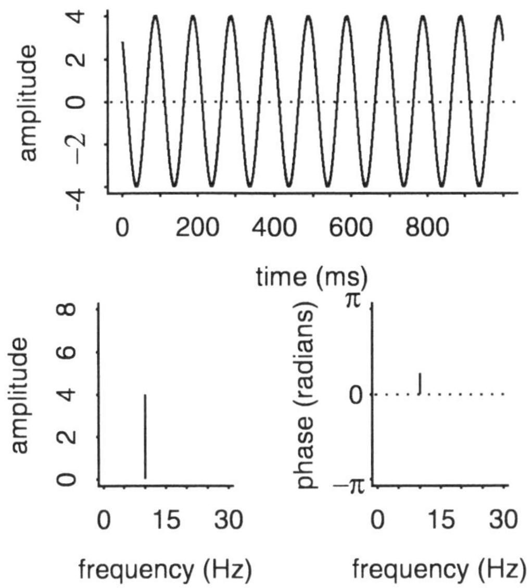

Figure 2.9: Waveform (*top*), the amplitude spectrum (*bottom left*), and the phase spectrum (*bottom right*) of a sinusoid.

is implied, and when we wish to discuss phase as a function of frequency we shall use the term *phase spectrum*.

2.2.3 The decibel scale

When displaying spectra of speech sounds, it is more common to convert the vertical axis of amplitude of air pressure variations into the logarithmic *decibel* scale. This is for two main reasons. First, the difference in sound pressure variation between the quietest sound and the loudest sound we can tolerate without causing pain is very large (a difference of more than a million times): a logarithmic scale converts these very large values to more manageable ones. The second motivation is that judgements of loudness correspond more closely to logarithmic, than to linear, changes in sound pressure variation.

When we measure speech sounds in *decibels*, we are usually asking: how much louder is a particular speech sound relative to the threshold of hearing (the quietest sound we can hear)? In order to make this comparison, we have to compare the *power* (rather than the air pressure) differences between the target the reference sound. Second, the reference sound is defined to have an intensity of 0 dB. The various steps in deriving an intensity value in decibels are summarised as follows:

1. Calculate the power of the sound by squaring the air pressure values;

2. Divide by the power of a reference sound to obtain the *power ratio* (a

dimensionless number);

3. Take the common logarithm (logarithm to the base 10 or $\log_{10}$) of the power ratio to obtain the intensity in Bels; and

4. Multiply the value in Bels by 10 to obtain an equivalent intensity value in decibels.

Steps 1 and 2 amount to calculating the *power intensity ratio* between the sound in question and the reference sound. The important point to note about this division (ratio calculation) is that it produces a dimensionless number that is relative to the reference sound. For example, if the calculations from 1 and 2 produce a power intensity ratio of 1000, this means that the power of the sound is 1000 times greater than the power of the threshold of hearing (assuming we have taken this to be the reference sound). The reference sound itself always has a power intensity ratio of 1.

Steps 3 and 4 are used to convert the power intensity ratio into the logarithmic Bel and decibel scales. For example, a (power) intensity ratio of 1000 is equivalent to $log_{10}1000 = 3$ Bels, which is 30 decibels (dB). Notice that since the logarithm of 1 is always zero, the reference sound has an intensity value of 0 Bels or 0 dB.

From the above discussion, we can see that any value in decibels is always relative to a reference sound. When the decibels are calculated relative to the threshold of hearing, the units are sometimes written as dB SPL where SPL stands for *sound pressure level.* Conversational speech usually falls in the range of 0 to 70 dB SPL; a sound that is at threshold of pain is in the region of 130 to 140 dB SPL.

2.2.4 Perceptually weighted frequency scales

One of the motivations for the decibel scale described above is that physical amplitude is not directly related to perceived loudness. Similarly, the physical frequency scale in Hertz is not representative of the way in which the ear resolves frequency. A simple example of this is that a doubling of frequency in Hertz at different parts of the scale (e.g., 500 Hz to 1000 Hz; 2000 Hz to 4000 Hz) does not correspond to a perceived doubling of pitch.

In some speech studies, the physical frequency scale is sometimes transformed into the *mel* scale, which was devised to be proportional to pitch (so a doubling of frequency in mels produces a sensation of a doubling in pitch). In Fant (1968), the formula used to convert Hz to mels is given by

$$f_{mel} = \frac{1000}{\log(2)} \log(1 + f_{Hz}/1000),$$

where f_{mel} and f_{Hz} are frequencies in mels and Hertz, respectively.

Another perceptual scale, known as the *critical band* or *Bark* scale, is more directly based on the view that the peripheral auditory system contains a bank of filters within which energy is summed (e.g., Plomp, 1975; Schroeder, Atal,

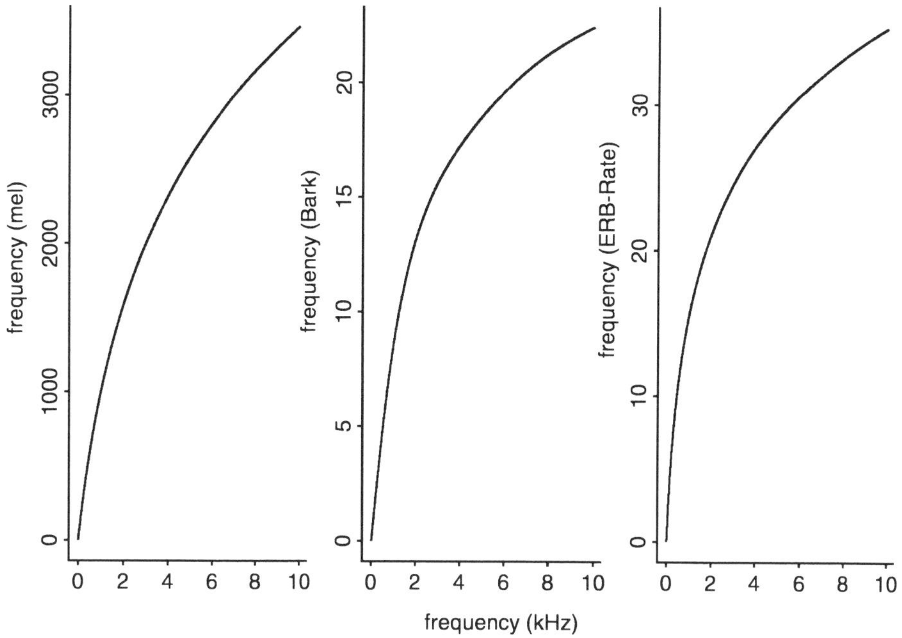

Figure 2.10: Relationship between Hertz and the mel, Bark, and ERB scales.

& Hall, 1979; Zwicker, 1961; Zwicker & Feldtkeller, 1967; Zwicker & Terhardt, 1980). The spacing between the centre frequencies of the filters is nearly linear up to 1000 Hz and quasi-logarithmic thereafter (Scharf, 1970; Zwicker, 1961). Just over twenty critical bands span the frequency range 0 to 10000 Hz as shown in Figure 2.10. The relationship between the physical Hertz and perceptual Bark scale is given by

$$f_{Bark} = 13\tan^{-1}(0.0076 f_{Hz}) + 3.5\tan^{-1}(\frac{f_{Hz}}{7500}^{2}), \qquad (2.1)$$

where $\tan^{-1}$ is the arctangent in radians.

A third perceptual scale, which is more directly based on the shape of the auditory filters (Moore & Glasberg, 1983), is known as the *ERB* scale (the *equivalent rectangular bandwidth* of the auditory filter). In Moore and Glasberg (1990), a formula relating the number of ERBs, E, to frequency in Hz, f_{Hz}, is given by

$$E = 21.4\log_{10}(4.37 f_{Hz}/1000 + 1).$$

The relationship between the physical Hertz scale and these three perceptual scales is shown in Figure 2.10.

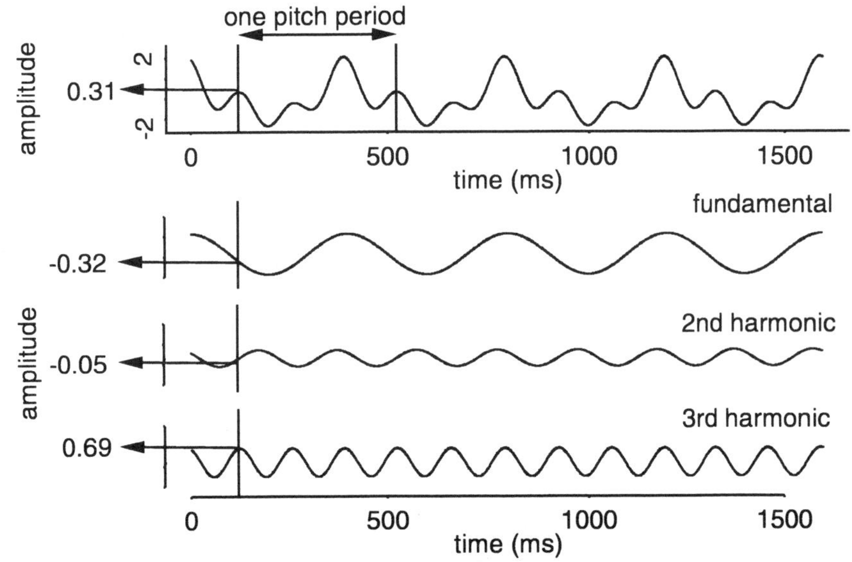

Figure 2.11: *Top panel*: Four periods of a waveform with a fundamental frequency of 2.5 Hz. *Bottom panel*: The three sinusoids below this are the result of applying a Fourier analysis to this waveform. When these sinusoids are summed, the waveform in the top panel is exactly reconstructed.

2.2.5 Fourier Analysis

Fourier was a nineteenth century mathematician who made the important discovery that any periodic waveform can be decomposed into a set of harmonically related sinusoids whose fundamental frequency is equal to that of the periodic waveform (Fourier, 1822). Figure 2.11 shows four periods of a periodic waveform with a fundamental frequency of 2.5 Hz (each period has a duration of 0.4 second or 400 ms) as well as the result of applying a Fourier analysis to this waveform. Consistently with Fourier's theorem, the resulting sinusoids are harmonically related at multiples of 2.5 Hz. Furthermore, if the sinusoids are summed at equal points in time (this process is known as *Fourier synthesis*), the original periodic waveform is reconstructed. For example, at the vertical line in Figure 2.11, the amplitude of the waveform that Fourier analysis is applied to is 0.31, which is almost the same as the sum of the amplitude values of the fundamental, second and third harmonics at the same time point (the sum of these values is $-0.32 - 0.05 + 0.69 = 0.32$; the discrepancy of 0.01 is caused by rounding errors).

The amplitude spectrum for the waveform in Figure 2.11 is shown in Figure 2.12. Notice that there are three frequency components at multiples of 2.5 Hz whose amplitudes are equal to those of the sinusoids in Figure 2.11. This type of spectrum is known as a *line spectrum* because in periodic waveforms,

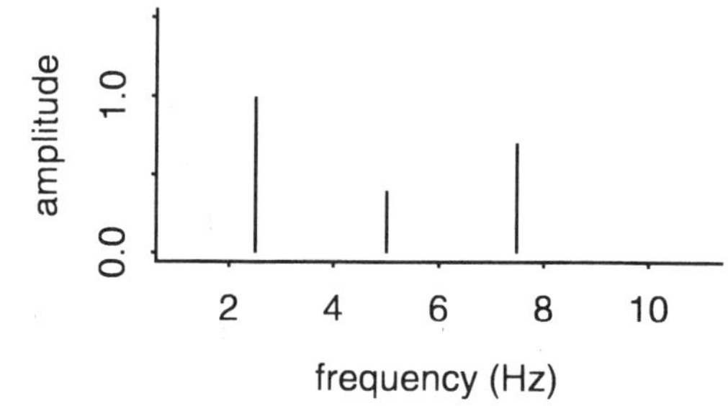

Figure 2.12: The amplitude spectrum for the waveform in Figure 2.11.

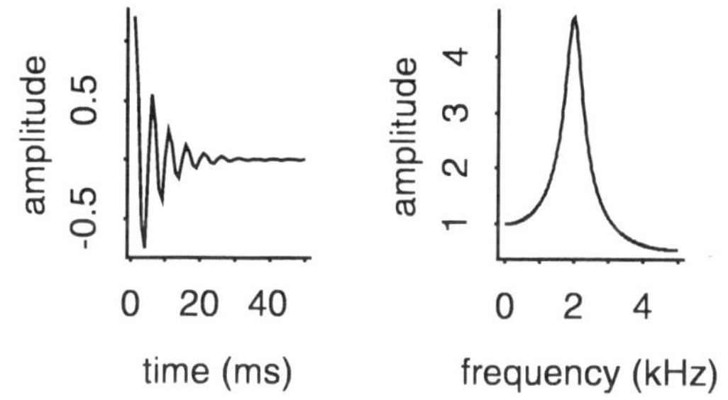

Figure 2.13: A non-repetitive waveform (*left*) and its spectrum (*right*).

frequency components occur only at multiples of the fundamental frequency.

We must now turn to a consideration of nonrepetitive waveforms. Fortunately, Fourier also demonstrated that these can be decomposed into a set of sinusoids in a similar way to periodic waves. One of the important differences is that the number of sinusoids needed to reconstruct a nonrepetitive waveform is infinite which means that the resulting spectrum is continuous (Figure 2.13). Although the spectrum is continuous, the same principle of summation applies: if the infinite number of sinusoids that is represented by the continuous spectrum is summed, the non-repetitive waveform on the left is exactly reconstructed.

2.2.6 Spectral characteristics of voiced and voiceless speech sounds

In order to make a spectrum of a speech sound, it is almost always necessary to extract the waveform that is to be Fourier analysed from a recording of a longer speech utterance. For example, in order to make a spectrum of the [æ] in "stamp", Fourier analysis must be applied to part of the [æ] waveform, rather than the entire word. There are then further aspects to consider. For example, do we convert all of the [æ] waveform into a frequency representation or just part of it? And if only part of it, then how we decide in a principled way on the beginning and end of the section that is to be Fourier transformed? These

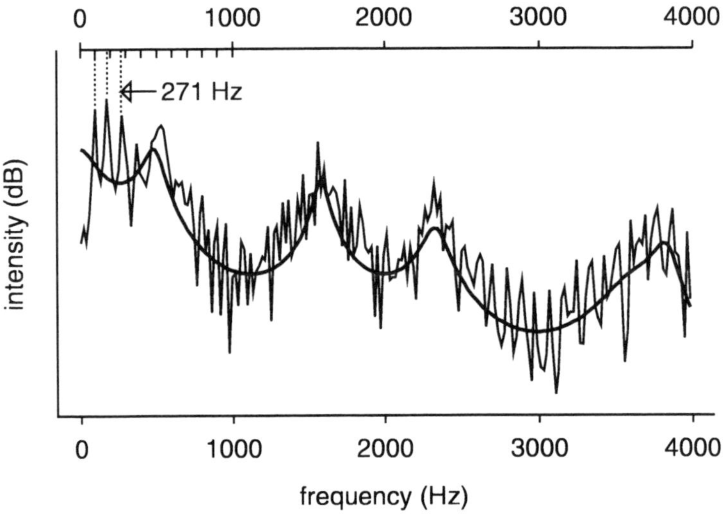

Figure 2.14: The spectrum of a section of the [æ] waveform in Figure 2.3. The display also shows a smoothed spectrum from which the spectral peaks, corresponding to formant frequencies, are more clearly discernible.

issues will be considered in detail at a later stage; for the present, we will note that in almost all kinds of frequency analysis of speech, the waveform that is to be Fourier transformed should at least be relatively unchanging or *steady-state*. In producing speech, the greatest change in the vocal tract shape usually occurs at the transitions between sounds: between the [t] and [æ] or between the [æ] and [m] of "stamp" for example. It is usually better therefore to apply the Fourier transform nearer the middle of the speech sound to reduce the confounding influences due to coarticulatory effects of neighbouring segments.

For the present example, we shall consider a Fourier transform of the section of the periodic waveform in Figure 2.3 that has been cut out near the central part of this vowel. As before, the purpose of the Fourier transform is to decompose the waveform into a number of sinusoids and thereby obtain a frequency representation of the speech sound. Furthermore, since the waveform is periodic, we should expect to see harmonics close to multiple frequencies of the fundamental. Using the formula given earlier, the fundamental frequency values for this section of the waveform are estimated at 102 Hz, 92 Hz, 90 Hz, and 87 Hz, which gives an average of just under 93 Hz; we might therefore expect to find peaks in the spectrum that are at multiples of a value somewhere between 90 Hz and 100 Hz.

The spectrum for this section of the waveform is shown in Figure 2.14. Only the spectral range from 0 to 4000 Hz is shown, since this contains most of the information that is relevant to vowel distinctions. There are peaks in the spectrum that tend to occur at intervals of between 90 to 110 Hz, as predicted

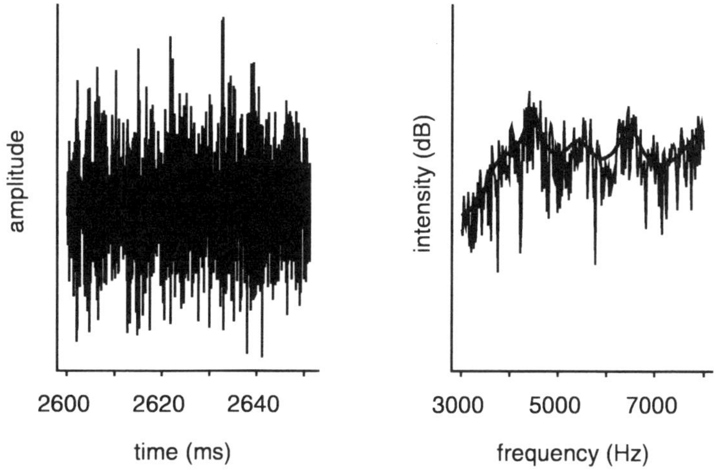

Figure 2.15: Waveform (*left*) and spectrum (*right*) of a section of [s] in "stamp". A smoothed spectral trend is superimposed on the spectrum showing three principal peaks that occur in the 4 to 7 kHz range.

from the earlier analysis of fundamental frequency. The first three of these are shown by vertical dotted lines, and they are separated by approximately 90 Hz.

If the spectrum is examined from another point of view, there are clear rises and falls in the spectral shape that span a wider frequency range. This is illustrated by the smoother "trend line" in Figure 2.14, which shows that the spectrum of this vowel has four main broad peaks and troughs. These peaks are estimates of the resonances of the vocal tract and in speech analysis they are called *formants*. Formants are the main determiners of phonetic vowel quality. For the present (Australian English) [æ] vowel, the three formants are those of the first three peaks of the solid line at approximately 500 Hz, 1620 Hz, and 2350 Hz respectively. They are labelled in numerical order and are abbreviated as follows: F1=500 Hz, F2=1620 Hz, F3=2350 Hz. Vowels that differ in phonetic quality also usually have different formant patterns (or F-patterns): for example, an [i] vowel tends to have a much lower F1 and higher F2 value. The F-pattern is the acoustic correlate of the vocal tract shape: thus since [æ] and [i] are produced with differently shaped vocal tracts, their F-patterns differ, and the different F-patterns are also the main cues to vowel quality distinctions.

The analysis of the periodic speech waveform has shown that the spectrum contains very narrow peaks that are spaced at almost equal intervals throughout the frequency range that are due to the vibrating vocal folds and a number of much wider troughs and peaks that depend on the shape of the vocal tract. In nonrepetitive waveforms that are characteristic of voiceless sounds, the vocal tract makes a similar contribution to the spectral shape in terms of broad peaks and dips, but these are superimposed on short-term random fluctuations in the spectrum, rather than on the series of narrow peaks that are due to vocal fold vibration (Figure 2.15). The random short-term fluctuation in the spectrum is the acoustic correlate of a turbulent airstream that is produced when the air is channelled through a narrow constriction in fricative-like sounds. The spectrum in Figure 2.15, which is of a section taken from the [s] of "stamp", is displayed above 3 kHz, since most of the acoustic information for identifying this sound is contained in this frequency region.

2.2.7 Spectrographic analysis

The spectra discussed so far have all been derived by applying a Fourier transform to a section of speech data. This type of analysis is *static* because the spectrum is a display of the frequency content of a *single* time section. While static spectral slices can provide much information about the characteristics of speech sounds, for many kinds of speech analysis we will need to consider how the frequency content of the waveform *changes in time*. Essentially, we need to change the two-dimensional spectral representation into a three-dimensional one whose axes are amplitude, frequency, and time.

This raises various issues to do with how we are going to partition the waveform in order to convert it into a frequency representation: we clearly do not want to calculate a single spectrum for an entire waveform such as the one shown in Figure 2.1 (of the word "stamp") because this would represent an averaged spectrum that does not preserve any of the important differences between the separate speech sounds. Instead, the waveform must be divided into a number of equal portions in order to calculate a spectrum for each of these. This means decisions must be made about how many portions the waveform is to be divided into and also whether the sections will overlap and if so by what amount.

There are two sorts of three-dimensional displays that can be produced from this kind of analysis. The first is sometimes known as a *waterfall display* in which each of the spectral sections is arranged in a three-dimensional perspective that allows the user to look at the changes in spectral shape as a function of time (Figure 2.16). The second is a *spectrogram*, which is far more common in speech research. This display encodes the three-dimensional information as two dimensions by representing intensity by the darkness of markings (an increasing darkness represents a greater intensity). Part of the reason for the popularity of spectrograms is that they succinctly encode many of the known acoustic cues to different speech sounds; in particular, a spectrogram displays very effectively the formant frequencies of vowels. Another reason for the popularity of the spectrogram is historical. Since the invention of the spectrograph in the 1940s, much of our knowledge about the acoustics of speech sounds has been based on spectrographic data. Nowadays a spectrogram can be efficiently calculated on a computer, and there are also many other ways in which speech data can be rapidly parameterised which were not easily available even fifteen years ago. Nevertheless, spectrograms have had such an influence on speech research that every student of speech science should be acquainted with the kind of information that they contain.

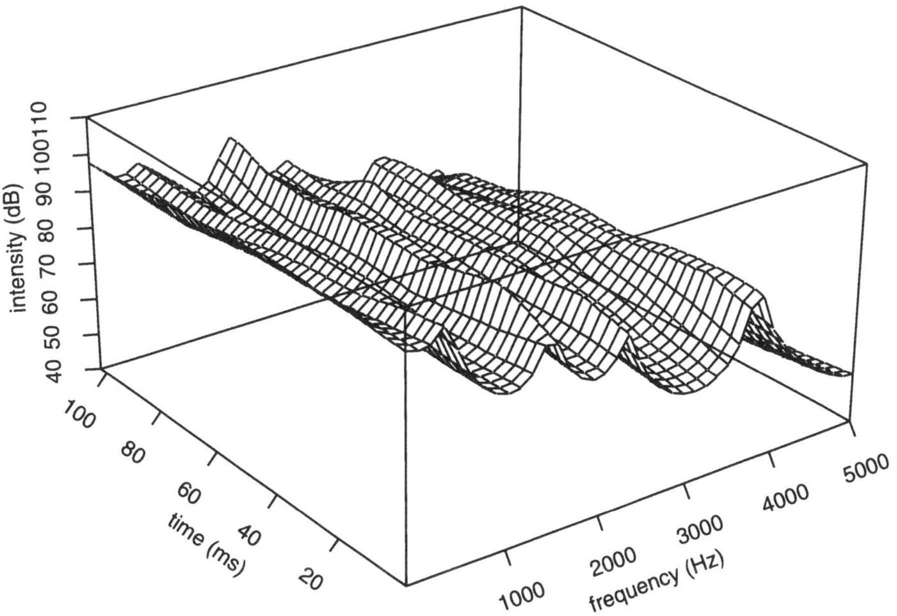

Figure 2.16: A "waterfall" display of *frequency*, *amplitude* and *time* for a vowel sound.

The most common kind of spectrogram is a *wideband spectrogram* in which the frequency resolution is often set to around 300 Hz for speech analysis. This means that a wideband spectrogram does not differentiate two frequencies that are less than 300 Hz apart: so two tones of frequencies 50 Hz and 150 Hz (interval = 100 Hz, which is less than 300 Hz) would be undifferentiated and represented by a single dark line in the 0 to 300 Hz band. Relatedly, since the fundamental frequency for most male and many female talkers is less than 300 Hz, the separate harmonics, which occur at multiples of the fundamental, are not differentiated. Therefore, a wideband spectrogram cannot usually be used to track individual harmonics in the speech of most adult, male talkers.

One of the properties of spectrographic analysis is that frequency and time resolution are approximately inversely proportional: in other words, the coarser the frequency resolution (that is, the wider the analysing filter), the better the time resolution and vice-versa. For example, when the frequency resolution is 300 Hz (as for most wideband spectrograms), the time resolution is approximately 1/300 seconds or just over 3 milliseconds: this means that two events that are temporally separated by more than this value are distinguished on a wideband spectrogram. The reason that the wideband spectrogram in Figure 2.17 has a series of vertical striations during the voiced speech is that these are the successive closures of the vocal folds. Since the fundamental frequency is usually around 100 Hz for this talker, the duration between closures (the du-

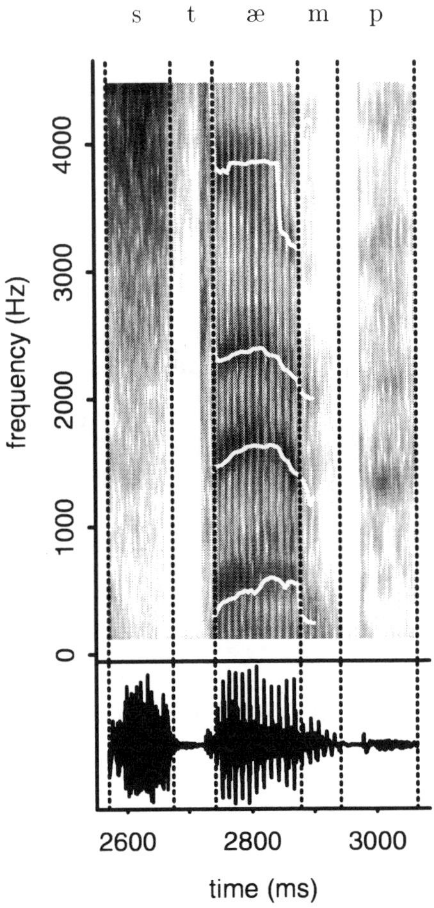

Figure 2.17: A wideband spectrogram of "stamp" calculated with a 300 Hz analysing filter and showing the first four automatically tracked formant frequencies.

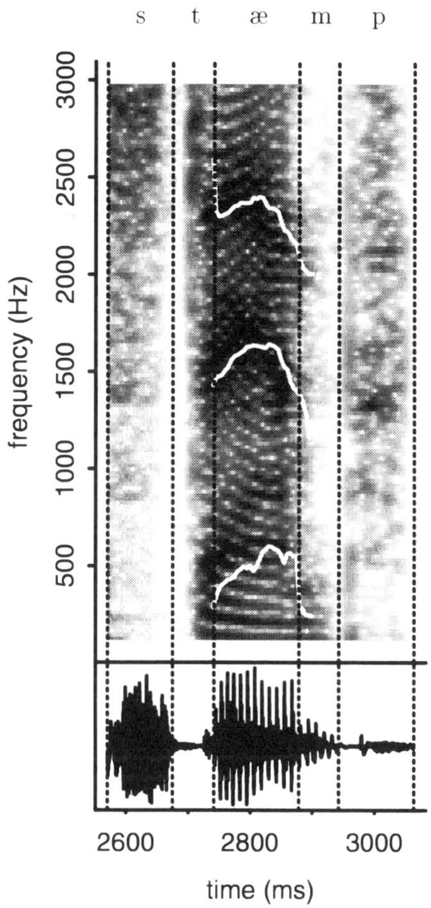

Figure 2.18: A narrowband spectrogram of "stamp" calculated with an analysing filter of 45 Hz and showing automatically tracked formant frequencies.

ration of the pitch period) is 1/100 seconds = 10 milliseconds. Another way of estimating the fundamental frequency of voiced speech is therefore to determine how many vertical striations occur per unit of time.

In another kind of spectrogram, known as a *narrowband* spectrogram (Figure 2.18), there is a much finer frequency resolution. Typically, the analysing filter in narrowband spectrograms is set to around 50 Hz. Since the talker's fundamental frequency is normally greater than this value at 100 Hz, the harmonics are separately resolved. Yet another way of estimating fundamental frequency is therefore to track the frequency of one of the harmonics. Higher harmonics are often used to estimate the fundamental frequency (for example, by tracking the tenth harmonic and dividing by ten) since changes in pitch produce much more dramatic excursions in them than in the fundamental.

When the frequency resolution is 50 Hz, the temporal resolution is approximately 1/50 seconds or 20 milliseconds. Since this exceeds the duration of a pitch period at 100 Hz, the striations due to separate vocal fold closures are not visible for a fundamental frequency at this value.

Wideband spectrograms are generally preferred for formant analysis because the visible presence of harmonics can make the formant more difficult to see. Because of their efficient temporal resolution, they are also used in preference to narrowband spectrograms for calculating the duration of speech sounds.

Further reading

The following books discuss in further detail some of the aspects of the physics of speech and the fundamentals of acoustic phonetics dealt with in this Chapter (see also the references under *acoustic phonetics* at the end of Chapter 1): Borden et al. (1994), Clark and Yallop (1995), Fry (1979), Kent and Read (1992), Ladefoged (1962), and Pickett (1980).

3
The Acoustic Theory of Speech Production

The acoustic characteristics of any speech sound are determined by the whole complex of the movement and configurations of the speech production process. We have seen that some aspects of speech production have a fairly predictable effect on the acoustic speech signal. For example, periodicity in the acoustic waveform is the acoustic consequence of vocal fold vibration that characterises voiced sounds, while a nearly random fluctuation in air pressure variation results from a turbulent airstream in the production of most voiceless sounds.

These observations form part of a general theory about how the production of speech is related to the acoustic speech signal. What might we expect a more detailed elaboration of such a theory to include? And why is it so important to speech acoustics?

The goal of an acoustic theory of speech production must be to explain as many of the acoustic attributes of a speech sound as possible in terms of its speech production characteristics. It must explain why broad phonetic categories, such as nasal and oral, have quite different acoustic consequences, and it must explain how formants are related to the expected vocal tract shape that produced them. There are other effects that are less directly related to the phonetic quality of sounds that must be explained as well. We might want to know why formants of females are higher in frequency than those of their male counterparts or why the spectrum of voiced sounds usually decreases in amplitude with increasing frequency. Since all these acoustic effects are caused by the production of speech, it must be possible to explain each of them individually in terms of some aspect of the speech production process. This componential analysis of the acoustic speech signal in terms of articulatory effects is the core of an acoustic theory of speech production.

The overarching reason that modelling the acoustics of speech sounds in terms of production effects is so important to speech acoustics has already been mentioned: we will be in a better position to understand the relationship between sounds as phonetic units and the speech waveform if we can model the logically prior step of how the production of speech is translated into an acoustic signal. An understanding of articulatory-acoustic relationships can provide an initial hypothesis about the most likely acoustic cues to speech sounds. For example, a knowledge of articulatory-to-acoustic mappings would tell us that the [s]/[ʃ] distinction is more than likely to be based on the frequency location of the major energy concentration in the 2 to 6 kHz range. Still another reason for the importance of research in this field is that digital speech signal processing is to a large extent built on a model of speech production and its relationship to the acoustic speech signal.

There are four main processes in the production of speech whose acoustic effects can be considered independently. The first of these is the *sound source*, which is due either to the vibrating vocal folds (in voiced sounds), a turbulent airstream (in voiceless sounds), or a combination of the two (in voiced fricatives). Second, the shape of the vocal tract, which is acoustic terms is modelled as a *vocal tract filter*, can be analysed to a large extent independently of the source. Third, there are *energy losses* to consider that make various contributions to a speech sound's acoustic structure. Finally, the way in which sound as a speech pressure waveform *radiates from the mouth* needs to be taken into account in decomposing the sound's acoustic signal into its basic components.

Each of these four main processes will be considered in turn. As a preliminary, it will be necessary to review the theoretical basis for the independence of the acoustic source and vocal tract filter.

3.1 The source-filter decomposition of speech

In the preceding chapter, we had alluded to the idea that at least two major processes of speech production are distinguishable in the spectra of speech sounds. One of these can be attributed to the shape of the vocal tract above the larynx or supralaryngeal vocal tract shape; the other is caused either by the vibrating vocal folds (in voiced sounds) or a turbulent airstream (in voiceless sounds). The supralaryngeal shape of the vocal tract is manifested as a slowly varying spectral trend consisting of peaks and troughs (see the trend line in Figure 2.14 and Figure 2.15 of Chapter 2). In speech acoustics, this spectral trend is said to be attributable to the filtering effects of the supralaryngeal vocal tract shape, or the vocal-tract filter, while the short-term fluctuations in the spectrum are caused either by a periodic source (in the case of vocal fold vibration) or an aperiodic noise source (in the case of sounds produced with a turbulent airstream).

The terms source and filter are directly applicable to simple experiments that serve to illustrate the properties of *resonance*. Consider as an example an experiment involving an audio oscillator for generating pure tone sine waves of the same amplitude and a set of thin glass jars open at one end and closed at the other. The experiment involves holding the oscillator up to each jar separately and varying the frequency of the generated sine wave starting with a low frequency (just above 0 Hz) and then increasing the frequency up to a maximum value at equal frequency intervals (of, for example, 100 Hz). At these frequency intervals, the sound that is produced by holding the oscillator close to the jar is recorded. In this experiment, we would notice that the resulting sound is louder at some frequencies than at others; furthermore, the measurements of the sound's amplitude would show clear peaks at certain frequencies. If the experiment were repeated but with a jar of a different length, the spectrum would

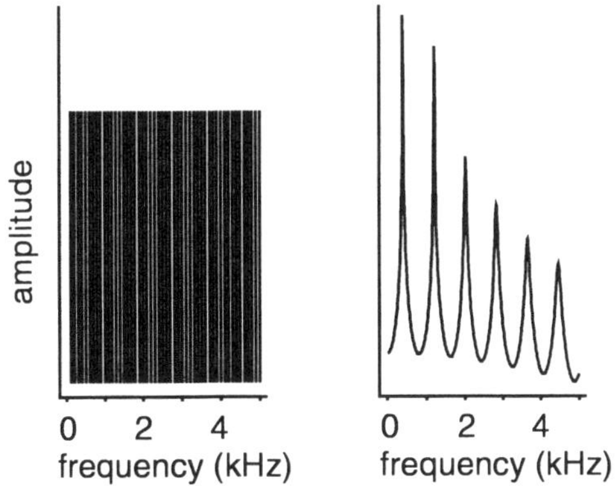

Figure 3.1: A source spectrum (*left*) with a fundamental frequency of 100 Hz and the spectrum of the filter (*right*).

once again show amplitude peaks, but they would be at different frequencies. The reason that there are peaks in the spectrum is because the oscillator makes the column of air in the jar vibrate with high amplitude at the jar's *natural frequencies of vibration*. These natural frequencies of vibration are called *resonances* and they depend on the jar's shape. For example, when two jars have the same shape but are of different lengths, the resonances of the longer jar are lower in frequency.

The experiment just described could be summarised for each jar separately in terms of three spectra: one for the source (the oscillator), one for the filter (the jar we have selected) and one for the combined sound output (holding the oscillator up to the jar as it is stepped through increasing frequency values).

The spectra of a hypothetical source and filter are shown in Figure 3.1. The source spectrum shows single lines at multiples of 100 Hz because the oscillator was made to generate sine waves at a frequency interval of 100 Hz: these sine waves are harmonically related with a fundamental frequency of 100 Hz. The spectrum of the filter is also sometimes known as a *resonance curve*. The peaks in the spectrum are the *resonance centre frequencies*, and the width of the peaks are the *resonance bandwidths*.

The *resonance bandwidth* defines a range of frequencies on either side of the peak which are within 3 dB of the peak. For example, if a resonance has a centre frequency of 2000 Hz and a bandwidth of 400 Hz, then the frequencies at $2000 - 400/2 = 1800$ Hz and $2000 + 400/2 = 2200$ Hz should be approximately 3 dB down from the dB level of the resonance centre frequency (Figure 3.2). The motivation for the choice of 3 dB is that a reduction in amplitude of 3 dB is approximately equal to halving the sound's power.

When the oscillator generates a sine wave close to one of the resonances, the resulting sound output is high because the air in the jar vibrates with high amplitude at this frequency. On the other hand, if a sine wave is generated close

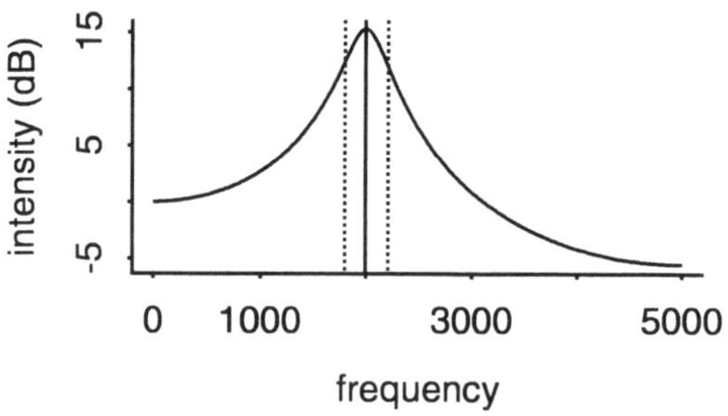

Figure 3.2: A resonance curve showing the centre frequency (solid line) and bandwidth (the interval between the vertical dotted lines).

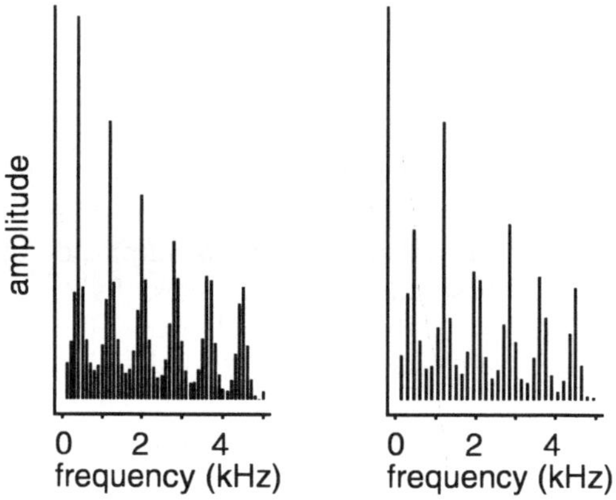

Figure 3.3: The effects of combining the source and filter for two different fundamental frequencies.

to one of the troughs in the filter spectrum, then the resulting sound output is comparatively low — the term *filter* is applicable because the sound source is attenuated at these frequencies.

The spectrum of the combined sound output (left panel, Figure 3.3) has the characteristics of the filter 'imprinted on' the source: the frequency components at 100 Hz intervals are the result of the source; the amplitudes of these frequency components rise and fall in accordance with the resonating properties of the filter. Notice in particular that the harmonics that are closest to the resonance frequencies of the filter have the highest amplitude(s) in the combined spectrum.

In the production of speech, the resonances of a given vocal tract shape are known as *formants*. The idea that the source and filter make an (almost) *independent* contribution to the spectrum of the combined sound output in the production of speech (Stevens, Kasowski, & Fant, 1953; Fant, 1960; Stevens & House, 1963a) is a very important part of the acoustic theory of speech production. This means that if there is a change to the source but a minimal change to the filter, the formant centre frequencies do not change and there is a minimal change in the long-term spectral trend of the combined sound output.

Consider as an analogy of this, the right panel of Figure 3.3 in which the fundamental frequency is at 150 Hz (i.e., the oscillator was stepped through frequencies of 150 Hz, 300 Hz, 450 Hz ...). Although the source characteristics have changed, the filter is the same, and consequently the two spectra of the combined sound output in Figure 3.3 have a very similar spectral trend. There are of course differences as well. One that is immediately noticeable is that the harmonics are spaced at greater intervals in the combined spectrum in the right panel in which the fundamental frequency is higher. Another is that the frequencies of the harmonics that have *highest* amplitude are slightly different in the two spectra. This comes about because, as stated earlier, the resulting spectra are a combination of the acoustic properties of the source and those of the filter which are imprinted upon it. Therefore, when the fundamental frequency is 100 Hz and the first resonance of the filter (first formant in the production of speech) is at 500 Hz, the fifth harmonic at 500 Hz has the highest amplitude. However when the source fundamental frequency is changed to 150 Hz, it is the third harmonic at 450 Hz that has the highest amplitude, a shift of 50 Hz. Although the combination of two different source signals with the same filter produces some perturbations in the exact location of the frequency peaks in the combined spectrum, listeners' judgements of vowel quality (or of speech sound quality in general) are to a large extent based on the filtering properties of the vocal tract independently of the source signal. Therefore, when the source fundamental frequency changes but the vocal tract shape stays in the same configuration for an [i] vowel (implying a constant filter and therefore constant formant frequencies), listeners judge the quality of the vowel as having the same phonetic [i] value. Conversely, if the shape of the vocal tract changes the acoustic filter changes and there is a shift in the phonetic quality of the vowel (for example, from [i] to [ɛ]).

The independence of the acoustic source and filter has direct correlates in the production and perception of speech. In speech production, the rate of vocal fold vibration (the source) can be varied independently of the shape of the supralaryngeal tract. In perceptual terms, this implies that pitch can to a large extent be varied independently of phonetic quality. We know this must be so since it is possible to vary the pitch while producing a sound of the same phonetic quality (e.g., a prolonged [i] produced on a continually varying pitch). As listeners, we hear the same phonetic quality despite the change in pitch because a change in the fundamental frequency does not imply a change in the formant frequencies and perturbs minimally the peak frequencies in the combined spectrum of the source and filter.

3.2 The acoustic source in speech production

3.2.1 Glottal source

In normal vocal fold vibration, the vocal folds are initially drawn together, or adducted, so that the subglottal chamber is separated from the rest of the vocal tract. The deflation of the lungs produces a continual increase in subglottal air pressure and when this force overcomes the adductive resistance offered by the vocal folds, they are blown apart. Under the aerodynamic-myoelastic theory of vocal fold vibration (van den Berg, 1958), once the folds have been opened, their closure is assisted by the *Bernoulli effect* (van den Berg, Zantema, & Doornebal, 1957; Lieberman, 1968) in which a local drop in air pressure is produced between the vocal folds as the lung air is channelled through the narrow glottal opening. Thus the vocal folds close once again and the cycle is repeated.

The description of the acoustic characteristics of the cycles of vocal fold vibration is complicated by the difficulty of obtaining precise measurements. One technique might be to photograph the vocal folds to obtain the glottal area as a function of time (e.g., Farnsworth, 1940; Moore & van Leden, 1958; Hirano, Kakita, Kawasaki, Gould, & Lambiase, 1981) using a transillumination technique usually by means of a fiberscope inserted through the nasal cavity (Sawashima & Hirose, 1968; Sonesson, 1959, 1960) and then to estimate from that the glottal volume velocity waveform (Flanagan, 1958); however, since the glottal area is not always directly proportional to glottal airflow (Flanagan, 1972), and since these glottal photographic techniques produce in the first instance qualitative rather than quantitative data, they are of limited value in determining the acoustics of vocal fold vibration. Another way of estimating the acoustic waveform of glottal vibration is to attempt to cancel out the effects of the vocal tract resonances: this can be done by attaching a long tube to the subject's lips (a reflectionless tube) and recording the resulting speech production of a neutral vowel (see, for example, Monsen & Engebretson, 1977; Sondhi & Resnick, 1983). Another technique that attempts a reconstruction of the glottal waveform is *inverse filtering* in which the effects of the formants (specifically of the vocal tract transfer function) are removed using mathematical models of the source-filter decomposition of speech (Millar, 1959; Rothenberg, 1973, 1981; see also Fant, 1993 and Fant, 1995 for a review of this technique). The volume velocity of glottal flow can also be reconstructed by inverse filtering the volume velocity of airflow (rather than the speech pressure waveform) at the lips (Rothenberg, 1973, 1977).

The waveform shown in Figure 3.4 provides a very simple model of the volume-velocity of airflow through the glottis that is approximated in some text-to-speech systems (e.g., Klatt, 1980). This glottal waveform was derived from perception experiments by Rosenberg (1971), who replaced the glottal waveform of natural utterances with various kinds of synthesised glottal waveforms; the waveform that was most preferred by subjects was found to be similar to the one shown in Figure 3.4 (see, for example, Ananthaphadmanabha, 1984, for a summary of further details and developments of this model; and Chapter 7 on

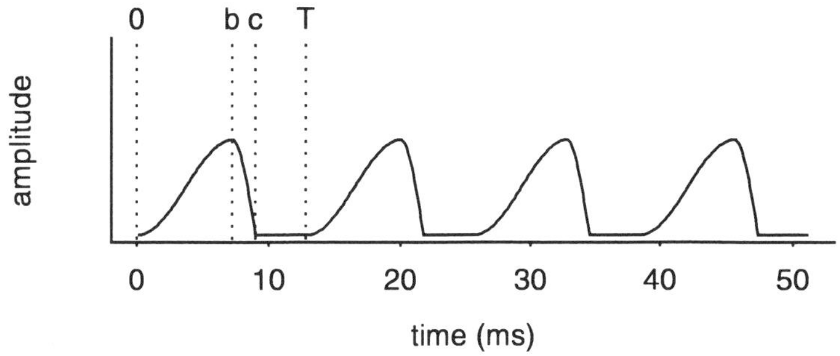

Figure 3.4: A glottal pulse modelled after Rosenberg (1971). 0–b: opening phase; b–c closing phase: c–T: closure. One glottal cycle corresponds to 0–T.

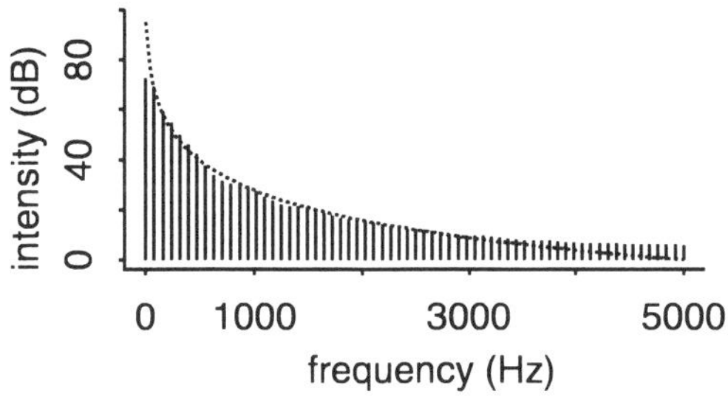

Figure 3.5: The spectrum of the glottal pulse shown in Figure 3.4. The dotted line shows a -12 dB/octave slope.

digital formant synthesis for other parameters that are included to model different kinds of vocal quality). In this waveform, each cycle (from 0 to T) includes an opening phase (0 to b), from the time when the vocal folds are closed to their extent of opening, and a closing phase (b to c), from their maximum extent of opening to closure. When the vocal folds are completely closed (c to T), there is no glottal sound output. A pitch period extends from the beginning of the opening phase until the end of the closure phase; as stated earlier, the duration of the pitch period is inversely proportional to the fundamental frequency of vocal fold vibration.

Since the glottal waveform is periodic, we should expect to see harmonics at the pitch frequency in the resulting spectrum: the spectrum for the glottal waveform in Figure 3.4 is shown in Figure 3.5 and the lines are spaced at just over 78 Hz, corresponding to the pitch period duration of 12.8 ms ($1000/12.8 \approx 78$). Another important spectral characteristic of the glottal waveform that is captured by Rosenberg's model is that the spectrum falls at a rate of about 12 dB per doubling of frequency, or 12 dB/octave (for example, the amplitude of the glottal spectrum in Figure 3.5 at 1 kHz is roughly 27 dB; the amplitude of the spectrum at 2 kHz, which is double this frequency, is approximately 15 dB

= 27 – 12 dB).

The glottal waveform and spectrum shown in Figure 3.4 and Figure 3.5 represent a highly simplified model of the glottal airflow and glottal spectrum that was found to be adequate in synthesising normal adult male vocal fold vibration in early speech synthesis systems. There are of course many other possible phonatory settings (see, for example, Kaplan, 1960; Laver, 1980; and Zemlin, 1981, for an analysis of the various possibilities) that cause differences in the glottal waveform and its corresponding spectrum. For example, in a breathy articulation, in which there may be almost continuous airflow throughout the glottal cycle, the opening phase is likely to be increased at the expense of the rate and duration of the glottal closure. Some typical spectral characteristics of breathiness include an increase in noise particularly in the higher harmonics and an increase in the amplitude of the first harmonic possibly resulting in a greater spectral tilt (Klatt & Klatt, 1990). There may also be an increase in the formant bandwidth and some indication of the presence of *subglottal* formants and antiformants due to the increased coupling to the trachea (Ishizaka, Matsudaira, & Kaneko, 1976; Fant & Lin, 1988; Karlsson & Neovius, 1993; Fant & Kruckenberg, 1995).

On the other hand, in a more forceful articulation that might occur while shouting, the duration of the closure (from b to c) increases at the expense of the opening phase (from 0 to c); the acoustic consequence is greater energy at higher frequencies so that the spectrum falls off at a lower rate than 12 dB/octave (Fant, 1959).

Research in the last twenty years has also shown that female glottal pulses cannot simply be considered as linearly rescaled male ones (see, for example, Klatt & Klatt, 1990), and there has been extensive research both on characterising female glottal pulses more accurately and on improving the naturalness of a female voice quality in speech synthesis (e.g., Fant & Lin, 1988; Karlsson, 1990; Karlsson, 1991; Karlsson, 1992). There is evidence from these studies that some female talkers may have a slightly breathier voice qualities than male talkers, which is compatible with earlier research (Monsen & Engebretson, 1977) that suggests a proportionally greater opening phase in female glottal pulses.

In recent years, a much more sophisticated model of the glottal flow (specifically of the rate of glottal flow) has been developed by Fant and colleagues known as the LF-model (Fant, Liljencrants, & Lin, 1985; Fant, 1993). This model, which has been used to synthesise some of the voice quality difference discussed above, as well as male-female differences, has four main parameters: the time of peak glottal flow, the time at which the glottal closure rate is a maximum, a time interval that is used to control the abruptness of the glottal closure, and the time of the glottal closure itself. A useful recent summary of the LF-model is given in Childers and Ahn (1995) (see also Childers & Hu, 1994 and Chasaide and Gobl (1997) for a comparison with other glottal source models).

3.2.2 Noise source

The position of the noise source in the production of speech, in contrast to that of the glottal source, varies with place of articulation ranging in English from a labiodental to a glottal place of articulation. The acoustic characteristics of a noise source are far less well researched than those of a glottal source. Part of the reason for this is that both voice quality differences as well as differences between male and female voice qualities are much more dependent on variations in the glottal waveform than in the fricative noise source. Another is that, as various studies have demonstrated (Meyer-Eppler, 1953; Heinz & Stevens, 1961; Fant, 1960; Shadle, 1991; Badin, 1989), the predicted relationship between the noise source and its spectrum is complicated by many factors including estimating precisely where the source is located in fricative production (see Shadle, 1990, 1991 for a further discussion).

Sounds that are produced with a noise source include voiceless and voiced fricatives in English, as well as the release and aspiration stage of voiceless stops. The noise source is caused by a turbulent airstream that is the result of a jet of air being channelled at high speed through a narrow constriction. As discussed in further detail in Stevens (1972a) and Catford (1977), the random nature of the acoustic noise source can be attributed to some of the random circulation effects and eddies that are caused by the airstream: these can be intensified when the jet of air strikes an object such as the upper front teeth in the production of alveolar and dental sounds.

The source in the production of fricatives has been classified in two different ways. Firstly, in the production of [s] and [ʃ], the jet strikes the teeth that are located approximately at right angles to it. Secondly, in the production of palatal and velar fricatives, the obstacle is formed by the hard-palate which is at a more oblique angle to the jet than in [s] and [ʃ]. Shadle (1997) also notes that noise is generated along the lips in the production of [f] and [θ].

As discussed in Stevens (1997), the amplitude and spectral shape of the noise source are dependent on the pressure drop across the constriction, the cross-sectional area of the constriction, and the locations of any obstacles in the airstream downstream from the constriction. There is some agreement that that the amplitude of the noise source is greatest when the airstream directly strikes an obstacle, which is particularly the case in the production of the sibilant fricatives [s] and [ʃ] which are higher in amplitude than the other fricatives.

The acoustic correlate of a noise source is aperiodicity, which, as already indicated, has no clearly identifiable repetitive pattern. The corresponding spectrum has been described as relatively "flat" that is, there is energy at all frequencies of the spectrum and there are no clearly identifiable peaks and troughs. In some fricatives, notably [f] and [x] (German "ach"), the spectrum of the noise source may slope downwards at roughly 6 dB/octave (Fant, 1960; see also the synthesis of the noise source in Klatt, 1980). A recent discussion of the spectra of noise sources is given in Stevens (1997).

Many phonologically voiced fricatives, in particular the voiced sibilants [z], [ʒ] (and the fricated components of the affricate [ʤ]), are realised with a simultaneous glottal and noise source corresponding to simultaneous vocal fold vibration and a turbulent airstream respectively. The waveform and spectrum share many features of both the glottal and noise sources: in the waveform, random fluctuations are likely to be superimposed on detectable periodicity, while in the spectrum the presence of harmonics caused by the glottal source are obscured by the random nature of the flat spectrum that is the consequence of the noise source.

3.3 The acoustic filter in speech production

In the production of a voiced oral sound (for example, a vowel), the vocal tract can be considered as a tube which is closed at one end (during the closure phase of vocal fold vibration), open at the other (the lip end) and with a constantly varying cross-sectional area between the glottis and the lips. In estimating the corresponding spectrum and resonances, certain assumptions have to made about the shape of the tube and how the airstream propagates through the tube from the glottis to the lips.

First, the directional change in the shape of the vocal tract — specifically the fact that the tract makes a near right-angled bend at the junction between the pharynx and the oral cavity — is assumed to be unimportant as far as estimating the spectrum is concerned. This simplification is justifiable since the airstream propagates through the vocal tract in a planar fashion for frequencies below about 4000 Hz — that is perpendicularly to the vocal tract itself.

Second, if we took two slices out of the tube at different points and compared them, it is more than likely that there would be gross differences in both their areas and shapes. For example, if one of the slices were taken near the glottis, it might have a nearly ellipsoidal shape, whereas if the other slice were taken at the uvula, the top part of the section might be U-shaped due to the uvula itself. As far as estimating the spectrum is concerned, the differences in shape are far less important than the *cross-sectional area* of the slices. Consequently, the further simplifying assumption is often made that all such slices taken out of the tube are circular – which means that the tube is reduced to a cylindrical shape with a continuously varying diameter.

Third, an examination of this cylindrical tube in speech sound production would show that the diameter along certain parts of the tube changes minimally. For example, at the place where the tongue makes a constriction in the vocal tract for producing an [i] vowel, the diameter is not likely to change a great deal for the length of the constriction; similarly, the diameter of the tube may be more or less constant for the whole length of the pharynx in producing certain sounds. Consequently, the further simplifying assumption is made that the diameters are equal over large lengths of the vocal tract: this effectively means that the vocal tract is now being modelled as a set of interconnecting cylinders, where each cylinder represents a length of the vocal tract for which the diameter change is negligible.

Usually, rather than representing the vocal tract in cylindrical form, the more common procedure is to specify it in terms of an *area function* that shows

Figure 3.6: A vocal tract area function for a Russian [i] vowel given in Fant (1960). The area is based on thirty-three cylinders extending from the glottis (*left*) to the lips (*right*).

the cross-sectional area and length of each cylinder that is used to represent the vocal tract (Figure 3.6). Clearly, the more cylinders that are used in the model, the closer the approximation will be to the shape of the vocal tract itself.

Having reduced the vocal tract shape to an area function, the acoustic theory of one-dimensional wave-propagation can be used to calculate the *transfer-function*, from which a resonance curve of the interconnecting cylindrical tubes can be derived. However, these calculations can be further simplified if the tubes are assumed to be *lossless*, which implies that there is no energy loss as the airstream passes through the cylindrical sections. As far as the vocal tract is concerned, this is a gross simplification, since energy losses are introduced at various stages in speech production due to a number of factors, including vibration of the vocal tract walls and lip-radiation effects to mention but two. Nevertheless, a lossless model can provide a fairly good approximation for the purpose of estimating resonances of certain sounds; moreover, the calculations, which initially assume a lossless model, can be modified in various ways to account for some of the known losses (see in particular Fant, 1972, and also Badin & Fant, 1984; Wakita & Fant, 1978).

3.3.1 The acoustic filter of vowels

Single-tube model

The simplest model to be considered is the approximation of the vocal tract by a single cylinder of constant cross-sectional area. The tube is furthermore assumed to be closed at the glottal end and open at the lips. Although this is obviously a highly simplified model of speech production, it nevertheless provides a useful approximation to the central vowel that occurs in productions of "bird" in many non-rhotic English accents (e.g., Southern British English, Australian English).

In order to gain some understanding of how the resonance frequencies are dependent on tube length, we must first consider what is meant by a *wavelength*. When a tuning fork vibrates, it creates variations in air pressure (compressions and rarefactions). These air pressure variations occur both in the vicinity of the tuning fork itself and are also propagated outwards in all directions. It is possible to measure the distance between any two propagated compressions (or rarefactions) of a sinusoid: this is known as the sinusoid's wavelength. The wavelength is dependent both on the speed of sound in air and the frequency with which the tuning fork vibrates. The relationship is

$$\lambda = \frac{c}{f} \quad \text{cm}, \tag{3.1}$$

where λ is the wavelength (in centimetres), c is the speed of sound in air (in cm/s) and f is the frequency of frequency in Hz.

Resonance in the straight-sided tube comes about because of the occurrence of *standing waves* that are produced when air pressure waves cancel and reinforce each other at various points along the tube itself. As discussed earlier, a vibrating sound source causes a succession of air-pressure compressions and rarefactions, and when the source is placed at the open end of a tube, the air-pressure wave that is propagated along the tube is reflected at the closed end causing a set of compressions and rarefactions to be propagated back in the direction of the source. The summation of the air pressure wave from the source (the *incident wave*) and the reflected wave produces a standing wave that has air pressure reinforcements and cancellations at various points along the tube: the reinforcements, or air pressure peaks, are sometimes called *anti-nodes*, and the cancellations, at which the air pressure is atmospheric are the *nodes* of the standing wave.

Resonance occurs when the air inside the tube vibrates with maximum amplitude. For the air inside the tube to vibrate with maximum amplitude, it can be shown that the standing wave has to have an air pressure maximum at the closed end of the tube (because at the closed end, the air molecules are not freely displaced), and atmospheric pressure at the open end of the tube (because at the open end, the air molecules distribute themselves much as they do outside the tube). In other words, a resonance occurs when there is an air pressure antinode at the closed end of the tube, and an air pressure node at the open end.

The lowest frequency sinusoids whose wavelengths conform to this distribution of air pressure variations in this straight-sided tube open at one end and closed at the other are shown in Figure 3.7: these have wavelengths of 4 (top) and 4/3 (centre) times the tube length, which is assumed to be 17.5 cm. Therefore, for the tube of this length, the wavelengths are $4 \times 17.5 = 70$ cm and $4 \times 17.5/3 \approx 23.3$ cm. From Equation 3.1, the corresponding frequencies (in Hertz) are $35000/70 = 500$ Hz and $35000/23.3 \approx 1500$ Hz (assuming the speed of sound is 35000 cm/s). More generally, the resonances for a lossless straight-sided tube open at one end and closed at the other occur at multiples of quarter-wavelengths of the tube, that is, at:

$$F_n = \frac{(2n-1)c}{4l} \quad \text{Hz}, \tag{3.2}$$

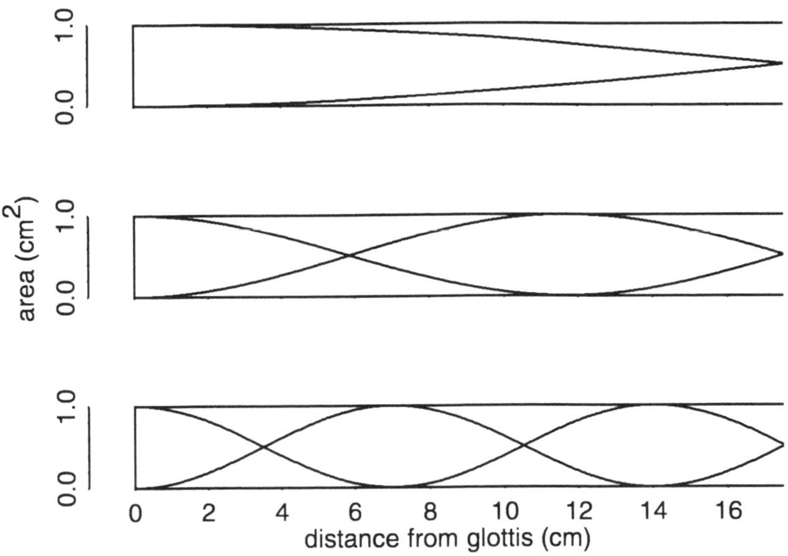

Figure 3.7: Distribution of the standing wave of air pressure in a straight-sided tube of length 17.5 cm used to model a central vowel. Resonances occur when there is a node, or atmospheric air-pressure at the open end of the tube and an antinode, or an air pressure peak at the closed end of the tube. This condition is met for sinusoids whose wavelengths are a 4 (*top*), 4/3 (*centre*) and 4/5 times the length of the vocal tract (adapted from Chiba & Kajiyama, 1941).

where F_n is the nth formant, c is the speed of sound in air, and l is the total length of the vocal tract (therefore the resonances occur at 500 Hz, 1500 Hz, 2500 Hz and at successive intervals of 1000 Hz).

The simple uniform tube can be used to model the effects on the formant frequencies of vocal tracts of different total lengths. Consider, for example, a typical female vocal tract of total length l = 14.5 cm. In this case, the first resonance is calculated as

$$\begin{aligned} F_1 &= \frac{(2-1) \times 35000}{4 \times 14.5} \\ &\approx 603 \quad \text{Hz.} \end{aligned}$$

For this vocal tract length, F_2 and F_3 are at approximately 1810 Hz and 3017 Hz: that is, all formant frequencies are raised compared with those produced from the longer vocal tract. This simple model shows that since female vocal tracts are usually shorter than those of males, their formant frequencies are almost always higher in the production of the same phonetic vowel.

It can also be shown from acoustic theory that when the straight-sided tube is constricted at a node, the corresponding resonance frequency is decreased, and when a constriction occurs at an antinode there is an increase in the associated

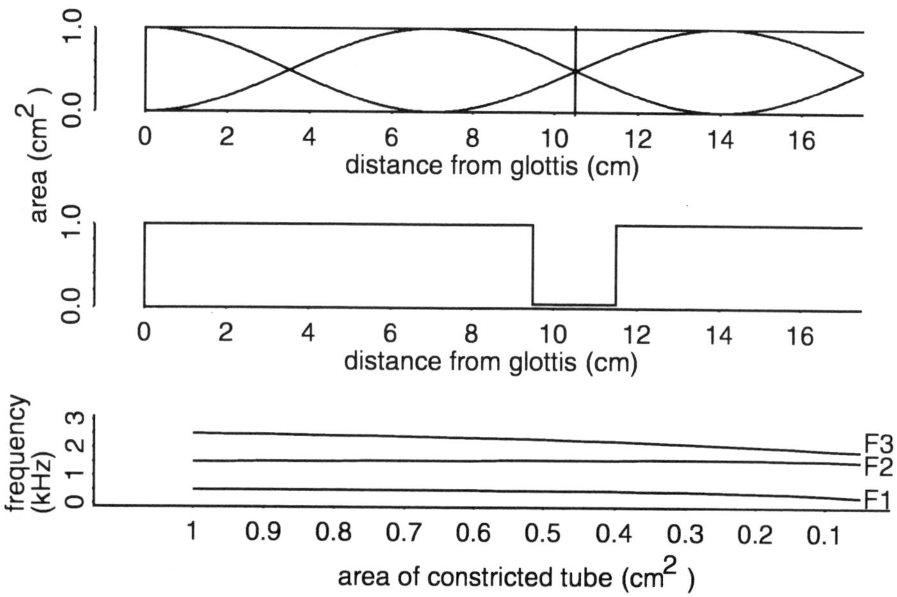

Figure 3.8: The figure shows the effects on formants of a progressive narrowing at the second node from the glottis associated with F3 — such a constriction often occurs in the production of [ɹ] or [ɝ]. *Top panel*: the distribution of the air pressure standing wave associated with F3. *Middle panel*: a constriction of length 2 cm and area 0.05 cm^2 at the second node. *Bottom panel*: the formants were calculated by progressively decreasing the area of the constricted tube in 10 equal steps from 1 cm^2 (that is, no constriction) to the maximum constriction shown in the middle panel. The largest shift in the formants occurs for F3 from its value for a straight-sided tube at 2500 Hz at the left of the plot to around 1800 Hz when the tube is maximally constricted.

resonance frequency (e.g., Chiba & Kajiyama, 1941; Fant, 1973). For example, since there must always be a node at the lip-end of the tube (Figure 3.7), a constriction at the lips caused by lip-rounding or the production of a bilabial consonant results in a lowering of all formant frequencies. A constriction in the velar region, as in the production of a back vowel such as [u], coincides approximately with the node associated with F2 in a straight-sided tube nearest the glottis (Figure 3.7). Consequently, F2 of back vowels is lowered relative to the 1500 Hz resonance in the model for the central vowel.

Figure 3.8 shows the effects on the third formant frequency of a constriction at a node associated with F3 that occurs about two-thirds of the way along the vocal tract from the glottis. This configuration is appropriate for modelling either [ɹ] in English "red" or [ɝ] in many American English productions of "bird" which are often produced with a constriction at the hard-palate. The bottom panel of Figure 3.8 shows the effect on the formants of a progressive narrowing of the constriction (of length 2 cm) which is centred at the node at just over 10.5 cm from the glottis associated with F3. As the bottom panel shows, as the area of this constricted tube is decreased, F3 is progressively lowered and moves closer to F2. Many acoustic studies have shown that a low F3 is indeed characteristic of [ɹ] or [ɝ] (for example, Lehiste, 1964).

Four-tube, three-parameter model for vowels

Once we go beyond a neutral vowel, we have to consider more complicated models than a straight-sided tube of uniform cross-sectional area. While two-tube models can be used for a fairly crude approximation to the formant frequencies of some peripheral vowels, the salient differences in the acoustic structure of vowels only emerge by considering more complicated articulatory-to-acoustic models of speech production. At the same time, X-ray tracings of the vocal tract show that there are small variations in cross-sectional area that are either specific to a particular talker or that are not relevant to the vowel's acoustic-phonetic description. A principal aim of research in this area has been to reduce the detailed shape of the vocal tract to a small number of articulatory parameters that model the essential characteristics of articulatory-to-acoustic relationships in vowels. One of the most influential models of this kind is Fant's (1960) three-parameter model of vowel production (see also the three-parameter model in Stevens & House, 1955) in which the vocal tract is represented by four interconnecting cylinders.

The first tube to be considered models the vowel's *constriction location*. The X-ray analysis of vowel production in Fant (1960), as well as various other studies (e.g., Wood, 1979), have shown that vowel production is accompanied by a narrowing or constriction in the vocal tract, in much the same way that the vocal tract is constricted at a place of articulation in consonant production. From the point of view of the articulatory-to-acoustic mapping, this constriction can be represented by a single tube of fixed area (assuming, therefore, that the diameter of the constriction does not vary appreciably).

Having identified the constriction, there will be a cavity behind it, extending from the glottis to the constriction, and a cavity in front of it extending from the front of the constriction to the lips. If we model the cavities with one tube each, the vocal tract has now been divided into three cylinders representing the back cavity, the constriction, and the front cavity. Finally, a fourth tube is needed to model the configuration of the lips, which is an important articulatory parameter in vowel production.

The principal articulatory variations in vowels of different quality can now be explained in terms of three parameters that have consequences for the areas and shapes of these four tubes. First, vowels differ in the *horizontal distance* of the constricted tube from the glottis. In [i] vowels, the constricted tube is nearer the front of the vocal tract; in [u] the constriction is close to the soft-palate; and the open vowel [ɑ] is produced with a constriction in the pharynx. Second, vowels differ in the diameter and therefore the *area of the constriction*. For example, in producing [i], the constriction is narrower than in the production

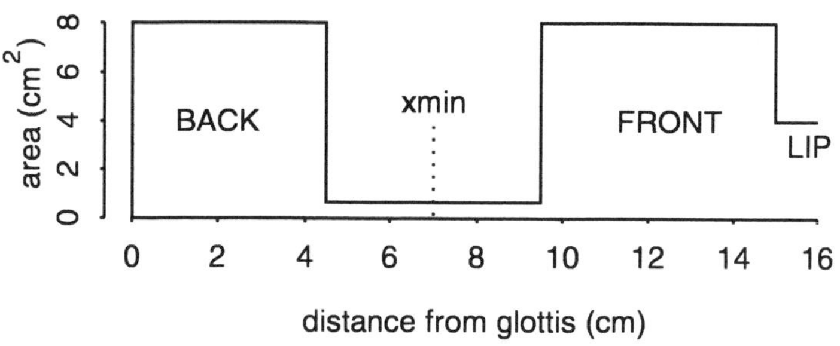

Figure 3.9: One of the models used in Fant (1960) to predict resonance frequencies for vowels. *xmin* is the constriction coordinate.

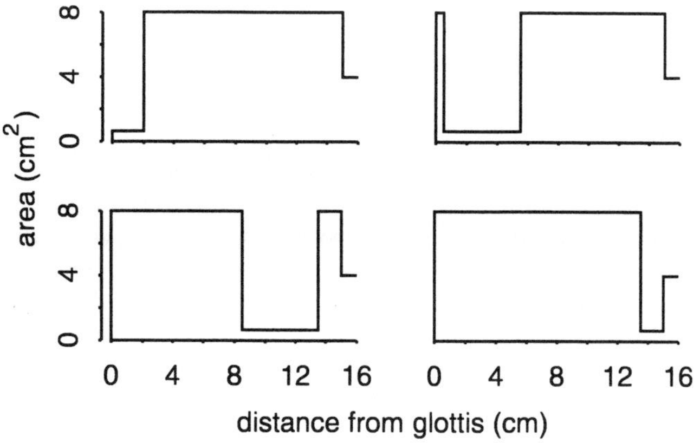

Figure 3.10: The four-tube model at four different values of *xmin*. At extreme *xmin* values (*top left* and *bottom right*), the back or front tubes disappear completely to maintain a constant overall tract length.

of [ɛ] (Southern British English, "head"). Third, vowels differ in the *extent of lip-opening* (compare, for example, [ɔ], which has rounded lips and therefore a narrow lip-opening with [æ] in which the lips are open and unrounded).

In Fant's (1960) *Acoustic Theory of Speech Production*, the effect of moving the constricted tube horizontally from the glottis to the lips (first parameter) is examined by calculating the resonances of the four-tube model at each stage. Specifically, in one type of analysis, the model shown in Figure 3.9 is used for this purpose. The areas of all four tubes are fixed at 8 cm^2, 0.65 cm^2 (the constricted tube), 8 cm^2, and 4 cm^2 (the lip tube). The lengths of the constricted tube and lip tube are fixed at 5 cm and 1 cm respectively while the lengths of the back and front tubes vary with the horizontal position to maintain a total vocal tract length of 16 cm.

The resulting plots are often arranged in the form of *nomograms* that show

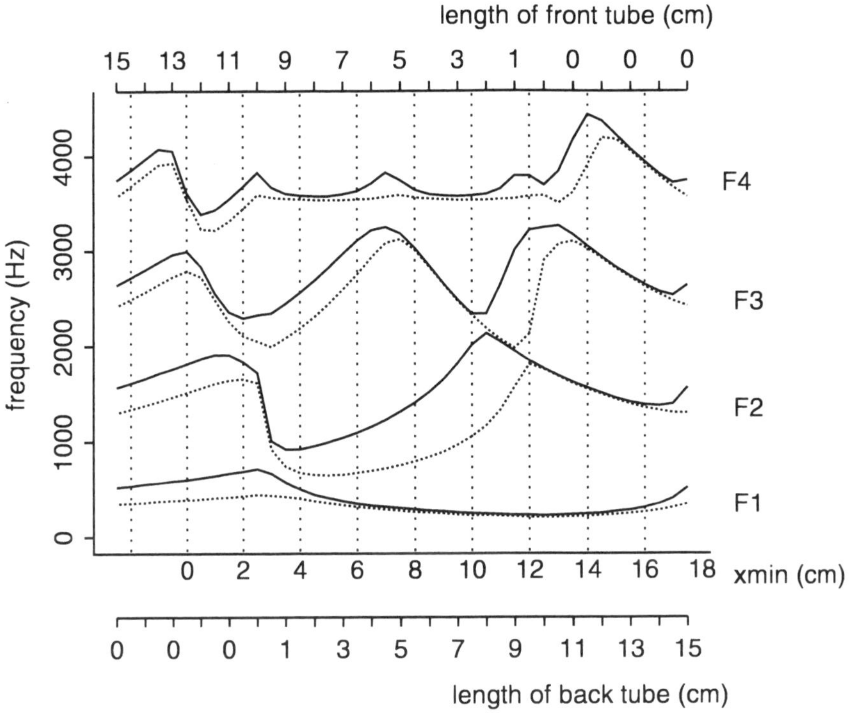

Figure 3.11: Nomograms for incremental values of *xmin* from the glottis to the lips. The top/bottom scales show the lengths of the front/back tubes for different values of *xmin*. *Solid line*: lip tube area of 4 cm^2 (unrounded vowels). *Dotted line*: lip tube area of 0.65 cm^2 (rounded vowels).

how the resonances change for different values of the constriction coordinate *xmin* and its associated front and back cavity lengths (that are predictable from *xmin*). A nomogram plot is shown in Figure 3.11. These resonances were calculated from a four-tube lossless model with the parameters set as in Figure 3.9 by incrementing *xmin* in steps of 0.5 cm between the glottis and the lips.

The nomogram shows some of the important effects on formant frequencies of varying the horizontal location of the constricted tube. The most systematic variation is in the first two formant frequencies. When *xmin* is just under 11 cm from the glottis, F1 and F2 are far apart and F2 is close to F3 — this corresponds approximately to the production of the front vowel [i]. In front of this constriction location — toward the lips, there is a progressive increase in the F3 centre frequency (up to about 13 cm) and then in F4 (from 13 cm to 15 cm) so that there is almost a continual rise first in F2, then F3, then F4 from an *xmin* value of about 10 cm to the front of the vocal tract. As discussed in Ladefoged (1985), this increase in frequency toward the lips can be attributed to the shortening of the front tube as the constriction moves forward in the

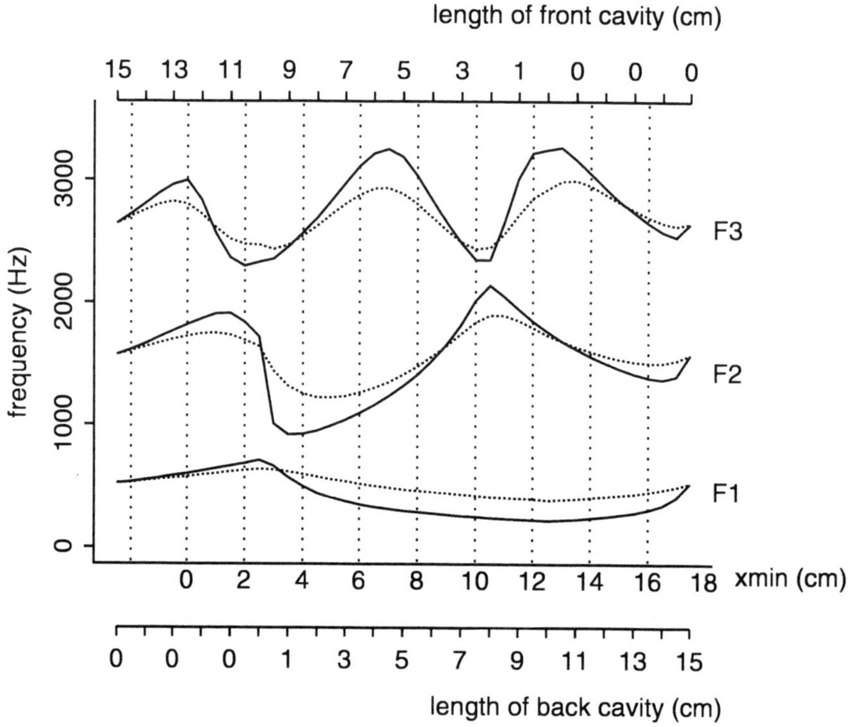

Figure 3.12: Solid line: constriction coordinate area of 0.65 cm^2; dotted line: constriction coordinate area of 2.65 cm^2.

mouth.

As *xmin* moves closer to the glottis, F1 and F2 converge while F2 and F3 diverge. Therefore, vowels that are produced with a constriction further back in the vocal tract (e.g., "hoard") have F1 and F2 closer together and F2 and F3 further apart.

The same chart also shows the effects of lip-rounding (dotted lines), which is simulated with the same parameters as before, but with a smaller lip area — specifically 0.65 cm^2 as opposed to 4 cm^2. The main effect of lip-rounding is to lower the second formant frequency for most constriction locations at which vowels are produced (between *xmin* values of 3 cm and 12 cm). Thus changing the lip configuration from an unrounded position (as for "heed") to a rounded position while keeping the tongue position constant (as for French "lune") is accompanied by a lowering of F2.

The final parameter to be considered is the area of the constriction itself. In the nomogram considered so far, the constriction was fixed at a narrow area of 0.65 cm^2. In Figure 3.12, the parameters are the same as those for Figure 3.11, but the area of the constriction is increased to 2.65 cm^2, corresponding to the production of a more open vowel. As Figure 3.12 shows, one of the main effects of increasing the constriction area is to raise the first formant frequency: it confirms therefore, that phonetically more open vowels (e.g., [æ] in "had") have a higher F1 value compared with that of their closer counterparts ([ɛ] in "head").

In a study by Ladefoged and Bladon (1982), a comparison was made between Fant's nomograms and real vowels in an attempt to copy the movement of the maximum point of constriction from the front to the back of the mouth. Their general conclusion was there was a good correspondence between nomograms derived from tube models and the formants of real vowel productions but with two main discrepancies. First, contrary to the data from the nomograms, there was little evidence that F2 decreases in frequency in real vowel productions in the region of [i] when $xmin$ is roughly 10 to 14 cm from the glottis. A possible explanation for this discrepancy is that the simulated configurations in the region appropriate for [i] have an unrealistically large lip-aperture; another is that it is unlikely that the back cavity behind the point of maximum constriction fills up the space as the constriction tube moves toward the alveolar ridge to maintain a constant vocal tract length. Second, they note some differences in the effects of rounding between the simulated data in the nomograms and the real vowels: in particular, in real [i] vowels (corresponding to $xmin \approx 11$ cm), there is a greater lowering of F3 than suggested by the nomograms (compare the solid with the dotted line), and in human back rounded vowels ($xmin \approx 4$ cm), there is a pronounced change in F2 but very little change in F3 relative to their unrounded counterparts. In their articulatory and acoustic study vowels, Lindblom and Sundberg (1971) also confirm that F3 lowering is most pronounced in front vowels and F2 lowering in back vowels when these are rounded.

The relationship between an articulatory tube model of the vocal tract and the acoustic signal are central to many research areas and theories in speech analysis including the influential *quantal theory of speech production* that has been developed by Stevens (Stevens, 1972b, 1989) over a number of years. Stevens has proposed that there are regions of the vocal tract for which a large change in constriction location results in relatively small change to the acoustic signal. One of these quantal regions is at the constriction location appropriate for [i] (see also Perkell & Nelson, 1985). Simulations using tube models of the vocal tract can be used to show that, if the constriction location is advanced or retracted slightly in a quantal region, the formants change minimally. This implies that even if talkers are imprecise in positioning their tongue horizontally in an [i] vowel (for example, if the tongue dorsum is advanced due to the coarticulatory effects of neighbouring alveolar consonants), there will nevertheless be a minimal perturbation to the formant frequencies and consequently to the listener's recognition of the vowel as [i] (see also Beckman et al., 1995, for a recent analysis of the quantal theory using speech movement data and Wood, 1979, for evidence from area functions derived from X-ray tracings of naturally produced vowels). Some evidence for the relative stability of the first two formant frequencies in back rounded vowels is shown in Figure 3.13. In this case, a high back rounded vowel is modelled using four-tubes with a narrow tongue-constriction and a narrow lip-aperture (Stevens, 1989). The formant frequencies

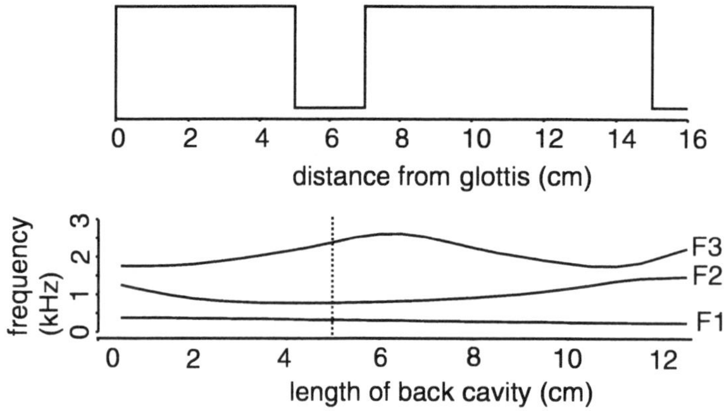

Figure 3.13: *Top*: A four-tube model in Stevens (1989) used to predict the formant frequencies of a high back rounded vowel such as [u]. *Bottom*: The corresponding F1 to F3 values for the above tube configuration are at the vertical dotted line. The formant trajectories were obtained by incrementing the distance of the constricted tube between the glottis and the lips (producing a progressive increase in the back cavity length). The plot shows that F1 and F2 are close together and low in frequency for a wide range of back cavity lengths (adapted from Stevens, 1989).

in this figure were obtained by moving the constriction from the glottis toward the lips in equal increments and then calculating the resonances at successive stages. Figure 3.13 shows that there is a considerable range of back cavity lengths (from almost 2 to 8 cm) for which F1 and F2 are low and close together in frequency. Therefore, quite large variations in the horizontal position of the constriction (and therefore of the length of the back cavity) for this modelled high back rounded vowel result in comparatively stable F1 and F2 frequencies.

There are good reasons for languages to prefer quantal vowels, not only because formants change minimally in quantal regions but also because there are very clear and marked acoustic differences as the constriction location moves between *different* quantal vowels. This is apparent in the formant changes in the nomogram in Figure 3.11 when the constriction location changes from a configuration appropriate for [i] at $xmin \approx 10.5$ cm to that of another quantal vowel [u] for which $xmin \approx 4$ cm. Therefore, since quantal vowels are distinctive acoustically, they are less likely to be confused by listeners (see also Liljencrants & Lindblom, 1972, and Boë, Schwartz, & Vallée, 1994, for evidence that [i u ɑ] are perceptually maximally contrastive and therefore preferred in languages).

3.3.2 The acoustic filter for some consonant classes

Vowels can be described as sounds in which there is a single path through the vocal tract from the sound source at the vocal folds to the opening at the lips.

As we have seen, an appropriate model for such sounds is a set of interconnecting cylinders that are closed at one end (the source) and open at the other (the lips). A physical property of systems in which there is a single acoustic path from the source to the opening is that they are uniquely characterised by resonances. Essentially, this means that if we know the resonance frequencies of a vowel filter, the rest of the shape of the spectrum, including the dips between the resonance peaks, is entirely predictable from the frequency location of their resonances (Fant, 1980).

However, this is not so for fricative and nasal sounds. The filter spectrum for these classes of sounds is characterised by both formants and antiformants. Anti-formants produce dips in the filter spectrum, and they are introduced whenever there is more than one acoustic path from the source to the mouth opening. In the case of nasal consonants, two acoustic paths are formed, from the glottis into the nasal cavity and from the glottis into the oral cavity as a result of the opening of the velar port. In fricatives, the source is not at the glottis but near the point of maximum constriction. There are therefore once again multiple acoustic paths, from the source to the lips and also from the source to the back cavity behind the point of closest constriction.

We shall consider briefly the acoustic characteristics of these two major classes in turn.

Fricatives

The relationship between vocal tract shapes for fricatives and the resulting speech waveform is much less well understood than for vowels. This is for at least three reasons. First, the fact that the noise source can be located in different parts of the vocal tract, as well as the presence of antiresonances in the spectrum are considerable complicating factors in predicting the articulatory-to-acoustic mappings. Second, there is a limited resource of available X-ray data to validate the models that are used to predict the acoustic consequences of modelled shapes of the vocal tract. Third, since fricatives can be synthesised adequately without explicitly modelling the effects of antiresonances and since fricatives can also be classified quite accurately from the frequency location and amplitudes of resonances alone (but see Fujisaki & Kunisaki, 1976), the association between cavities formed in fricative production and the resulting spectral shape has assumed much less importance than for vowels.

There is some agreement from early studies in the literature (Fant, 1960; Flanagan, 1972; Heinz & Stevens, 1961) that the *length of the front cavity* can account for many of the spectral differences between fricatives of different places of articulation. The production of [s] is often modelled using three cavities: a constriction channel formed in the narrow groove between the tongue-tip and the alveolar ridge that divides the rest of the vocal tract into cavities in front of (front cavity) and behind (back cavity) the constriction channel itself. Shorter front cavities result in higher resonance frequencies, which appropriately predicts that the dominant energy for [s] is at a higher frequency than that of [ʃ]: the resonance due to the front cavity has been estimated between 4000 Hz and

7000 Hz for [s] depending on the assumed length of the front cavity and generally around 2000 to 4000 Hz for [ʃ]. However, as stated earlier, it is difficult to predict accurately the resonance frequencies not only because of the lack of X-ray data, but also because there are likely to be marked speaker effects (in particular, since female talkers have shorter front cavities their front cavity resonances are appreciably higher than those of male talkers). Furthermore, as Badin and Fant (1987) and Badin (1991) show, the palato-alveolar [ʃ] represents a particularly complicated case because there is likely to be considerable coupling between the back and front cavities (that is, the resonances and antiresonances are probably determined by the *combined* effects of these cavities). With regard to the nonsibilants [f] and [θ], these have shorter front cavities, which would suggest that they have even higher main resonances than those of the sibilants. In fact, the extreme shortness of the front cavity, combined with relatively open lips produces such high energy losses that their spectra are relatively *diffuse* (that is, there are no appreciable resonant peaks) and this is one of the features than can sometimes be used to differentiate them acoustically from the sibilants (see Chapter 4).

Our discussion so far has focused on the influence of the front cavity partly because the back cavity is presumed to have much less influence on the spectrum. It can be shown (see Fant, 1960 and Heinz & Stevens, 1961) that because the noise source is located in the constriction channel, and because the channel is fairly narrow, the back cavity produces resonances and antiresonances below about 4000 Hz that are quite close together in frequency. Since they almost coincide, they tend to nullify each other which can be used to explain the lack of appreciable energy in the sibilant fricatives in the low part of the spectrum.

Finally, some of the first studies on articulatory-to-acoustic mappings of fricatives by Fant (1960) and Heinz and Stevens (1961) also point to the possibility that the constriction channel itself can contribute an antiresonance and a resonance at different frequencies in the production of [s]. The antiresonance is calculated from a quarter-wavelength resonator of the constriction channel and is estimated at 3500 Hz, while the constriction channel resonance is presumed to originate from a half-wavelength resonator and occurs at almost double this frequency. These features would in any case accentuate some of the acoustic characteristics of [s] that have already been discussed that is, a lack of energy in the low part of the spectrum and clear evidence of resonances above 4 to 5 kHz.

Nasal consonants

Nasal consonants and nasalised vowels are very difficult to model accurately using compound cylindrical models and their acoustic characteristics are perhaps the least well understood of all classes of speech sounds. One method of modelling nasal consonants is to assume there are three tubes, one for each of the nasal cavity, oral cavity (closed at the place of articulation of the nasal consonant), and the pharyngeal cavity (Figure 3.14). The tube of the oral cavity is longest in the production of the bilabial [m] but then progressively shortens as

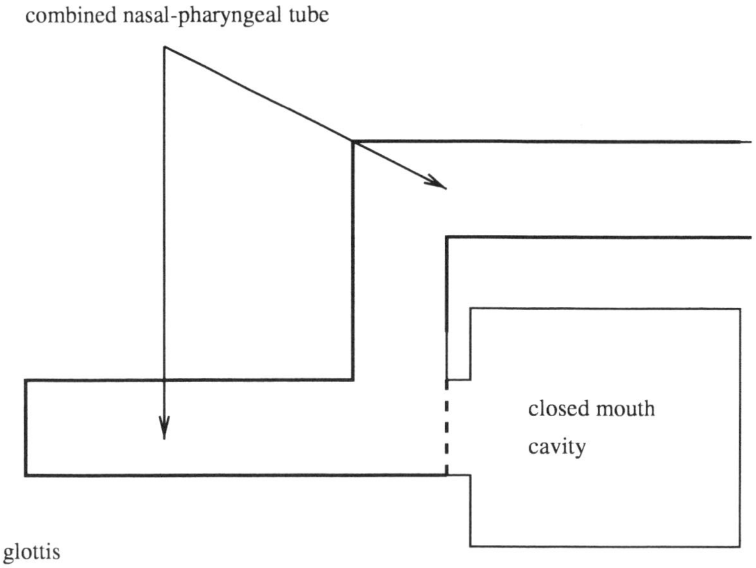

Figure 3.14: Three-cylinder model for nasal consonants. The combined nasal-pharyngeal tube contributes the nasal formants to the spectrum, while the (oral) antiformants are the result of the sealed mouth cavity, which acts as a side-branching resonator to the main tube.

the place of articulation moves towards the uvula.

The spectra of nasal consonants are characterised by *nasal formants* (labelled N1, N2, N3,...) which are due to the combined nasal-pharyngeal tube. For nasal and pharyngeal tubes with lengths typical of those of an adult male vocal tract, the first nasal formant, N1, is calculated to occur in the 300 to 400 Hz region; higher nasal formants occur approximately at 800 Hz intervals (Fant, 1960; Flanagan, 1972).

In uvular and post-velar articulations, the mouth cavity is effectively cut off from the nasal-pharyngeal tube and therefore has little effect on the resulting spectrum: for nasals produced with the tongue at the far back of the mouth, the spectrum is determined almost entirely by the nasal formants discussed above. However, when the tongue articulation is further forward in the mouth, as in the production of palatal, alveolar, or bilabial nasals, the oral cavity acts as a *side-branching resonator* to the main nasal-pharyngeal tube and introduces (oral) *antiformants* into the spectrum. The first antiformant frequency occurs at a quarter-wavelength of the mouth cavity tube that is, at $c/(4l_m)$ Hz, where l_m is the length of the side-branching mouth cavity. This implies that the frequency of the first anti-formant varies inversely with the length of the oral tube, being lowest for bilabial [m], higher for [n], and highest for [ŋ]. Assuming a mouth cavity length of 6.5 cm in bilabial productions, the first antiformant for

[m] is calculated at 35200/(4 × 6.5) Hz or approximately 1350 Hz: it therefore almost coincides with the second nasal formant due to the combined nasal-pharyngeal tube at 1200 Hz. In general, the effect of the introduction of oral antiformants is both to "flatten" the spectrum, particularly if nasal resonances and oral anti-formants coincide, and to lower its amplitude. In spectrograms, nasal consonants typically show overall amplitude dips and (nasal) formants which are very low in amplitude.

A further complication that must be considered in the production of nasal consonants is that *oral* formants, as well as antiformants, can be introduced into the spectrum. However, since these are likely to be close to the oral anti-formants in frequency, they are usually very low in amplitude.

3.4 Vocal tract losses

The calculations of resonances have been made under the assumption that the cylinders used to model the vocal tract are lossless. In reality, there are considerable losses to the acoustic energy in speech sound production that are caused by various factors (Badin & Fant, 1984; Fant, 1972; Flanagan, 1972; Wakita & Fant, 1978). It is beyond the scope of this chapter to discuss these in detail, but a general indication of their effects can be given. In all cases, it should be noted that when resonances are calculated from a lossless model of the vocal tract, they have zero bandwidths (which implies that the resonances have infinite amplitude). One of the main consequences of including losses is to broaden the bandwidths and also to shift the values of the resonance centre frequencies. Three main types of losses can be considered.

First, the cylinders are modelled under the assumption that they are hard-walled and smooth. In the case of the vocal tract, however, energy losses occur due to the vibration of the cavity walls. This type of loss causes a slight raising of the formant centre frequencies and a broadening of the formant bandwidths and affects primarily the formants in the frequency range below 2 kHz (Wakita & Fant, 1978; see also Rabiner & Schafer, 1978). Fant (1985) proposes the following correction that can be applied to the resonances calculated from a lossless model:

$$R_c = \sqrt{R^2 + R_w^2}, \tag{3.3}$$

where R_c is the corrected resonance, R is the resonance calculated from the lossless (hard-walled) tube, and R_w is a constant equal to 190 Hz. A formula for modifying the bandwidths is proposed in the same paper.

Second, losses in the vocal tract also occur due to viscous friction and heat conduction as the air travels through the vocal tract. The effect of these losses is small however compared with the losses produced by wall vibration: for oral sounds they can be disregarded below 3 to 4 kHz. On the other hand, viscous friction and heat conduction losses are considerably greater in nasal sounds (Flanagan, 1972) due to the properties of the nasal tract resulting in greater bandwidths for nasal than oral resonances.

Third, energy losses result when the acoustic energy radiates from the lips

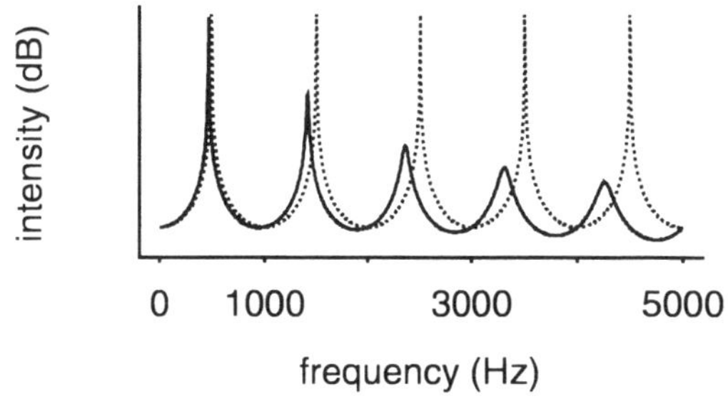

Figure 3.15: *Dotted resonance curve*: Resonances for a lossless single-tubed model of length 17.5 cm. *Solid resonance curve*: The same as the dotted one but including losses due to lip-radiation effects.

and nostrils in sound production. This type of loss is often estimated by modelling the mouth as a circular aperture in a sphere (the sphere represents the head) and it has the greatest effect of all types of losses on the resonance centre frequencies and bandwidths. Generally, this type of loss causes a lowering of the resonance centre frequencies and an increase in the bandwidths; furthermore these two effects are progressively more pronounced at higher frequencies. An example of how energy losses due to acoustic radiation from the mouth affects formant resonances and bandwidths is shown in Figure 3.15. In this figure, the resonance curve for a straight-sided tube of length 17.5 cm and constant cross-sectional area is calculated from a lossless model (this is the tube representing the central vowel discussed in Section 3.3.1); a second resonance curve is then calculated for the same vocal tract model but taking into account energy losses due to the radiation at the lips using the formula proposed by Wakita and Fant (1978). As the Figure shows, the bandwidths are progressively increased and the resonance centre frequencies are progressively lowered with increasing frequency in the model that includes losses due to lip radiation.

3.5 Radiated sound pressure

Although we have not shown the detailed mathematical analysis, the resonance curve for any vocal tract model represented as cylindrical sections is derived by calculating how the *volume-velocity of airflow* at the glottis is transformed at each cylindrical junction until the final transformation is obtained at the cylinder representing the lips. There is then a direct relationship between the transformed volume-velocity of airflow at the lips and the resonance curves of the vocal tract filter that we have shown in the preceding sections. In recording speech, what is measured however is not the volume velocity of airflow at the lips, but the acoustic speech pressure waveform at a distance from the lips (using a pressure-sensitive microphone). We must therefore consider the final correction that must be made to the resonance curve to take into account

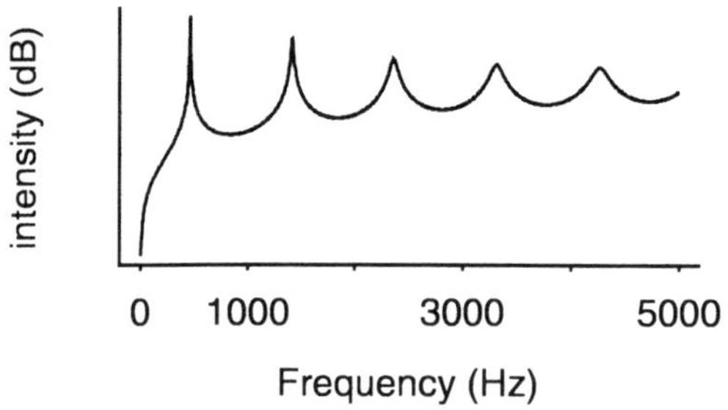

Figure 3.16: The 6 dB/octave rise when the spectrum is adjusted to include the effects of radiated sound pressure (compare with this the solid resonance curve in Figure 3.15).

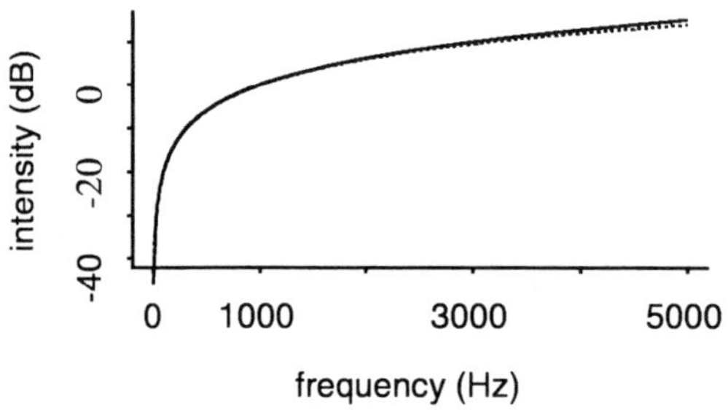

Figure 3.17: The difference between the two solid resonance curves in Figure 3.16 and Figure 3.15 is nearly 6 dB/octave (the dotted line is exactly a 6 dB/octave rise).

this transformation from volume-velocity to air pressure at a distance from the lips. Fortunately, the transformation is simple and produces approximately a 6 dB/octave rise to the spectrum: in other words, the amplitude of the resonance curve is progressively boosted at higher frequencies.

Figure 3.16 shows the modification to the solid resonance curve in Figure 3.15 when the effect of radiated sound pressure is taken into account. This effect can be most clearly demonstrated by subtracting the original solid line resonance curve in Figure 3.15 from the new resonance curve in Figure 3.16: the result of this subtraction, which is shown in Figure 3.17, confirms that the effect of taking account of radiated sound pressure is to boost the spectrum by approximately 6 dB for every doubling of frequency.

A further detailed investigation of the effects of sound radiation about the head is given in Flanagan (1972, pp.34–36).

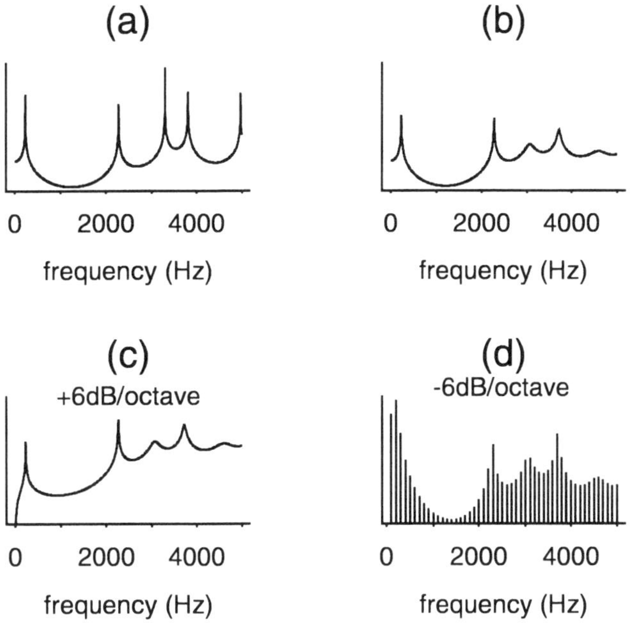

Figure 3.18: *(a)*: The filter spectrum based on a lossless tube model of the area function shown in Figure 3.6. *(b)*: The same, but including the effects of losses caused by radiation of sound from the lips. *(c)*: The further effect of including the 6 dB/octave rise as a result of radiated sound pressure. *(d)*: The combination of the filter with a -12 dB/octave source of fundamental frequency 100 Hz.

3.6 The composite model

In this chapter we have decomposed sounds into a number of speech production processes which, to a large extent, make an independent contribution to the shape of the sound's spectrum. Figure 3.18 is a summary of these processes and shows the final result of their combination for a spectrum modelled on the area function for a Russian [i] vowel derived from an X-ray photograph in Fant (1960). The area function is represented as 33 sections each of lengths 0.5 cm (see Fant, 1960, pp.115).

Figure 3.18(a) shows the calculated spectrum of the filter assuming no losses in the tubes. The filter spectrum shows the location of the formant peaks and, since the calculations are based on a lossless model, the bandwidths are zero (in theory, therefore, the formant peaks have infinite amplitude). The modification of the filter to take account of one type of loss, due to radiation of the acoustic energy from the lips, is shown in the top right panel. With the incorporation of this energy loss, the bandwidths are no longer zero, they also become increasingly broader with increasing frequency, and the formants are shifted slightly downwards in frequency. The second modification (bottom left panel) takes account of the radiated sound pressure waveform, which produces a 6 dB/octave boost to the spectrum.

The final panel shows the effect of combining this filter (with losses and the 6 dB/octave radiation effect included) with a source spectrum due to the vibrating vocal folds. If the vocal folds are vibrating at 100 Hz, then the harmonics are spaced at 100 Hz intervals, and it also a property of the source spectrum that it slopes downwards at approximately 12 dB/octave (see Figure 3.5). When the source spectrum is combined with the filter to produce the spectrum of the sound, peaks occur at those harmonics that are closest to the formant centre frequencies of the filter. The combination of the -12 dB trend caused by the source spectrum and the +6 dB trend due to the radiated speech pressure waveform produces a net downward sloping spectrum of -6 dB/octave.

The principal for obtaining the spectra of nasal and fricative consonants is very much the same that is, they can also be decomposed into the main processes discussed in this chapter. A key difference, however, is that, particularly in the case of nasal consonants, but also for many fricatives, the spectrum of the filter is characterised by both formants and antiformants.

Further reading

Much of the material reviewed in this chapter is based on Fant (1960) and Flanagan (1972). An extensive discussion on the quantal theory of speech production is in a special edition of volume 17 of the *Journal of Phonetics* (1989).

See also Lindblom and Sundberg (1971), Harshman, Ladefoged, and Goldstein (1977), Ladefoged, Harshman, Goldstein, and Rice (1978), and Maeda (1990) for different kinds of vocal tract parameterisations and their relationship to the acoustic speech signal.

Computer programs for calculating resonance frequencies from cylindrical models of the vocal tract are given in Fant (1985) and Lin (1992).

4
SEGMENTAL AND PROSODIC CUES

In this chapter, our concern is with the different kinds of cues that can be used to identify speech sounds from the acoustic waveform. As discussed Chapter 1, the majority of studies in this field were carried out after the invention of the sound spectrograph in the late 1940s and following the development of speech synthesis techniques and their application to speech perception studies.

Much of the research in this area models the relationship between acoustic waveform and speech sound units in speech perception research. Since sounds have often been analyzed acoustically using a fairly restricted speech corpus (citation-form speech produced by a small number of speakers), the acoustic cues carry over only indirectly to a system that is designed to recognise continuous speech from a large number of talkers. On the other hand, the knowledge that acoustic phonetic studies provides has been directly incorporated into many speech synthesis systems, including the successful MITtalk system, which is discussed in Chapter 7.

The review in this chapter is divided into three major sections: acoustic cues to vowels, consonants, and some prosodic units. In most cases, we begin by identifying the major characteristics of the sounds from spectrograms and then focusing on the different kinds of parameterisations of the signal that can enhance the separation of linguistic-phonetic units within these three major categories.

4.1 Vowels

Vowels can be identified from wideband spectrograms by the presence of a formant structure, which often shows transitions at the vowel margins, and by vertical striations characteristic of periodic vocal fold vibration. Figure 4.1 shows a spectrogram of the utterance "Is this seesaw safe?" in which the vowels are segmented from their neighbouring fricatives. Unstressed vowels in function words and schwa vowels can be a lot more difficult to segment in acoustic data both because they are inherently shorter in duration and because they are influenced a good deal more by the phonetic context effects.

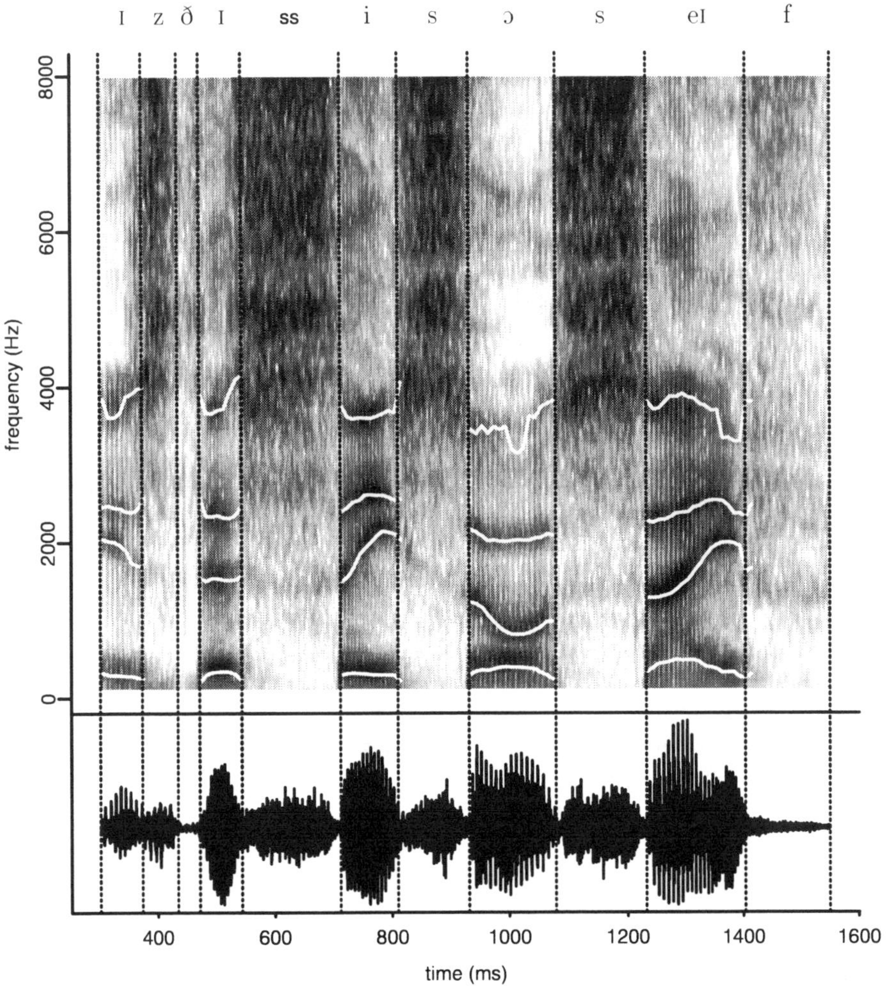

Figure 4.1: Wideband (300 Hz filter) spectrogram of the utterance "Is this seesaw safe?" showing automatically tracked formants.

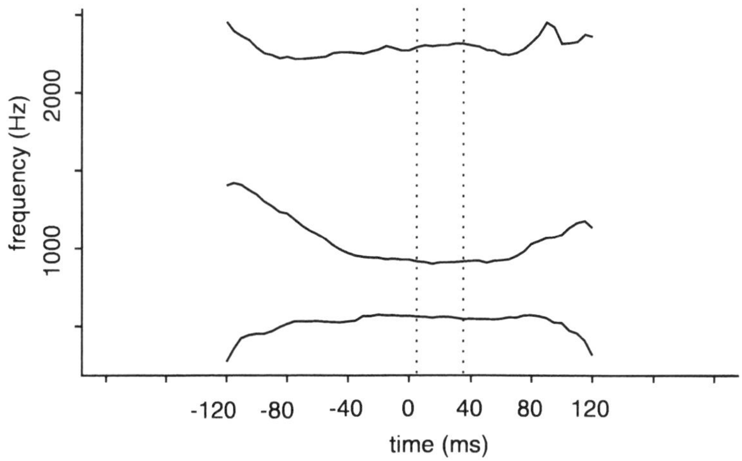

Figure 4.2: First three formant frequencies and the estimated vowel target (interval between the dotted lines) based on the section of the vowel for which there is least change to the formant frequencies.

4.1.1 Vowel targets

In classifying vowels from acoustic data, a common practice is to begin by locating a section of the vowel known as the *vowel target* or *acoustic vowel target*. The acoustic vowel target, which in monophthongs typically occurs near the vowel's temporal midpoint, is presumed both to be the section of the vowel that is influenced least by context effects and to be relatively steady-state (that is, unchanging). The acoustic vowel target is very often defined to be an entire *section* of the vowel that has either fixed (e.g., Stevens & House, 1963b) or variable (Lehiste & Peterson, 1961b) duration. In other studies, the vowel target is defined as the formant values at a *single time point* (e.g., Di Benedetto, 1989; Lindblom, 1963) rather than over an interval of time.

Figure 4.2 shows F1 to F3 traces for an [ɒ] vowel from "dod" spoken by a male talker. The vowel target is estimated in this case as the section of the vowel corresponding to 25 per cent of its total duration for which the formant movement is least. The algorithm for placing the vowel target is based on finding the section of the vowel for which the Euclidean distance between successive formant values is minimal as in Schouton and Pols (1979a) (see also Broad & Wakita, 1977, for a variation on this algorithm). As many authors have noted (see, e.g., the remarks by Gay, 1978; Nearey & Assmann, 1986; and Benguerel & McFadden, 1989), the concept of a steady-state vowel target is often not substantiated by the data either because many monophthongal vowels have no clearly identifiable steady-state section or else because the steady-state interval is different for each formant (e.g., Di Benedetto, 1989).

Faced with these difficulties, it is not surprising that vowel targets are often estimated using criteria other than the vowel's most steady-state interval, as judged either by eye or automatically from formant data. In some studies (e.g., Stevens & House, 1963b), target values are taken at the vowel's temporal midpoint on the (not unreasonable) assumption that this is usually the most steady-state part of the vowel. In other studies (e.g., Di Benedetto, 1989; Lisker, 1984), the vowel target is estimated from the time at which the first formant reaches its maximum value. The motivation for choosing maximum F1 is related to articulatory-to-acoustic modelling which shows that jaw lowering and a more open vocal tract cause F1 to rise (Lindblom & Sundberg, 1971). In many syllables, jaw lowering can be assumed to increase from the initial syllable margin to the vowel target and then to decrease from the target to the final syllable margin. Therefore, at least for most phonetically open and mid vowels, F1 should in general be in the shape of an inverted parabola whose maximum occurs at the vowel target (Stevens, House, & Paul, 1966; Di Benedetto, 1989). Finally, some studies also use the overall amplitude as an additional source of information for marking the vowel target (Schouten & Pols, 1979a). In many cases, the intensity of the vowel can be expected to increase as the jaw lowers and as the consonantal constriction is released but this parameter is again likely to be primarily characteristic of open and mid vowels because of their comparatively higher intrinsic amplitude (Lehiste & Peterson, 1959).

A detailed comparative study of the different ways of identifying the vowel target is presented in van Son and Pols (1990). In their study of Dutch vowels in continuous read speech, formant frequencies at vowel targets were obtained at the following time points in the vowel: (1) at the stationary (steady-state) section, which was defined as the interval with the least variance in the logarithm of the first three formants, (2) at the midpoint and (3) at formant maxima or minima depending on the vowel type (maxima for open vowels, minima for close vowels). Their results showed that it made little difference which technique was used, and they conclude that "When studying vowel targets, the method that is most convenient can be used" (p. 1692).

4.1.2 The F1/F2 space for vowel classification

There is a long line of research which shows that the first two formants are the most important acoustic parameters for vowel quality distinctions. The relationship between the first two resonances and vowel quality had already been documented in the nineteenth century and early part of the twentieth century by authors such as Willis (1829), Helmholtz (1863), Bell (1879), Paget (1923), and Stewart (1922) (see Ladefoged, 1967, and Traunmüller and Laçerda, 1987, for a historical account). Following the invention of the sound spectrograph in the late 1940s, various spectrographic analyses, one of the most notable being the study by Peterson and Barney (1952) of the vowels of men, women, and children, confirmed the importance of the first two formant frequencies as the basis for vowel quality distinctions (see also a recent extension of the Peterson & Barney, 1952, study by Hillendbrand, Getty, Clark, & Wheeler, 1995). Experiments using synthetic speech at the Haskins Laboratories (Delattre, Liberman,

Cooper, & Gerstman, 1952) demonstrated that vowels of different quality could be adequately synthesised using only the first two, and sometimes only the first formant frequency. The studies of articulatory-to-acoustic mappings discussed in Chapter 3 also show that changes in the (modelled) shape of the supralaryngeal vocal tract for vowels are accompanied by significant differences in the first two formant frequencies (Fant, 1960).

An analysis of listeners' confusions between vowels (e.g., Fox, 1985; Kewley-Port & Atal, 1989; Klein, Plomp, & Pols, 1970; Rakerd & Verbrugge, 1985; Shepard, 1972; Terbeek, 1977) provides evidence of a different kind that the first two formants provide the primary cues to vowel quality. In these experiments, listeners might be presented with three vowels at a time and then asked to judge whether the third vowel sounds more similar to the first or second. Alternatively, vowels might be presented in background noise, and listeners are asked to write down the vowel they think they heard. Experiments such as these produce a matrix of the number of confusions between different vowel types. Subsequently, the technique of *multidimensional scaling* (Carroll & Chang, 1970; Harshman, 1970; Kruskal, 1964; see also Berdan, 1978; Wish & Carroll, 1982) is used to transform the confusion matrix to a spatial representation. Although there are differences in the various scaling techniques, the general aim is to represent the distance between a pair of vowels in the derived space in such a way that it is directly related to the frequency with which they are confused (Shepard, 1972). For example, if listeners often confuse [i] and [ɪ], then these vowels would be represented by points that are close together in the derived space. Multidimensional scaling also gives an indication of how many dimensions are necessary to reflect adequately the confusion matrix and also of the most important dimensions for explaining the nature of the vowel confusions. The results of these experiments show that up to six dimensions can be necessary to explain the nature of the confusions adequately (e.g., Terbeek, 1977). But in almost all such studies, the two most important dimensions that underlie listeners' confusions are closely correlated with the vowels' first two formant frequencies.

Essner (1947) and Joos (1948) were the first to demonstrate the relationship between the F1/F2 space and the *vowel quadrilateral*, or *cardinal vowel space*, devised by Daniel Jones (1917) at University College London. These studies showed that F1 and F2 are negatively correlated with phonetic height and backness respectively.

The emergence of a quadrilateral-like space is shown for averaged tokens of Australian English and Southern British English vowels in Figure 4.3. Displays such as these can often be used to illustrate some of the principal phonetic differences between accent types, such as the more central vowels in "who'd" and "hard" in Australian English compared with their British English Received Pronunciation (RP) counterparts.

4.1.3 Higher formants and the fundamental frequency

The well-known experiments on vowel perception at the Haskins Laboratories (Delattre et al., 1952) mentioned earlier showed that vowel quality judgements

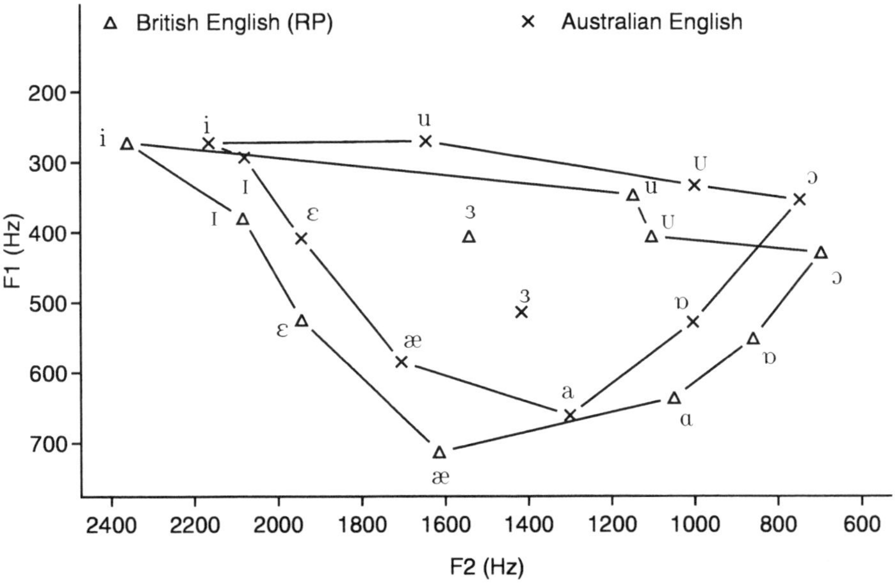

Figure 4.3: A comparison of Southern British English Received Pronunciation (RP) and Australian English vowels. The RP vowels, which are taken from Henton (1983), are averages of ten male talkers' productions of /hVd/ words. The Australian vowels are averages across five male talkers with approximately 22 tokens per talker per vowel type. The display reflects the front vowel raising and the more central vowel in "hard" of Australian English compared with RP (adapted from Harrington & Cassidy, 1994).

can be influenced by the third formant frequency. In their synthesis and labelling experiments, Delattre et al. (1952) found that many front vowels have to be synthesised with an F2 which is somewhat higher in frequency than the F2 in the corresponding natural vowels. Delattre et al. (1952) explain this finding in terms of the influence of F3 and speculate that listeners' perception of front vowels might depend on an effective upper formant that is based on an averaging of F2 and F3; in more recent studies, various formulae have been proposed for the effective upper formant (often known as F2') using F2 to F4 (see Carlson, Fant, & Granström, 1975; Bladon & Fant, 1978; Bladon, 1983; Paliwal, Lindsay, & Ainsworth, 1983).

Although these perception experiments demonstrate that F3 does have some influence on vowel quality, early acoustic studies showed that vowels of different quality have quite similar third formant frequency values (Peterson & Barney, 1952); additionally, the nomograms that are derived from vocal tract models for vowels show much less variation in F3 and F4 due to vowel quality differences compared with the first two formant frequencies. (The main exception to this is r-coloured, or rhotic, vowels as in American English "bird" which, like the consonant [ɹ], are distinguished by a very low F3).

Nevertheless, in some recent acoustic studies (Syrdal, 1985; Syrdal & Gopal, 1986; Sussman, 1990), the difference between the second and third formant frequencies, F3 − F2, has been suggested as an alternative to F2 as the main correlate of vowel backness. The basis for this distinction is twofold. First, F3 is very close to F2 for front vowels such as [i] (which therefore have a low F3 − F2 value) but far apart from F2 for back vowels such as [ɔ] (thus a high value on F3 − F2). Second, Syrdal and Gopal (1986) propose an absolute cutoff of 3.5 Bark as the threshold for the front (less than 3.5 Bark) from back (greater than 3.5 Bark) separation. This threshold is taken from a set of perceptual experiments (Chistovich & Lublinskaya, 1979; Chistovich, 1985) which suggest than when two formants are closer in frequency than 3.5 Bark, they are averaged by the listener (thus F3 and F2 would be averaged, under this interpretation for front vowels, but not for back vowels). In agreement with Syrdal (1985) and Syrdal and Gopal (1986), a study by Sussman (1990) of vowels in /bVt/, /dVt/, /gVt/ contexts produced by four male speakers has shown that the parameter F3 − F2 can be used for the front/back vowel separation.

Since the filter is usually modelled as independent of the source signal, it is perhaps surprising to learn that the fundamental frequency can also, under certain circumstances, provide cues to vowel quality (Johnson, 1990). Early experiments include those of Taylor (1933) and House and Fairbanks (1953), who demonstrated that, all things being equal, fundamental frequency varies positively with phonetic vowel height. Since, as discussed earlier, phonetic height is negatively correlated with the first formant frequency, the parameter $F1 - f_0$ should accentuate the difference between close vowels and other vowels: thus, this parameter should be low for vowels like [i] but considerably higher for open vowels like [æ]. Traunmüller (1981) has shown, from synthesis and labelling experiments, that listeners' perceptions of phonetic height are cued by the $F1 - f_0$ parameter in Bark (see also Traunmüller and Laçerda, 1987); and Syrdal (1985) and Syrdal and Gopal (1986) propose that $F1 - f_0$ is a more important correlate of phonetic vowel height than F1 on its own. In their reanalysis of the Peterson and Barney (1952) data, Syrdal and Gopal (1986) show that close vowels are well separated on the $F1 - f_0$ parameter (in Bark). However, in continuous speech, fundamental frequency is likely to be considerably less useful as a cue for vowel height because of the confounding influence of changes in prosody and intonation.

4.1.4 Length

In most English accents, a basic distinction can be made between tense/lax vowel pairs that are spectrally quite similar but that differ in duration. In Southern British English and Australian English, the lax vowels would normally be taken to include the nuclei of "hid", "hood", "head", "had", "bud", "hod") while other vowels, excluding schwa, are tense.

Other than total vowel duration, which interacts in continuous speech with many other prosodic variables (see, e.g., Umeda, 1975; Crystal & House, 1988),

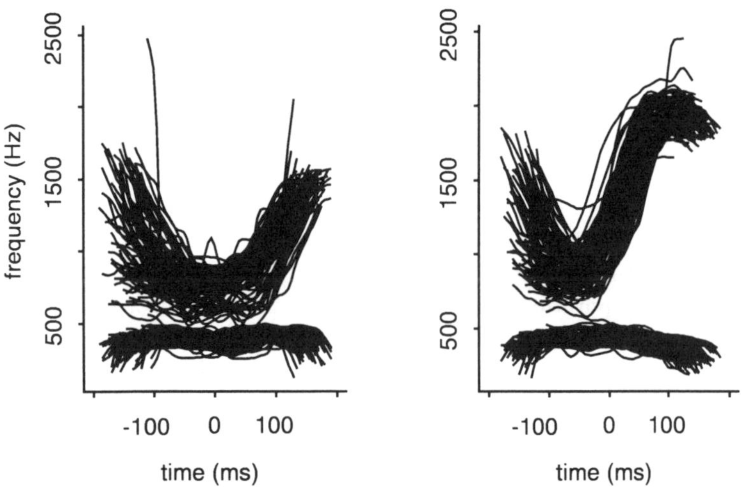

Figure 4.4: F1 and F2 trajectories of 100 [ɔ] (*left*) and 106 [ɔɪ] tokens (*right*) produced by 5 Australian English male talkers.

some studies suggest differences between tense/lax vowel pairs are cued by the *proportional duration of the acoustic vowel offglide* (the section of the vowel from the vowel target to the periodic offset). Lehiste and Peterson (1961b) were among the first to suggest that the offglide is longer in lax vowels than in their tense counterparts, and this is partially supported by more recent studies by Huang (1986), Rakerd and Verbrugge (1985), and Strange (1989a). Some lax vowels also have a more central quality: this applies in particular to the distinction between lax "hid" and tense "heed" in many English accents.

4.1.5 Monophthong/diphthong

The auditory distinction between monophthongs and diphthongs is quite salient (Bladon, 1985), and so the task of finding cues for distinguishing between these two groups would seem to be a reasonable undertaking. There is, however, little relevant acoustic data and the majority of more recent studies on diphthongs (Lindau-Webb, 1985; Peeters & Barry, 1989; Ren, 1986) have focused on cross-linguistic differences. The lack of interest in diphthongs has partly come about because, from a phonological point of view, diphthongs can be analyzed as tense monophthongs in English (Chomsky & Halle, 1968). Another reason may be that data from diphthongs are difficult to interpret in existing feature systems. As Disner (1983) points out, "They [diphthongs] must be represented by measures taken at several points in time and yet compact enough for use in feature specification".

Figure 4.4 shows F1 and F2 trajectories for the (Australian English) diphthong [ɔɪ] ("toy") and monophthong [ɔ] ("hoard"). There are approximately

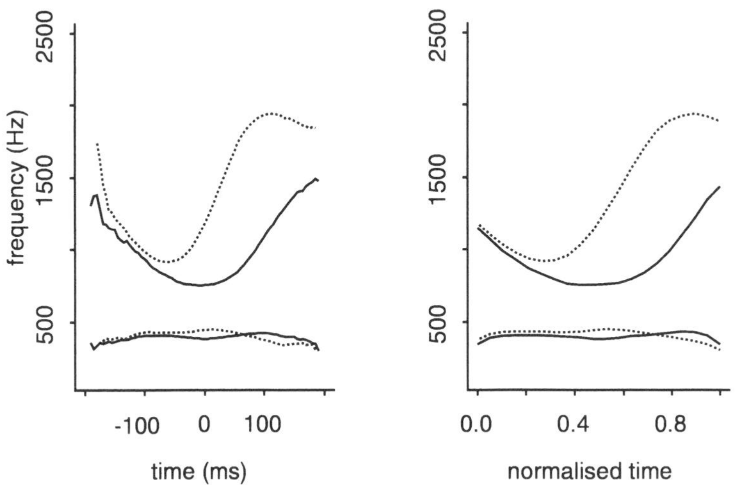

Figure 4.5: Averaged trajectories of the data in Figure 4.4 without (*left*) and with (*right*) linear time normalisation (*solid line*: [ɔ]; *dotted line*: [ɔɪ]). There is clear evidence of two targets for the diphthong [ɔɪ] and a single target for the monophthong [ɔ].

100 tokens in each panel produced by 5 male talkers (20 per talker) in a /CVd/ context, where C varies over a number of consonants. Averaged plots are shown in Figure 4.5.

Figure 4.4 and Figure 4.5 suggest that, whereas the [ɔ] tokens converge towards a single target near the vowel midpoint close to t = 0 ms), there are two principal targets for [ɔɪ], at approximately t = -60 ms and t = 90 ms, respectively.

There have been various studies in the literature concerning which components of a diphthong are most important for its identification. Following an acoustic analysis of diphthongs by Holbrook and Fairbanks (1962), various studies have investigated whether spectral change (between the two targets) is a characteristic feature of diphthongs (see, e.g., Bladon, 1985; Bond, 1982; Fox, 1983; Gay, 1970). Other studies have considered the relative importance of the first target compared with the second and have shown that, in general, the second target is less likely to be attained (undershot) due to factors such as tempo compared with the first (Gay, 1968, 1970; Jha, 1985; Nâbélek & Dagenais, 1986; Pols, 1977). This would suggest that the first target and the direction of formant movement are critical for diphthong identification, rather than whether the actual value of the second target is attained (Pols, 1977). One way of visualising this is to consider the diphthong trajectory in the F1 × F2 formant plane (on the assumption that only these parameters are being considered). In Figure 4.6, the time-normalised averaged formant display of Figure 4.5 for [ɔɪ] is redisplayed in the plane of F1/F2. Notice the loop at roughly F1=450 Hz,

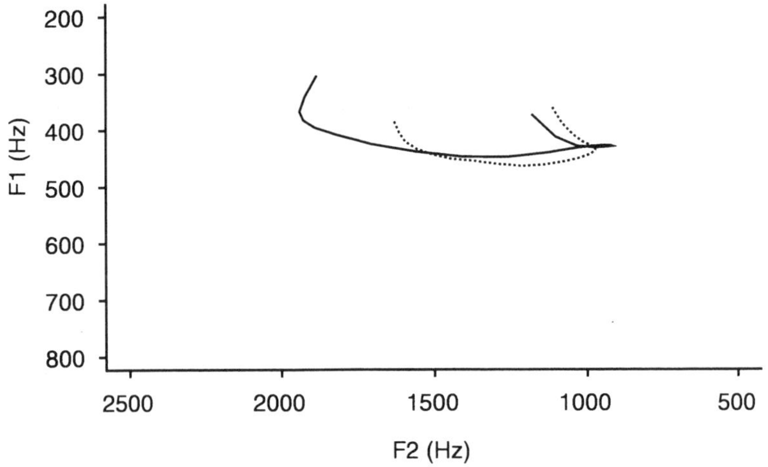

Figure 4.6: Trajectories from segment onset to segment offset in the formant plane of diphthongs in isolated (*solid line*) and continuous (*dotted line*) speech production. The trajectories are linearly time-normalised averages of 106 isolated and 134 continuous diphthongs respectively. Note the undershoot of the second diphthong target for the continuous speech data, as shown by the termination of the dotted line close to formant values for a central vowel.

F2=900 Hz corresponding to the first target and the "elbow" at approximately F1=400 Hz, F2=1900 Hz for the second target. Figure 4.6 also shows a time-normalised display for the same diphthong, averaged over 134 tokens produced by the same talkers but in continuous speech utterances (dotted line). There is clear evidence of an initial target for continuous speech [ɔɪ] at frequencies similar to those of isolated citation-form [ɔɪ], but the transition towards the second target is reduced in magnitude, terminating at a position close to schwa and there is also no distinct "elbow" that is indicative of a second vowel target. This data supports the position of Pols (1977) that the first target and the direction of the trajectory may be the critical cues for the phonetic identity of the diphthong, although it must be remembered that the display in Figure 4.6 represents both an averaging of tokens and an averaging of time differences.

In a more recent study by Gottfried, Miller, and Meyer (1993), three different hypotheses discussed in Pols (1977) (see also Nearey & Assmann, 1986) concerning the acoustic structure of diphthongs were analysed for a number of diphthongs in American English. The hypotheses were: *dual target* or *onset plus offset*, under which both diphthong targets are critical for their identification; *onset plus slope*, in which diphthong quality is presumed to depend on the first target and the rate of spectral change towards the second; and *onset plus direction* in which the first target and the direction of spectral movement were hypothesised as the main cues for diphthong identification (this third hypothesis is represented by the display in Figure 4.6; notice that it is different from the *onset plus slope* hypothesis, which explicitly incorporates time – that is, rate of spectral change – in its classification). The results of an analysis of 768 diphthongs in Gottfried et al. (1993) showed high classification scores for all three strategies (over 90% correct); the highest score was obtained from the *dual target* classification strategy.

4.1.6 Dynamic spectral cues

We have so far assumed that most of the important information for distinguishing between vowels of different quality is contained around the acoustic vowel target. There have been a number of experiments (reviewed in Strange, 1987, 1989b) that have focused on the extent to which *transitions* (that is, the onglide and offglide) might provide contributory information to phonetic vowel distinctions. In the various studies by Strange, transitions are considered to be as much an integral and indispensable part of the vowel as the closure and burst are of oral stops. It therefore makes no sense to try to identify the vowel from the target alone because to do so would be to throw away important contributory information that is contained in the transitions.

Two principal experiments have led Strange to such a conclusion. In the first, Strange, Verbrugge, Shankweiler, and Edman (1976) found that listeners identified vowels more accurately in CVC than isolated V syllables. Strange's (1987, 1989b) interpretation of this result is that CVC syllables contain transitions whereas isolated vowels do not, and so the transitions must provide listeners with additional information about the identity of the vowel. This interpretation is consistent with the results of the second principal experiment by Strange, Jenkins, and Johnson (1983), who presented listeners with various sorts of edited syllables. Two of these include silent centre (SC) syllables in which the vowel target is discarded leaving an onglide, a silent gap (where the target used to be) and an offglide; and variable centre (V) syllables in which the transitions are thrown away leaving only the vowel target. Their results show that SC syllables, containing only transitions, are identified as well as the unmodified, original CVC syllables. This experiment is contrary to the hypothesis that the most important information for cueing the vowel's identity is at the vowel target (because in SC syllables, the vowel target has been spliced out). Despite some methodological difficulties with these experiments (Assmann, Nearey, & Hogan, 1982; Diehl, McCusker, & Chapman, 1981; Macchi, 1980), many subsequent experiments have continued to show that vowels are as well identified from modified SC stimuli as from unmodified CVC syllables (Benguerel & McFadden, 1989; Fox, 1989; Jenkins, Strange, & Miranda, 1994; Parker & Diehl, 1984; Rakerd & Verbrugge, 1987; Strange, 1989a; Verbrugge & Rakerd, 1986), although the benefit that context provides to vowel identification has been shown to be affected by the phonetic identity of both the consonant and vowel (Gottfried & Strange, 1980; Rakerd, Verbrugge, & Shankweiler, 1984).

There are at least two main reasons why acoustic cues to vowels are presumed to be distributed throughout the vowel in these recent studies, rather than focused just at the vowel target. The first follows the reasoning developed in the action theory (e.g., Fowler, 1983, 1986) and task-dynamic (Saltzman & Munhall, 1989) theories of speech production that model phonetic segments in terms of *articulatory gestures that change in time.* As Fowler (1987, p. 581) comments:

> Evidently, vowels – and presumably phonetic segments more generally – must be ... patterns of articulatory activity that can be realized in any vocal tract or vocal-tract-like system. Just as the visible transformation, "rolling", for example, can be realized by any object of the appropriate sort in the appropriate environmental setting – regardless of its size and other properties – so, for example, /a/ may be seen as an articulatory transformation applicable to any vocal tract – indeed, to any physical system with appropriate physical characteristics – independently of its size.

The relationship, then, of vowels to articulation and the resulting acoustic waveform is rather like the relationship between the abstract concept "rolling" and the numerous bodies (for example, a football moving along the ground, a moving steam-roller, a gymnast producing a forward roll, a child rolling down a slope) in which rolling can be realised. If this is so, it makes no more sense to try to categorise vowels using single spectral sections taken from the vowel target than it would to explain adequately the concept of "rolling" from single snap-shot photographs. Instead, a motion picture is needed, and, applying the same logic, any acoustic-perceptual model of vowels needs to take account of the vowel gesture as it changes in time.

A second possible interpretation is that even "monophthongal" vowels are characterised by a changing acoustic pattern that is not attributable to context effects, but that forms part of the inherent structure of the vowel, in much the same way that the transition between two targets is an inherent part of a diphthong. This interpretation is consistent with studies by Nearey and colleagues, who show that at least some (so-called) monophthongal vowels in Canadian English are characterised by inherent formant movement that may also be important for their perceptual identification (Nearey & Assmann, 1986; Nearey, 1989; Andruski & Nearey, 1992).

Two kinds of *acoustic* studies have some bearing on the question of whether transitions provide contributory information to the vowel's phonetic identity. In the first type, Suomi (1987) and Sussman (1990) have shown that vowels are almost as well separated based on spectral or formant information taken from one of the transitions as from the vowel target (this is compatible with the dynamic view of vowels which emphasises that transitions contain valuable information for vowel distinctions). In the second type of study, which is a closer test of whether transitions provide contributory information to the vowel's phonetic identity, the extent to which vowels are separated in an acoustic space formed from *both* the vowel target and one or more time slices taken in the transitions is examined. The reasoning that underlies these experiments is that vowels should be more differentiated in this combined transition-target space compared with an acoustic space formed from the vowel target alone, if transitions provide contributory information to vowel quality. While Huang (1992), and Zahorian and Jagharghi (1993) show that this is the case (a better separation is obtained from a space that includes the transitions), Harrington and Cassidy (1994) show that for Australian English at least, the inclusion of transitions benefits only diphthongs but not monophthongs (which are as adequately classified from a single spectral slice at the vowel target as from multiple spectral slices taken at both the target and transitions).

4.1.7 Vowel reduction

In this section, we will consider some of the different factors that contribute to vowel reduction and analyse the effects of these factors primarily on duration and formant frequencies. We will begin with some general comments on the difference between phonological and phonetic vowel reduction, and then discuss some experimental data concerning the relationship between phonetic vowel reduction and tempo, stress, context effects, and clear speech.

We can begin by making a distinction between *phonological* or *lexical vowel reduction* and *phonetic vowel reduction*. Phonological reduction is the process whereby full vowels become unstressed weak vowels due to morphological and phonological processes of affixation (e.g., "major"/"majority"; "telegraphy"/"telegraphic"). This is an obligatory process (the nucleus of the second syllable of "major" is always weaker and more central than that of "majority") and is not the concern of the present discussion. The other kind is *phonetic vowel reduction* in which the quality of vowels changes because of segmental and prosodic context effects.

The articulatory processes that underlie phonetic vowel reduction are often described as *target undershoot*. Essentially, undershoot implies that, because of the influences of segmental and prosodic contexts, the vowel target cannot be attained. A *target* in this context can be thought of as either a spatial or auditory representation of the vowel produced in a citation and context-free form (for example, [i] produced in isolation). Vowel reduction is usually accompanied by a *decrease in the acoustic duration of the vowel.*

When examining the changes to formant frequencies from their theoretical target values caused by context effects, the term *formant undershoot* is often used. An example of formant undershoot (due in this case to the coarticulatory effect of a preceding consonant) is shown in Figure 4.7. The main observable effect of undershoot is an *increase in the variability* of vowels. This is evident in Figure 4.8, which shows the F1/F2 plane of vowels produced by the same talkers in a citation-form and continuous speech contexts. (In continuous speech, there are many more different kinds of segmental and prosodic variation and so phonetic vowel reduction is much more probable). There are likely to be two different, but related, kinds of undershoot: the first is *centralisation* (Joos, 1948; Tiffany, 1959), which implies that the vowel is displaced towards a more central schwa-like position (so towards the centre of the formant plane); the second is *contextual assimilation* (Stevens & House, 1963b; Lindblom, 1963), in which the quality of the vowel changes in the direction of the context that is influencing

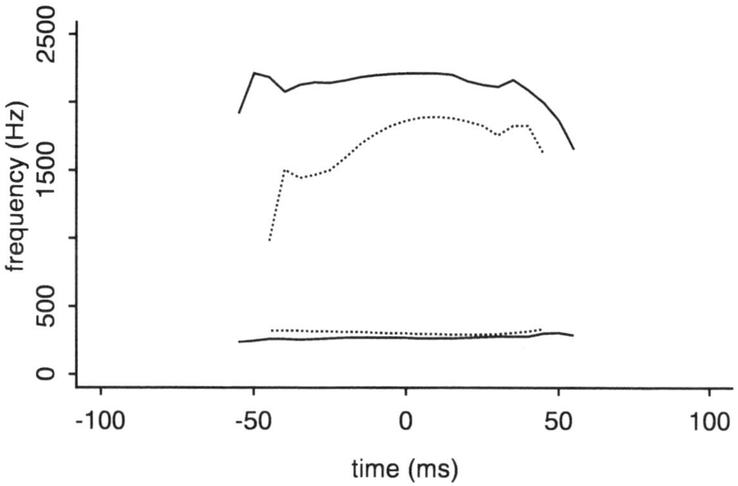

Figure 4.7: Averaged F1 and F2 trajectories of [i] vowels in "he" (solid lines) and "we" (dotted lines) words in 460 continuous speech utterances and a read passage produced by one male talker of Australian English. Note the lowered F2 and slightly raised F1 of the trajectories in "we" compared with "he" due to the coarticulatory influence of [w] on the [i] vowel target.

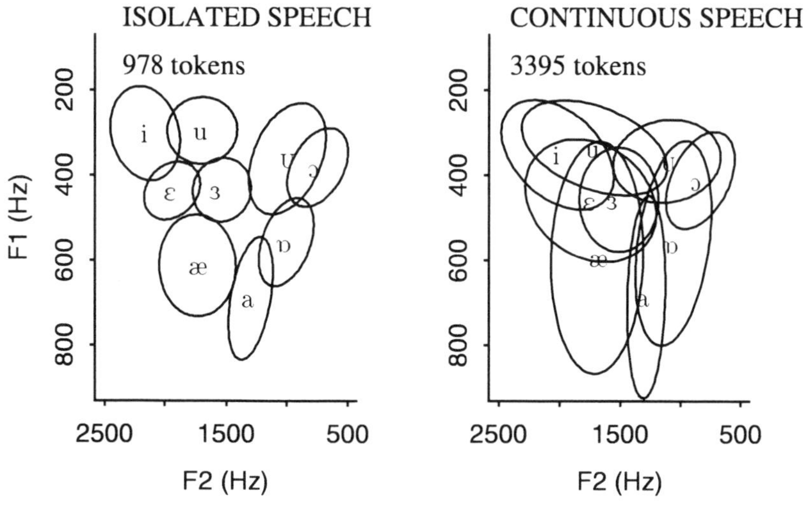

Figure 4.8: The formant plane for isolated and continuous speech vowels produced by five male talkers of Australian English.

it (as in Figure 4.7). There is some evidence of centralisation in the continuous speech data in Figure 4.8, but this is more marked for some vowels (e.g., [i] and [ɛ] that are closer to the centre of the F1/F2 plot in the continuous speech data) than for others (e.g., [ɔ], which shows very little evidence of centralisation).

In summary, when vowels reduce, their formant values may move towards those of a more central vowel, there is likely to be a greater degree of overlap between different vowel types in a formant space, and their duration usually decreases. Below, we briefly consider some of the main factors that contribute to vowel reduction.

Tempo

Continuous speech vowels are likely to be spoken at a faster rate, or tempo, than citation-form vowels in isolated words. One of the consequences of increased tempo is a decrease in vowel duration: since there is less time for the vowel to be produced, the articulators also have less time to attain the vowel's target position and as a result, the target may be "undershot". For example, at a faster rate of speech, there may be insufficient time for the tongue dorsum to be as raised as it would be in citation-form speech production of [i] or for the jaw to be as low as it would be in citation-form [a].

This is essentially the line of reasoning developed in Lindblom's (1963, 1964) classic study of vowel reduction in symmetrical CVC (e.g., [bib]) contexts spoken at different levels of tempo and stress. By fitting a mathematical model to the formant data, Lindblom explains the effects of tempo in terms of the effects of the reduced time to produce the vowel resulting in *formant undershoot* (the target position of the formants are not attained).

There are certainly a number of studies in speech perception that show that listeners' perception of vowel quality can be influenced by the tempo of the utterance (Miller, 1981, 1987) and that listeners compensate perceptually for a more central vowel quality at faster rates (Johnson & Strange, 1982; Verbrugge, Strange, Shankweiler, & Edman, 1976; Verbrugge & Shankweiler, 1977), but what is less clear is whether increasing tempo *necessarily* results in undershoot. Many studies in the production of speech point to the possibility of some form of articulatory reorganisation to compensate for the effects of the decrease in vowel duration caused by increasing tempo. Two main strategies are suggested. First, articulatory velocity can increase with increasing tempo (Beckman, Edwards, & Fletcher, 1992; Kelso, Vatikiotis-Bateson, Saltzman, & Kay, 1985; Kuehn & Moll, 1976; Sonoda, 1987): if the speed of articulation increases, then there should be sufficient time for the vowel target to be attained i.e. vowel undershoot need not necessarily occur. Second, other studies have shown that the extent of coarticulatory overlap between the consonant and a following vowel can increase with increasing tempo (Engstrand, 1988; Gay, 1981; Giles & Moll, 1975; Harris, 1978). If there is an increase in the extent of coarticulatory overlap, there should also be sufficient time for the vowel target to be attained (because the vowel effectively starts earlier during the preceding consonant), and so again undershoot need not necessarily occur with increasing tempo. The possibility that articulatory strategies are reorganised in these ways may explain why some acoustic studies have failed to find much evidence of undershoot with increasing tempo (e.g., Gay, 1978; Fourakis, 1991; van Son & Pols, 1990, 1992; Pols & van Son, 1993).

Stress

There is once again conflicting evidence concerning the relationship between stress and phonetic vowel reduction. As discussed more fully in Section 4.6.1, at least part of the reason for this is that researchers have not always been careful to distinguish between the different possible kinds of stress distinctions that are not necessarily cued by the same sets of articulatory or acoustic features.

The model of vowel reduction in Lindblom (1963) discussed earlier explained vowel reduction in terms of a decrease in vowel duration irrespective of the source of the reduction: that is, since both a decrease in stress and an increase in tempo result in a decrease in acoustic vowel duration, the formants are displaced from their targets accordingly because there is less time for the target to be attained. Since then, Lindblom has revised this model (Moon & Lindblom, 1994) to include an additional parameter that is more closely related to articulatory velocity i.e. to allow for the fact that talkers can speed up their articulations to compensate for a decrease in duration as discussed earlier.

Irrespective of how vowel reduction effects are actually modelled, recent research suggests that *accented* vowels – i.e. vowels with sentence level stress in the sense defined in Section 4.6.1 – are closer to the edge of the vowel quadrilateral than their unaccented counterparts. Thus accented open vowels are often produced with a lower jaw position and a greater vocal tract opening than unaccented vowels, which leads to a raising of F1, thus increasing the distance to the centre of the acoustic-phonetic vowel space (Summers, 1987; Beckman et al., 1992). Similarly, the tongue body is often raised to a greater extent in accented high vowels like [i], resulting in a raising of either F2 or F3 and thereby once again making the accented vowels more peripheral in an acoustic phonetic space (Harrington, Fletcher, & Beckman, in press; see also Engstrand, 1988, and de Jong, 1995, for other articulatory studies and van Bergem, 1993, for an acoustic study of accentuation effects in a large corpus of Dutch sentences).

Context effects

A well-known perception experiment by Lindblom and Studdert-Kennedy (1967) has shown that listeners' classifications of vowels are affected by consonantal context. In their experiments, a continuum from /i/ to /u/ was synthesised and presented to listeners in /j–j/ and /w–w/ contexts. Their results showed that a token from the middle of the vowel continuum (i.e. ambiguous between /i/ and /u/) is more likely to be perceived as /i/ in the /w–w/ context than in the /j–j/ context. This finding can be explained in terms of a perceptual compensation for target undershoot. In a /w–w/ context, listeners attribute a certain part of the /u/-colouring of the vowel to the coarticulatory effects of the adjacent /w/ consonants: listeners compensate for this coarticulatory effect and bias their responses towards /i/. The important conclusion from this experiment is that a vowel that has the same formant frequencies is perceived differently in different contexts because of listeners' compensations for the effects of the neighbouring consonants on the vowel target.

This perception experiment is consistent with many studies that have shown that vowel formant frequencies are influenced by the surrounding consonantal context (Broad & Fertig, 1970; Joos, 1948; Delattre, 1969; Fourakis, 1991; Lindblom, 1963; Moon & Lindblom, 1994; Stevens & House, 1963b; Stevens et al., 1966; van Bergem, 1993) although Schouten and Pols (1979b, 1979a), referring to a study by Pols (1977), show that the effect of consonants on vowel targets produces even less systematic variation in vowel targets than do different speakers. More recent studies have emphasised that it is not possible to generalise the influence of context on vowel reduction across all possible consonant-vowel combinations. Instead, there is some agreement (Moon & Lindblom, 1994; Pols & van Son, 1993; van Bergem, 1993) that context-effects are most pronounced for CV sequences which have a large distance from the consonant locus to the vowel target (the consonant locus is discussed in detail in Section 4.2.2).

As discussed in Moon and Lindblom (1994), the coarticulatory influence of neighbouring consonants on vowel targets need not necessarily imply centralisation i.e. a tendency for the vowel to become more schwa-like. For example, for an [i] vowel to become more centralised, F1 would have to be raised from around 200 Hz to a value nearer 500 Hz, which is more characteristic of a schwa vowel. However, as Moon and Lindblom (1994) show, F1 of [i] is *lowered* rather than raised in a [w–l] context because of the influence of the low F1 value of the initial labial consonant (see also van Bergem, 1993, 1994).

Clear speech and redundancy

We can conclude this section with a few remarks on speaker clarity, which has been shown to have some influence on vowel reduction. When talkers are instructed to produce words with maximum clarity, vowels are both lengthened and become more peripheral compared with their citation-form speech counterparts (Moon & Lindblom, 1994; Palethorpe, 1992; but see Pols & van Son, 1993, for different conclusions). This finding is relevant to spontaneous speech because the clarity of speech has been shown to vary with the extent to which words are predictable from context (Lieberman, 1963; Hunnicutt, 1985, 1987). For example, when a talker is communicating new information (see Chafe, 1974; Halliday, 1967, for a discussion of "new" as opposed to "old" or "given" information), such as mentioning a street name for the first time, a listener cannot predict the words the talker is about to say either from the utterance or from the discourse: therefore, the talker must produce the utterance with a high degree of clarity if the information is to be successfully conveyed to the listener. On the other hand, when the information is given (e.g., a person's name has already been mentioned several times before in a dialogue), the talker need not produce the information as clearly because the listener is better able to predict it from a variety of discourse cues. Fowler and Housum (1987) suggest that clarity is used in a dynamic way to enable talkers to signal to listeners whether or not the upcoming information introduces a new topic by varying the clarity of their speech. A detailed examination of formant values by Koopmans-van Beinum and van Bergem (1989) using experimental materials that were similar to those of Fowler and Housum (1987) confirmed that the vowels of more predictable (or "old") words are accompanied by a greater degree of phonetic vowel reduction.

4.1.8 Vowel normalisation

We have so far been discussing acoustic cues and major features for dividing the vowel space without explicitly acknowledging the fact that the acoustic structure of phonetically identical vowels varies a good deal across different speakers. The classic study by Peterson and Barney (1952) had already demonstrated that formant frequencies of women are around 20% higher than those of men, while children's formants are even higher than those of women. These differences between speakers can obscure the phonetic differences between vowels in a formant space. For example, [ɔ] has a lower F1 and F2 than [a] produced by the same speaker, but when [ɔ] and [a] are produced by a child and man, respectively, this distinction is often cancelled because of the child's higher average formant frequencies (Peterson, 1961). This kind of overlap between vowel categories even occurs for vowels produced by speakers of the same sex and regional accent, as the studies by Ladefoged (1967) and Pols, Tromp, and Plomp (1973) amply demonstrate.

It is clear enough that listeners can somehow disregard the acoustic variability that is due to differences between speakers. Thus despite marked acoustic differences in the productions of [i] by a male and female speaker, listeners can nevertheless identify the "same" vowel in both productions. From the listener's point of view, we can say that there must be a transformation, or set of transformations to the speech signal that makes the acoustic differences between these two versions of [i] irrelevant and that also separates the acoustically similar child [ɔ] and adult male [a] vowels considered earlier. Strategies of *vowel normalisation* have a similar aim of transforming the acoustic vowel space such that phonetically identical vowels, which might be indistinguishable at an acoustic level due to speaker-effects, are grouped and such that phonetically different vowels are separated. In some cases, a vowel normalisation technique is designed to copy the kinds of transformations that are made by the ear (e.g., Bladon, Henton, & Pickering, 1984), but this need not necessarily be the case (e.g., Gerstman, 1968).

There have been many vowel normalisation studies in the last twenty years or so that can be grouped in various ways. The principal distinction is between *speaker-independent* strategies, which are usually based on auditory theories of speech processing, and *speaker-dependent* strategies, which tend to make use of a statistical procedure to eliminate speaker differences.

Speaker-independent strategies

We can begin by making the (not unreasonable) assumption that the acoustic signal of a vowel contains cues to both the identity of the speaker and the phonetic identity of the vowel. Speaker-independent normalisation can be considered to be a transformation of the acoustic signal that both reduces the potency of the cues to the speaker and enhances the cues to the vowel identity at one fell swoop: it is therefore a single transformation that is double-edged.

One of the simplest kinds of speaker-independent normalisation is based on the formant-ratio theory, which can be traced back to Lloyd (1890), receiving renewed attention in Peterson (1952, 1961) and Potter and Steinberg (1950); more recently the formant ratio theory has been discussed at some length by Millar (1989). The basis of the formant ratio theory is as follows: when several speakers produce a vowel such as [i], the *absolute* values of the formants may well vary considerably from speaker to speaker, but the *ratio* between the formants is assumed to be more or less constant. However, Peterson (1952, 1961) has shown that talker-differences of back vowels are poorly modelled in terms of formant ratios (i.e. F1/F2 is not constant across different talkers in a back vowel such as [ʊ]). Furthermore, vowels of different phonetic quality, such as [ɑ] and [ɔ], may have very similar formant ratios (Potter & Steinberg, 1950), which implies that, although there may be a reduction in the variability across speakers when formants are normalised using formant-ratios, the confusion between some phonetically different vowels may actually increase. A more serious problem is that the formant ratio theory predicts a constant change across different formant numbers. For example, if F1 of speaker A is 20% greater than F1 of speaker B for a given vowel, then, under the formant ratio theory, F2 of speaker A must also be 20% greater than F2 of speaker B in the same vowel (otherwise the ratio F2/F1 would not be constant). However, considerations of articulatory-to-acoustic mappings in Fant (1966) data suggest otherwise. For example, in front vowels such as [i], the percentage increase in mapping male onto female vowels is considerably greater for F2 than F3 (because of the different ratio of the mouth to pharynx length in females).

Other speaker-independent normalisation techniques are based on the theory that the apparent acoustic dissimilarities in different speakers' productions of (phonetically) the same vowel are eliminated by virtue of the transformations that are made to the acoustic signal in the ear and auditory nerve. Auditory theories of vowel normalisation such as those discussed in Bladon et al. (1984) and Syrdal and Gopal (1986) have been inspired by a hypothesis advanced by Potter and Steinberg (1950) (see also Chiba & Kajiyama, 1941) that phonetically equivalent vowels produced by different speakers result in the same spatial pattern of motion along the basilar membrane, although the actual position of the pattern on the membrane may vary. Since displacement along the basilar membrane is directly related to the frequency of a sound in Bark, many of the differences between speakers are assumed to be filtered out by a transformation of the acoustic signal in Hertz to the Bark scale. In the study by Syrdal and Gopal (1986), various Bark-scaled parameters are tested for their effectiveness in reducing the speaker-variability in the Peterson and Barney (1952) data. In one classification, the parameters included the fundamental frequency and the first three formant frequencies in Hertz. In a second, the parameters were the same but in Bark. In the third, there were three parameters: F1-f_0, F2-F1 and F3-F2 all in Bark. The third set of parameters, based on Bark difference measures, produced the highest classification scores (although significant, the differences are small: 85.7% and 81.8% correct for Bark-difference and Hertz classifications, respectively). They also show that the differences between the three speaker groups are more effectively eliminated using the third set of parameters compared with the other two.

Bladon et al. (1984) have devised a model of male/female vowel normalisation which is based on sliding an entire auditory spectrum along the frequency axis. Normalisation is presumed to come about by shifting the auditorily transformed spectra of female talkers downwards by 1 Bark (or conversely the male spectra upwards by 1 Bark). The choice of a 1 Bark displacement is based partly on Bladon *et al.*'s (1984) reanalysis of formant data in six languages from Fant (1975) and also on the fact that the fundamental frequency of males is usually around 100 Hz, or 1 Bark lower, than that of females. However, as Bladon et al. show, the simple 1 Bark displacement is usually insufficient to model male/female differences in the lower part of the spectrum. This is because F1 tends to shift upwards in frequency when it is close to the fundamental (i.e. when the harmonic spacing is wide in relation to the first formant frequency value). A consequence of this upward shift is that the male/female differences can be modelled accurately only in the lower part of the spectrum under two conditions: first, if it is known whether an f_0-induced shift in F1 has taken place; second, if it can be determined by how much the first formant should be adjusted to compensate for the displacement caused by the fundamental. Formulae are tentatively proposed in Bladon et al. (1984), but they are not tested on any data (see also Holmes, 1986, for further remarks on the f_0 interaction with F1 in this model).

Speaker-dependent strategies

Most speaker-dependent normalisation strategies consist of two principal stages: first, those aspects of the speech signal that depend on the characteristics of the speaker are explicitly identified and second, once identified, they are factored out. Since a knowledge of those attributes of the acoustic signal that are specific to a speaker can usually be obtained only from a large sample of the same speaker's vowels, speaker-dependent strategies are almost always *extrinsic* i.e. the normalisation is accomplished using information beyond the vowel that is to be normalised. Such extrinsic information might include a range, mean, or other statistical measure of a large sample of the same talker's vowels.

Joos (1948) was one of the first to suggest that vowel normalisation might be speaker-dependent and implied that the vowels of a given speaker are perceived in relation to the same speaker's "point" vowels [i a u]. Thus, although the formant frequencies of the vowel [ɛ] may be different for a given male and a female speaker, the distance of the male [ɛ] to the male's point vowels is assumed to be about the same as the distance of the female [ɛ] to the female's point vowels. An early speech perception experiment to test the relationship between speaker-dependence and vowel quality was carried out by Ladefoged and Broadbent (1957). In their experiment, six versions of the phrase "Please say what the word is" were synthesised. The synthetic versions were identical in every respect except that the formant frequencies of the vowels were shifted up or down. This had the effect that the sentences sounded as if they were produced by different speakers of the same accent. Additionally, four /bVt/ words, which subjects were instructed to identify, were synthesised and appended to these phrases. Ladefoged and Broadbent (1957) found that the same test word could be perceived in different ways depending on which of the phrases preceded it. For example, shifting the first formant frequency of the introductory phrase downwards changed listeners' perceptions of one of the test words from "bit" to "bet". The experiment demonstrates quite clearly that listeners' judgements of vowel quality are influenced by extrinsic cues. From this, we can infer, as Ladefoged (1967) does, that vowel quality is partly dependent on the relationship between the frequencies of the first two formants for that vowel, and the frequencies of the first two formants of other vowels pronounced by the same speaker (p. 105). The influence of extrinsic cues has also been demonstrated in similar experiments by Ainsworth (1975), van Bergem, Pols, and Koopmans-van Beinum (1988), Dechovitz (1977), and Nearey (1989) and in reaction-time studies by Summerfield and Haggard (1975) and Mullenix, Pisoni, and Martin (1989).

Certainly, the evidence suggests that extrinsic cues can *influence* vowel quality, but what is less clear is whether extrinsic information is *necessary* to judge vowel quality accurately. Experiments by Verbrugge et al. (1976) and Assmann et al. (1982) suggest that it is not. In Assmann et al. (1982), listeners were presented with isolated vowels under two conditions: either they were all produced by the same speaker (blocked), or the speaker varied from vowel to vowel (mixed). Assmann et al. (1982) found that vowel identification was marginally better in the blocked condition, but also that very few errors were made in the mixed condition. Similar results are presented in Verbrugge et al. (1976) who also show that listeners' identifications of vowels are not improved when they have been previously exposed to the point vowels produced by the same speaker.

A well-known speaker-dependent normalisation strategy is set out in Gerstman (1968). In this strategy, all vowel formant frequencies are rescaled for each speaker separately using the speaker's highest and lowest formant frequency values. Where $F_{n.min}$ and $F_{n.max}$ are a speaker's minimum and maximum values for formant n (across a suitably large sample of vowels), the relationship between the speaker's formant value, F_n, and its normalised equivalent, $F_{n.norm}$, is given by

$$F_{n.norm} = (F_n - F_{n.min})/(F_{n.max} - F_{n.min}). \tag{4.1}$$

This formula should be compared with a similar one proposed by Lobanov (1971) in which normalisation is based on a speaker's formant mean ($F_{n.mean}$) and standard deviation ($F_{n.sd}$) rather than the formant endpoints

$$F_{n.norm} = (F_n - F_{n.mean})/F_{n.sd}. \tag{4.2}$$

The extrinsic normalisation strategy in Nearey (1977) is related to the formant-ratio theory discussed earlier that rests on two main assumptions: first, a speaker's vowel formant is normalised when it is multiplied by a constant that is likely to be different for each speaker; second, different formant numbers are multiplied by the same constant. Thus

$$F_{n.norm} = F_n k, \tag{4.3}$$

where $F_{n.norm}$ and F_n are the talker's original and normalised nth formant frequencies and k is the speaker-dependent constant. Nearey (1977) attempts to estimate k first by representing Equation 4.3 in its equivalent logarithmic form (which converts multiplication to an addition):

$$G_{n.norm} = G_n + k \tag{4.4}$$

where $G_{n.norm}$ and G_n are the logarithms of $F_{n.norm}$ and F_n respectively. The speaker-dependent constant, k, is then estimated in the following way. First, all the speaker's F1 and F2 values across a large sample of vowels are converted to logarithms that we can denote by $G1$ and $G2$. Second, the mean of all G1 values, $G_{1.mean}$, and the mean of all $G2$ values, $G_{2.mean}$, are determined. The final step involves the calculation of G_{mean}, the average of $G_{1.mean}$ and $G_{2.mean}$. Normalisation is assumed to consist of subtracting G_{mean} from each of the speaker's $G1$ and $G2$ values (i.e. $k = -G_{mean}$):

$$G_{n.norm} = G_n - G_{mean} \qquad n = 1 \text{ or } 2 \tag{4.5}$$

Other versions of Nearey's normalisation technique are set out in Assmann et al. (1982) (see also Nearey, 1989).

Figure 4.9 shows the application of three of the above extrinsic normalisation strategies to Australian English vowels produced by five male and four female talkers. The top row shows the unnormalised data; the bottom row shows the vowel normalisations of the combined male and female data — i.e. of the data in the top right panel — according to three different techniques. The resulting overlap between the vowel ellipses is least for the Lobanov technique, although the overlap between the vowel types is substantially reduced for all three techniques.

4.2 Oral stops

The acoustic characteristics of oral stops preceding stressed vowels can be characterised by three principal stages. First, during the occlusion phase of voiceless stops, there is either no, or very little, acoustic energy corresponding to complete vocal tract closure: this is shown in the first [p] of "pepper" preceding

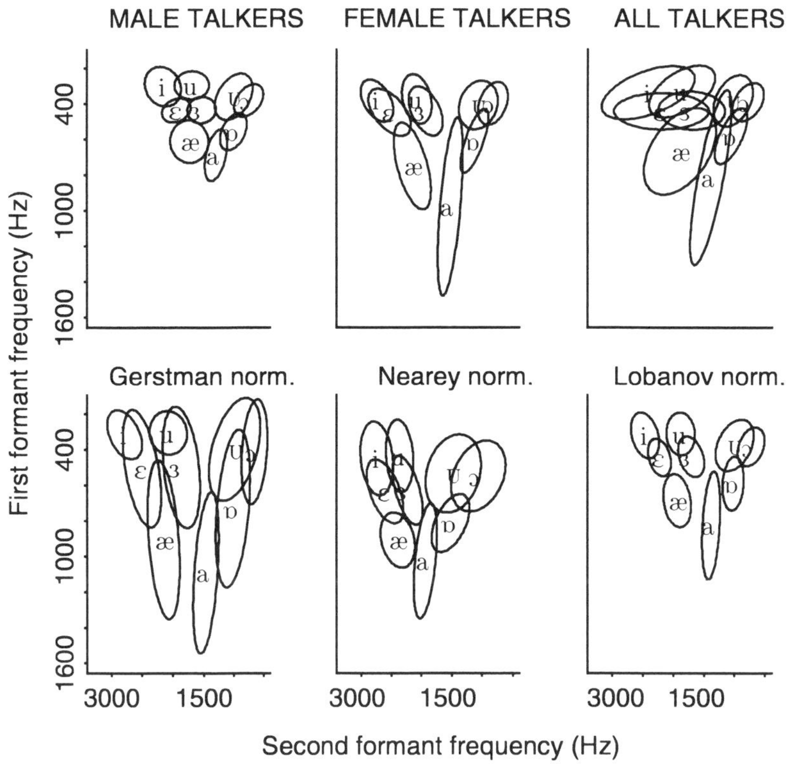

Figure 4.9: The bottom three panels show three different extrinsic vowel normalisation techniques applied to the raw data in the top right panel which is a combination of the male and female in the top left and centre panels. The normalised data was rescaled to ranges more typical for the first two formant frequencies (see Disner, 1980), and all axes are drawn to the same scale.

the stressed vowel [ɛ] in Figure 4.10. The occlusion of voiced stops may additionally be characterised by low frequency (0 to 400 Hz) periodic energy that corresponds to vocal fold vibration during the closure: this is clearly evident, particularly from the waveform, of the [g] of "gallon" in Figure 4.11. In continuous speech utterances, the boundaries of the occlusion may not always be clearly defined because of the greater likelihood that sounds overlap or are *coarticulated* with each other. Thus, while the release of the word-initial [p] of "pepper" in Figure 4.10 is quite clearly demarcated in the spectrogram, the exact acoustic boundary between the [ə] of "pick *a*" and the closure of the first [p] of "pepper" is less obvious: the difficulty is that there is fricative energy in the 1 to 2 kHz range in the first part of the closure which occurs either because the schwa is devoiced or (more probably) because the occlusion of [p] is gradually formed resulting in a voiceless bilabial fricative before the stop is completely closed.

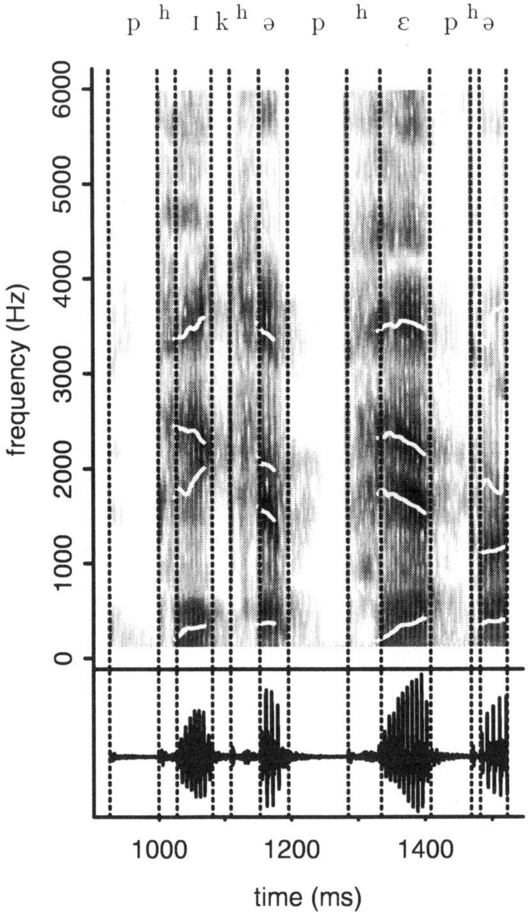

Figure 4.10: Spectrogram of the utterance "pick a pepper". There is some ambiguity in defining the precise point of segmentation between "a" in "pick *a*" and the closure of the initial [p] in "pepper" because of the presence of the progressively weakening fricative energy in the 1 to 2 kHz range in the first part of the closure.

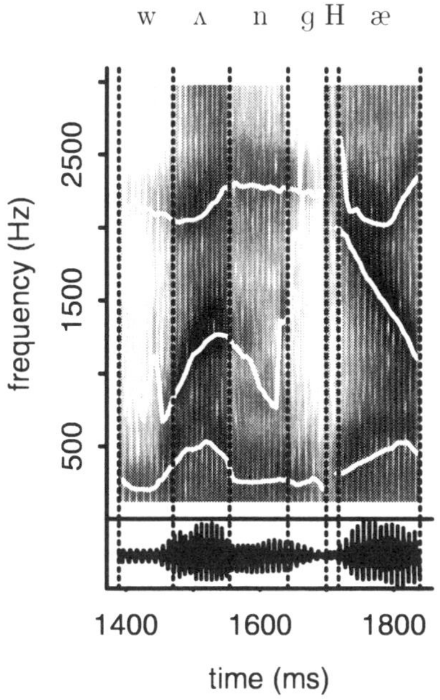

Figure 4.11: Spectrogram of the first syllable of "gallon" preceded by "one" (the burst of [g] is denoted by "H"). Note the periodicity throughout most of the closure.

When stops occur in unstressed syllables, the occlusion may be even less clearly defined and is sometimes incomplete, as in the realisation of the second /t/ of "thirty" in Figure 4.12.

The second principal stage is the *release burst* or *burst* that extends from the offset of the occlusion to the periodic onset of the following vowel: the duration of the burst is also known as *voice onset time*, which is the principal acoustic parameter for distinguishing between voiced stops and their voiceless counterparts in syllable-initial position in English. The burst is also sometimes subdivided into *transient*, *frication*, and *aspiration* stages (Fant, 1968). The transient corresponds to the release of the closure following a sudden rise of intraoral air pressure (during the closure) and is often detectable as a vertical "spike" on spectrograms. Frication results from a combination of high intra-oral pressure and a narrow channel at the point of release: in general, the frication is spectrally similar to a fricative at the same place of articulation (e.g., the frication stage in an alveolar stop is spectrally similar to that of [s]). Aspiration is caused by a noise source at the glottis which, particularly in voiceless stops, may result both in low frequency energy (below 1 kHz) and in a formant-like structure which is continuous with the voiced formants of the following vowel

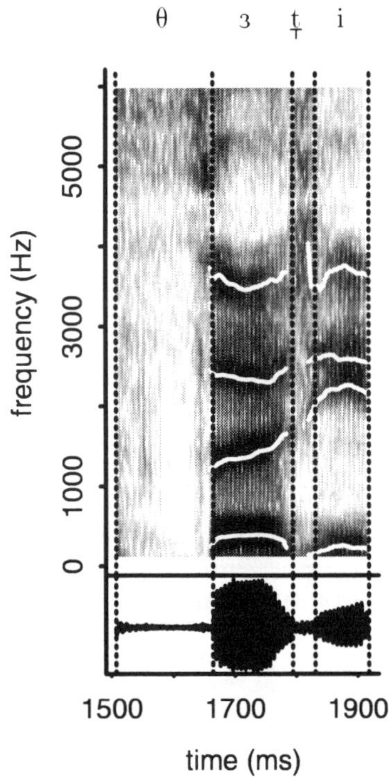

Figure 4.12: Spectrogram from [ɜ] to the offset of "thirty". The /t/ is realised without a closure and is strongly fricated as evidenced by the high-frequency energy in the spectrogram. The acoustic vowel onglide for [i] is shown by the rising F2 transition from roughly 1800 Hz, which is close to the "locus" frequency for alveolars.

(this happens because the shape of supralaryngeal vocal tract for the following vowel is "anticipated" during the final stages of the burst). For most purposes of discussion in this chapter, it will be sufficient to refer to the burst without further subdividing it into these three stages. Some examples of the burst stage are shown in Figure 4.13.

The third and final stage is the *transition* (Fant, 1968) or *acoustic vowel onglide* (Lehiste & Peterson, 1961b). At the periodic onset of the following vowel, the formants often show extensive movement to the vowel target (corresponding to the change in vocal tract shape from the stop to the vowel). Sometimes, the onglide can extend back into the aspiration and frication stages of the burst (particularly in voiceless stops). As we shall see in a subsequent section on formant loci, the formant transitions in the onglide provide important cues to the stop's place of articulation. A clear example of the F2-onglide

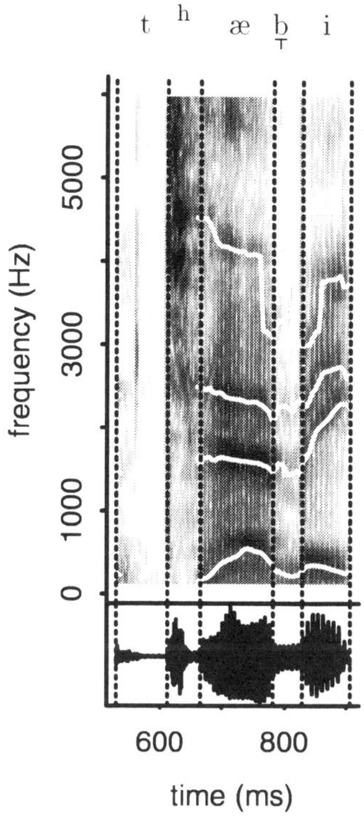

Figure 4.13: Spectrogram of "tabby" showing a clearly released [tʰ] burst.

in the second syllable of "thirty" is shown in Figure 4.12.

The above remarks on the spectrographic characteristics of stops apply primarily when they occur in syllable-initial position in stressed syllables: the various components identified above are present in different degrees depending on the context and speaking-style. For example, in continuous speech, oral stops are often doubly articulated with a following stop; consequently, the stops in "act" may be realised as a long closure (corresponding to the doubly articulated velar-alveolar stop) followed by a single release burst. Phonetics reference texts such as Gimson and Cruttenden (1994) provide an excellent overview of the major different phonetic realisations of oral stops and other sounds in English and many of the spectrographic characteristics of stops in these different contexts are discussed in detail in Olive et al. (1993).

4.2.1 Place of articulation: burst spectra

Studies of both stop consonant perception (Liberman et al., 1952) and of the acoustic characteristics of stops in natural speech (Fischer-Jørgensen, 1954;

Halle, Hughes, & Radley, 1957) have demonstrated that the spectrum of the burst carries primary cues for the place of articulation distinction in oral stops. The importance of the stop burst for the perception of place of articulation in stop consonants has been confirmed in several subsequent studies (e.g., Cole & Scott, 1974; Dorman, Studdert-Kennedy, & Raphael, 1977; Just, Suslick, Michaels, & Shockey, 1978; Ohde & Sharf, 1977; Repp & Lin, 1989; Winitz, Scheib, & Reeds, 1972), while several acoustic studies (e.g., Edwards, 1981; Fant, 1973; Repp & Lin, 1989; Zue, 1976) suggest the following distinguishing characteristics for the distinction between place of articulation in oral stops. The burst spectra of [b] and [p] are characterised by an even distribution of energy throughout the spectrum that is either *flat* (i.e. relatively unchanging) or *falling* with increasing frequency. The burst spectra of [t] and [d] also lack significant concentrations of energy in a particular frequency region, but in this case the spectrum *rises* with increasing frequency, particularly in the 2 to 5 kHz range; there may additionally be a peak in the spectrum at around 1.8 kHz corresponding to the alveolar "locus" frequency (Blumstein & Stevens, 1979), which is discussed in the next section. Velar stops are more difficult to characterise because their well-known context-dependent variation has such a marked effect on the burst spectrum. Before front vowels (and /j/), the tongue position for /k/ and /g/ is pre-velar or post-palatal (e.g., "key", "geese"), and in this case their spectra can be quite similar to those alveolar stops (Halle et al., 1957); before back vowels (e.g., "caught", "Gaul") their place of articulation is usually retracted to a post-velar position and their burst spectra may resemble those of bilabials (and it is on this basis, that Halle et al., 1957, propose an initial distinction between *acute*, which includes both alveolars and fronted velars and *grave*, which includes bilabials and retracted velars). The distinguishing characteristic of velar bursts is claimed to be a concentration of energy in the mid-frequency range (on a 0 to 5 kHz scale) which can vary between just under 2 kHz in the context of back vowels to over 3 kHz for velars before front vowels (Fant, 1973; Zue, 1976).

In the studies by Blumstein and Stevens (1979, 1980), which have received much attention in the literature, partly because they have suggested that the stop burst encodes *invariant* acoustic cues to place of articulation, the spectral differences are represented by three distinguishing templates, one for each place of articulation: the spectral burst template of bilabials is described as *diffuse-falling* or *diffuse-flat* (*diffuse* because the energy is presumed to be evenly distributed throughout the spectrum); the template for alveolars is *diffuse-rising*, while the velar template is *compact* and designed to model the concentration of energy in a central frequency range of the spectrum. In matching their templates to 900 CV and VC stops (C=/b d g p t k/), place of articulation could be identified by eye from the spectra 85% of the time.

The distinguishing spectral characteristics discussed above often emerge most clearly when spectra from a large number of stops are averaged (see, e.g., Cassidy & Harrington, 1995; Nossair & Zahorian, 1991; Repp & Lin, 1989). Figure 4.14 shows averaged spectra of the burst of [p^h t^h k^h] when they occur before the front vowel [i] ("key"), the central vowel [a] (as in Australian En-

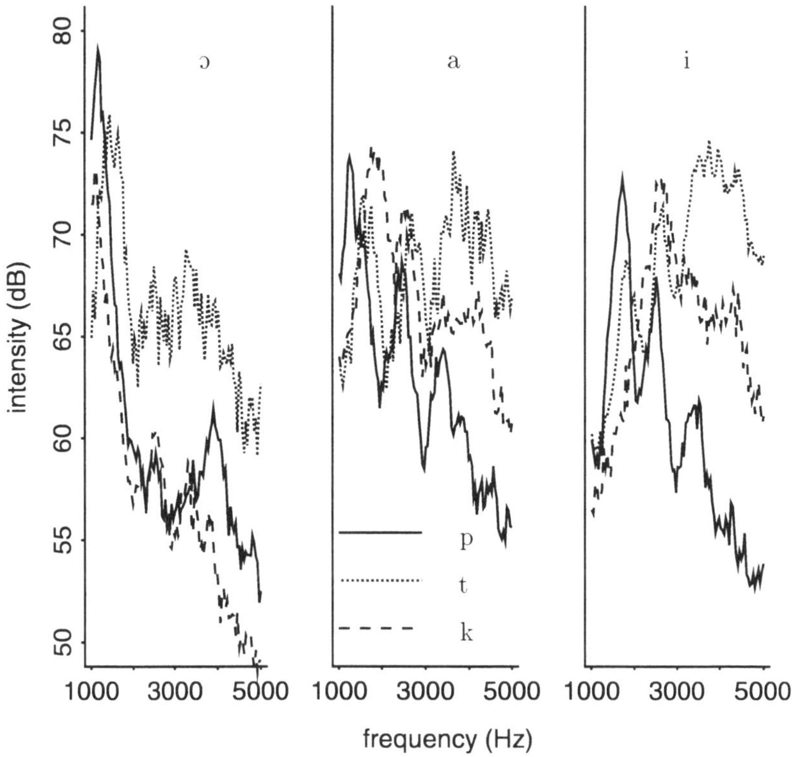

Figure 4.14: Averaged spectra centred at the bursts of the three voiceless stop categories preceding three different vowels. The spectra were centred at the burst and calculated from a 25.6 ms Hamming window. The stops are from both isolated and continuous speech data produced by five male Australian English talkers and there are between 26 ([tʰa]) and 147 ([tʰi]) tokens in each stop-vowel category.

glish "far" – which has an open central vowel nucleus in Australian English) and before the back vowel [ɔ]. The distinction between the burst in [pʰ] and [tʰ] is consistent with the analysis given earlier, at least to the extent that the bilabial burst exhibits a falling spectrum with increasing frequency, whereas the alveolar burst spectrum rises in the 2 to 4 kHz range, particularly before [i]. There is little consistency among the velar burst spectra, although there is some suggestion of a mid-frequency peak around 2.5 kHz (however, the bilabial and alveolar burst spectra also have peaks in this frequency region). In Figure 4.15, three-dimensional spectra are shown for the same data before [a]: in this case, the displays represent three averaged spectra that are centred at the bursts' 20%, 50% (midpoint), and 80% time points. The falling/rising spectral distinctions are once again in evidence for the bilabial/alveolar burst spectra,

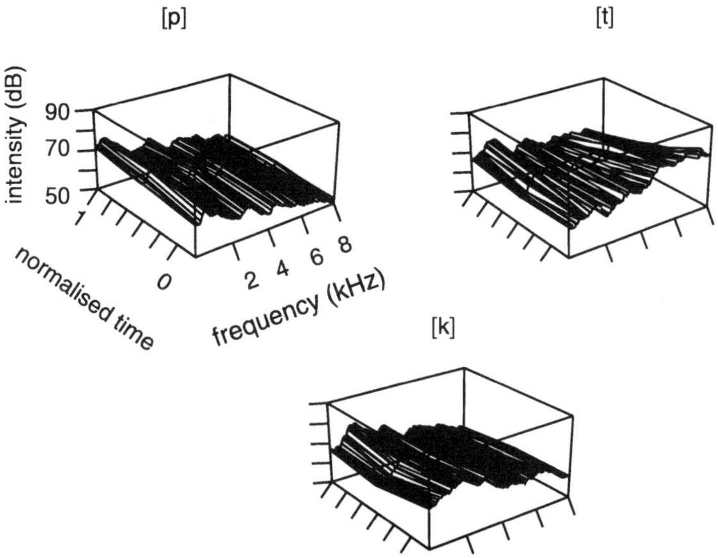

Figure 4.15: Running spectral displays of the burst in [pʰa tʰa kʰa]. The sections in the three-dimensional display at normalised time point 0.5 (centred at the burst midpoint) is the same as the spectral slices before [a] in Figure 4.14. This section is preceded by a spectral slice at time point 0.2 (spectra centred 20% of the duration into the burst) and followed by a spectral slice at time point 0.8 (80% of the duration into the burst).

and there is some suggestion of a mid-frequency peak at around 2.5 kHz, which extends throughout the velar burst spectra (Kewley-Port, 1983).

Since the influential studies by Blumstein and Stevens (1979, 1980), much research on the acoustic characteristics of stops has focused on *static* as opposed to *dynamic* cues to place of articulation in burst spectra (e.g., Blumstein, Isaacs, & Mertus, 1982; Kewley-Port, Pisoni, & Studdert-Kennedy, 1983; Nossair & Zahorian, 1991; Ohde & Stevens, 1983; Suomi, 1985; Walley & Carrell, 1983). The templates of Blumstein and Stevens (1979, 1980) are static because the spectra, which are calculated from a single long (25.6 ms) time window, represent a spectral averaging over a large part of the burst that can sometimes extend into the periodic vowel onset in voiced stops. Kewley-Port (1983) is critical of this approach and suggests that it fails to preserve the spectral *changes* during the burst that are particularly important for velar stop identification (see also Ohde, 1988). In her model, a succession of spectra are calculated for each stop from the onset to the offset of the burst resulting in a three-dimensional display which is not dissimilar to the one shown in Figure 4.15: velars are then identified in this model if spectral compactness persists throughout the running spectrum. In yet another formulation, Lahiri, Gewirth, and Blumstein (1984) advocate *difference spectra* as the basis for place articulation i.e. in their metric,

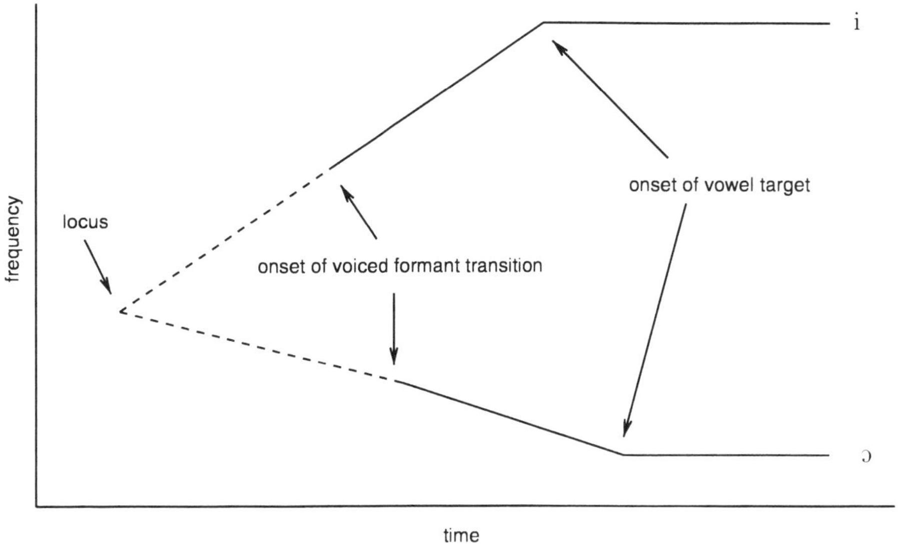

Figure 4.16: Relationship between the locus frequency and vowel target assuming straight-line transitions. The *solid lines* represent the formant transitions and targets; the *dotted lines* represent the theoretical extrapolation of the formant transitions back to a common locus frequency.

a spectrum taken in the burst is subtracted from one taken in the vowel onset.

4.2.2 Place of articulation: formant transitions

The idea that formant transitions provide cues for place of articulation can be traced back to both the work by Potter, Kopp, and Green (1947) on *Visible Speech* and to the various synthesis and labelling experiments using hand-painted spectrograms at the Haskins Laboratories (Delattre, Liberman, & Cooper, 1955a; Liberman, Delattre, Cooper, & Gerstman, 1954). These perception experiments showed that, in order to synthesise acceptable bilabial consonants, the second formant frequency had to "point to" a locus of just over 700 Hz. For alveolars, the locus frequency was shown to be nearer 1800 Hz. A locus of 3 kHz was found for velars preceding vowels that had a second formant target greater than 1.2 kHz (i.e. predominantly front vowels), but none could be found for velars preceding back vowels. A summary of the relationship between the locus frequency, transitions and vowel targets is shown in Figure 4.16. Notice that, given a locus frequency and the F2 frequency value at the vowel target, the slope of the formant is entirely predictable (assuming a straight-line transition): thus, assuming a locus for [d] at 1800 Hz, F2 should fall towards the vowel target in the context of most back vowels (for which the F2 targets are much less than 1800 Hz), whereas [di] and [dɪ] should have a rising F2 transition

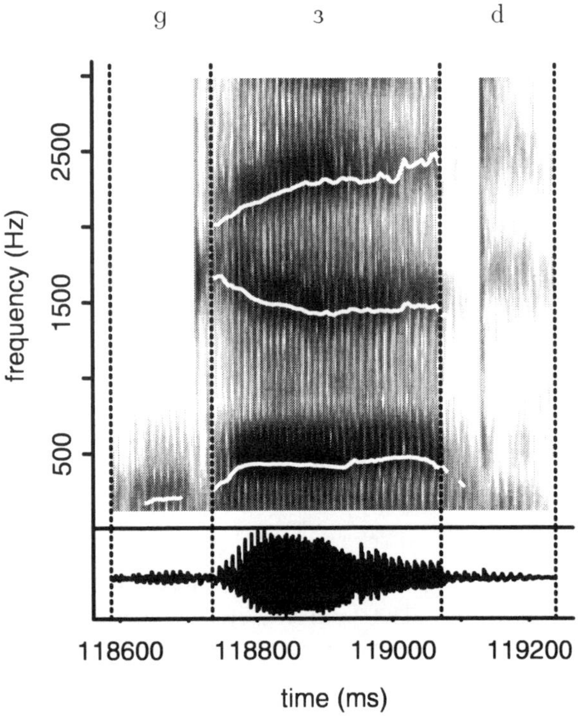

Figure 4.17: Formant transitions in the word "gird" produced by an adult Australian English male talker. There is some indication of the so-called "velar pinch" in which the F2 and F3 transitions start relatively close together in frequency (at the right boundary of [g]) and then split apart towards the vowel target.

(because the F2 target of these vowels is typically greater than 2000 Hz).

There have been various detailed acoustic studies that were, in part, designed to examine the extent to which the concept of an F2 locus was also manifested in natural acoustic speech data. Some of the most significant of these include Lehiste and Peterson's (1961b) spectrographic study of 1263 monosyllables, Fant's (1973) study of Swedish stops, Öhman's (1966) spectrographic analysis of stops in VCV syllables, and Kewley-Port's (1982) formant analysis of voiced stops in CV syllables. The following general points have emerged from these studies. First, the formant transitions are considerably more variable than predicted by the locus theory. Second, alveolar stops show the clearest evidence of an F2 locus close to 1800 Hz (for adult male talkers). Third, the variation in F2 transitions in the onglide is sufficiently great that there is little indication of a single locus frequency for bilabials (Kewley-Port, 1982, has suggested

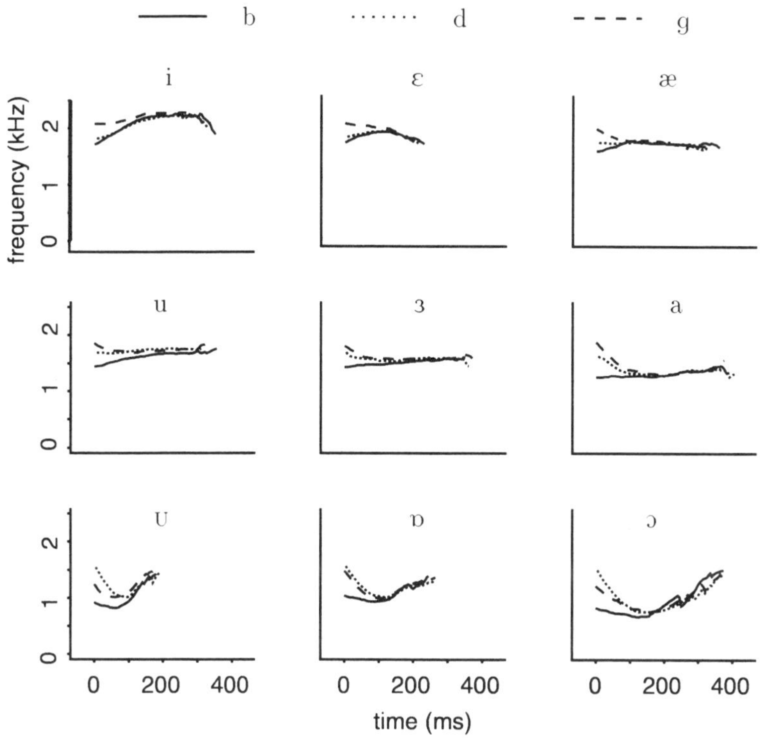

Figure 4.18: Averaged second formant frequency trajectories for [b d g] preceding various vowels in citation-form /CVd/ words produced by five male Australian talkers. The trajectories were aligned at the segment onset (at $t = 0$ ms) before averaging. Each trajectory represents an average of ten tokens (two tokens per talker) and talker-normalisation was carried out using Lobanov's (1971) technique described earlier (see Cassidy & Harrington, 1995 for further details).

there are at least two loci for bilabials, one before front vowels, one before back vowels). Finally, since velar stops are affected to such an extent by context, there is least evidence of a locus frequency for this stop class. A characteristic feature of formant transitions in velar consonants which is sometimes reported is the so-called "velar pinch" in which the F2 and F3 transitions are presumed to start close together in frequency and then split apart towards the target of the following vowel: an example of this is shown in Figure 4.17; an even more dramatic example of the velar pinch is evident at the stop-vowel transition in "gallon" in Figure 4.11 shown earlier. However, this cue in particular is very variable and may also occasionally be observed in stop-vowel transitions of the other two place categories.

Once again, the evidence for a locus frequency emerges most clearly for data that is averaged across a large number of tokens. Figure 4.18 shows averaged second formant frequency transitions from [b d g] to various vowels in isolated monosyllables. In all cases, the F2 trajectories were aligned at the periodic vowel onset and then averaged for each of three stop categories (each trajectory represents an average of approximately thirty tokens produced by five male talkers of Australian English). Consistently with the locus theory, the ordering of F2 frequency at vowel onset is highest for [g], intermediate for [d] and lowest for [b]; however, before the back rounded vowels [ʊ] [ɒ] and [ɔ], the ordering of F2 for [d] and [g] is reversed. The figure also shows that alveolars have the most consistent locus as reflected by the averaged F2 transition, which is close to 1800 Hz at vowel onset in most contexts.

Formant loci can also be estimated from acoustic data by calculating a *locus equation* in the plane of formant onset × formant target (Sussman, Hoemeke, & Ahmed, 1993). This method of analysing formant loci has been used by Sussman (e.g., Sussman, McCaffrey, & Matthews, 1991; Sussman et al., 1993; Sussman, Fruchter, & Cable, 1995) to argue that place of articulation in stop consonants can be reliably recovered from F2 and F3 transitions, even though there is a good deal of variability due to context (as Figure 4.18 shows). In order to estimate the locus frequency, a regression line is fitted to the data points in the F2 onset x F2 target plane for each stop category separately. It can then be shown that the best estimate of the locus frequency is the point at which this regression line bisects the line $y = x$ i.e. F2 onset = F2 target.[1] The locus frequencies estimated from these plots in Figure 4.19 agree quite well with those that are predicted from the averaged trajectories in Figure 4.18. The slope of the regression line is also indicative of the extent to which the formant transitions converge to a single locus (Krull, 1988, 1989): flatter lines (i.e. lower values for the slope) indicate a relatively stable locus i.e. that the straight-line trajectory from the onset to the target is strongly dependent on the phonetic identity of the preceding stop (thus note the flatter slope of the regression line for [d] compared with [b]; the slope values for the velar stops are only low because the regression lines were calculated separately in the context of the two major vowel classes that influence velar stop place of articulation).

4.2.3 Voiced/voiceless distinction: voice onset time

Voice onset time, which can be defined as the interval from the release of the stop to the periodic onset in the following vowel, is the principal parameter for distinguishing between voiced and voiceless oral stops when they occur in syllable-initial position in stressed syllables (Lisker & Abramson, 1964, 1967). VOT can be estimated from spectrograms in conjunction with waveform displays of stop consonants. Figure 4.20 shows VOT values for three different contexts: syllable-initial oral stops in isolated words; syllable-initial stops in stressed syllables in continuous speech; and syllable-initial stops in unstressed syllables in continuous speech. The results show trends that are similar to those reported in earlier studies (e.g., Lisker & Abramson, 1967; Zue, 1976): the distinction between the voiced/voiceless pairs is most clearly marked in isolated

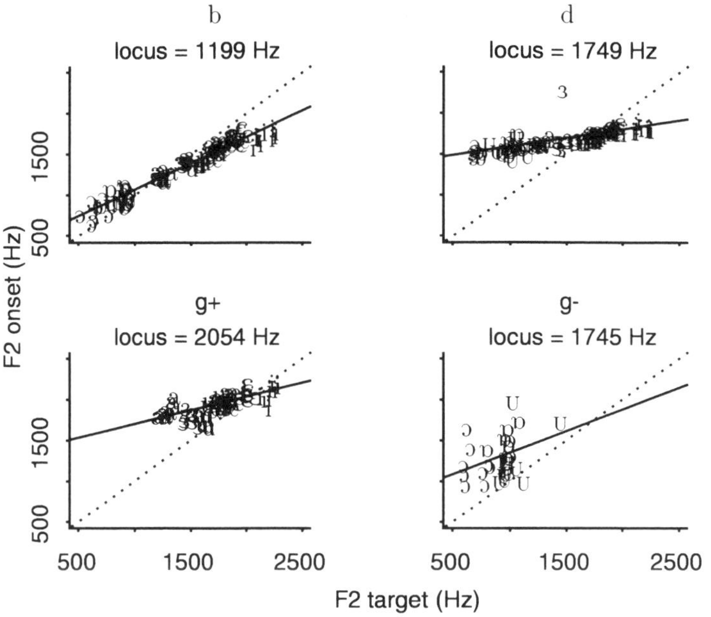

Figure 4.19: Locus equations for the same data displayed in Figure 4.18. The estimated locus, assuming straight-line transitions, is at the intersection of the dotted line $y = x$ and the solid regression line through the scatter. The labels represent the following vowel context for each stop token. The displays for [ɡ+] and [ɡ-] are for /ɡ/ before non-back, and back rounded, vowels respectively.

words and the overlap between categories at the same place of articulation is greatest in unstressed, continuous speech syllables.

The isolated word data illustrates another finding that, within any voicing category, velar stops have longer VOTs than alveolars that have longer VOTs than bilabials (Edwards, 1981; Lehiste & Peterson, 1961b; Lisker & Abramson, 1964; Kewley-Port, 1982; Zue, 1976). However, the data in Figure 4.20 and Table 4.1 show that, while this trend is clearly evident in isolated-word, citation-form speech, it is much less so in continuous speech (e.g., the VOT for [t^h] and [k^h] in continuous speech stressed syllables are very similar). The data would suggest, therefore, that VOT is of limited value in distinguishing between place of articulation in continuous speech.

4.2.4 Voiced/voiceless distinction: F1 transition

During the closure of a stop consonant, the jaw is raised, the vocal tract is occluded and the first formant frequency is at its theoretically lowest value: as the stop is released, the vocal tract opening increases and F1 rises (Delattre, 1951; Stevens & House, 1956; Fant, 1960). While a rising F1 transition is predicted to

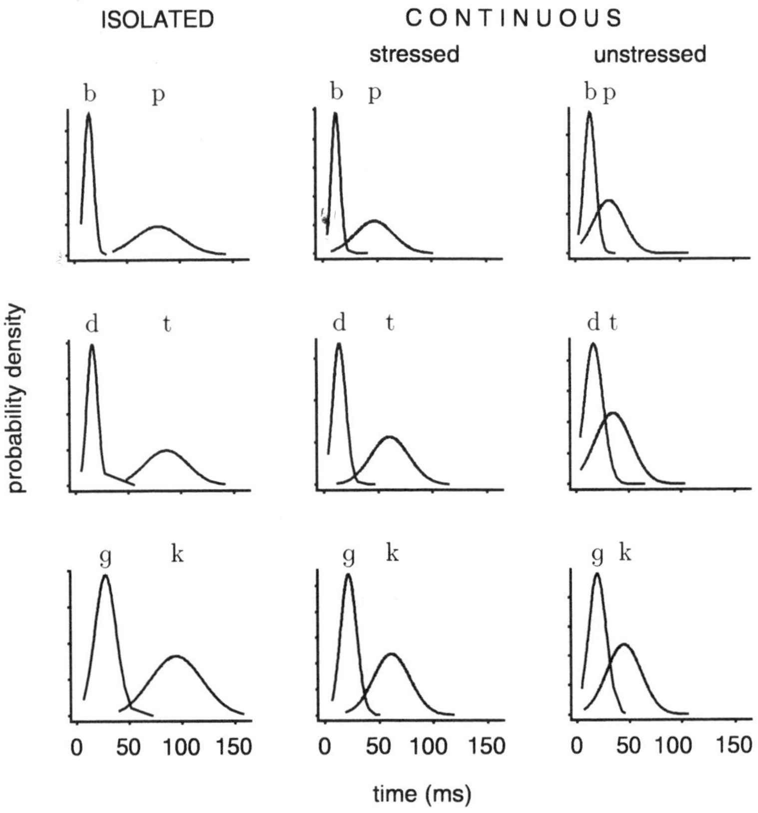

Figure 4.20: Fitted normal curves showing voice onset time durations (in milliseconds) for stops in three different types of context (see also Table 4.1)

characterise both voiced and voiceless stops in CV syllables, an essential difference is that, since periodicity starts earlier in voiced stops relative to the stop release, a greater part of the transition should be periodic: consequently, the onset frequency of voiced (pulse-excited) F1 is predicted to be lower in voiced stops (Figure 4.21). The lack of a clearly rising F1 transition in voiceless stops occurs not only because periodicity starts later, but also because of the presence of aspiration which occurs before voicing onset: aspiration can be produced only when the glottis is open and the coupling of the large resonance chamber below the glottis to the supralaryngeal tract is thought to attenuate considerably the first formant (Fant, 1960, 1980).

There is considerable evidence from perception experiments to show that a rising F1 transition (e.g., Delattre et al., 1952; Liberman, Delattre, & Cooper, 1958; Liberman, Harris, Kinney, & Lane, 1961; Stevens & Klatt, 1974; Walsh, Parker, & Miller, 1987) and lower F1 onset frequency (Lisker, 1975; Summerfield & Haggard, 1977; Revoile, Pickett, & Holden, 1982; Wolf, 1978) are cues for

	Isolated Speech			Continuous Speech					
				(stressed)			(unstressed)		
	Mean	s.d.	n	Mean	s.d.	n	Mean	s.d.	n
p	79.6	21.8	180	47.3	17.6	417	32.1	14.8	433
b	13.6	4.4	163	12.0	4.1	314	14.0	5.6	360
t	86.2	20.3	180	60.7	17.7	578	35.1	17.4	1265
d	15.8	5.1	178	14.2	6.0	273	16.8	8.8	454
k	94.5	24.4.	180	61.4	17.0	584	44.7	16.6	692
g	27.3	10.3	179	21.7	7.3	313	19.5	8.3	90

Table 4.1: Means and standard deviations of VOT (in milliseconds) for the displays shown in Figure 4.20 (n is the number of tokens per category).

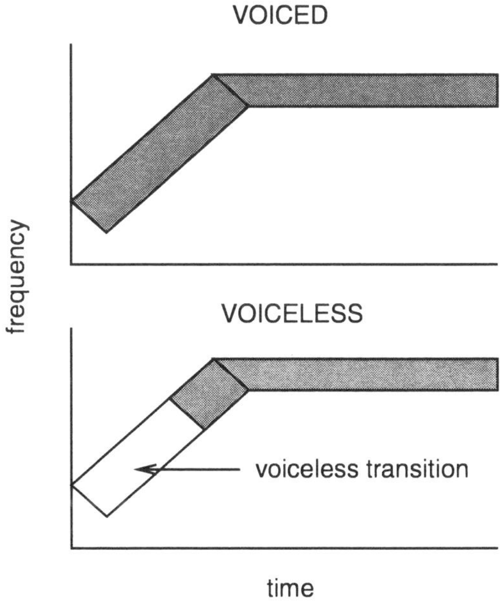

Figure 4.21: Schematic outline of F1 transitions in voiced and voiceless stops. The rectangles represent the movement of the first formant frequency from the burst release to the acoustic vowel target in a voiced (*top*) and voiceless (*bottom*) oral stop. The shading denotes the presence of voicing. Since periodicity starts earlier in voiced stops, a greater part of the F1 transition is periodic – consequently, voiced stops are characterised by a more extensive rising F1 transition.

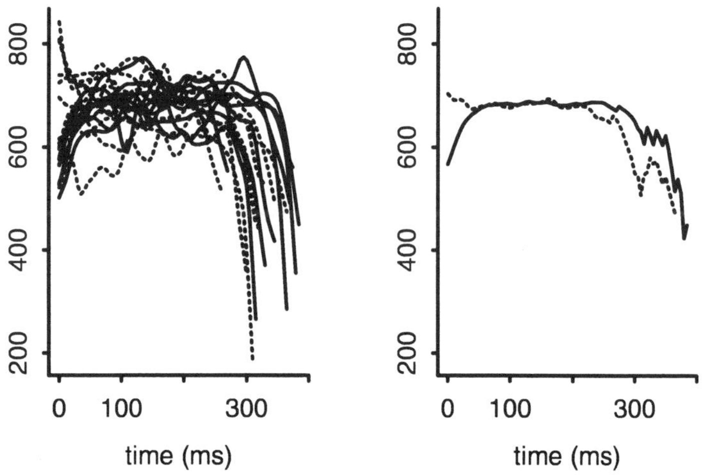

Figure 4.22: F1 transitions in [ba] (*solid lines*) and [p^{h}a] (*dotted lines*) syllables aligned at the onset of periodicity ($t = 0$ ms). When the trajectories are averaged (*right panel*), the difference in the direction and extent of the F1 transition emerges clearly.

voiced stops as opposed to their voiceless counterparts.

Some data on F1 transitions following voiced and voiceless stops is shown in Figure 4.22: the transitions are taken from the same data (five male talkers, citation-form speech) as used for the formant-loci plots, but only bilabials preceding [a] are considered (in which the extent of the F1 transition should be accentuated because the jaw moves between maximally raised and maximally lowered in [ba] syllables). The F1 traces are aligned at the onset of periodicity (left panel) and then averaged (right panel): the averaged plot shows evidence of a more extensive rising transition following [b].

4.2.5 Other cues for the voiced/voiceless distinction

The acoustic characteristics of voiced and voiceless stops have been intensively studied in the last forty years and a number of cues, in addition to VOT and F1 transition, have been suggested for their distinction. The most important of these are summarised below.

The *duration of the vowel that precedes syllable-final stops* is greater when the stop is voiced (e.g., "tab") that when it is voiceless ("tap"). This effect has been demonstrated by a number of researchers (Hogan & Rozsypal, 1980; House, 1961; House & Fairbanks, 1953; Klatt, 1973; Parker, 1974; Peterson &

Lehiste, 1960; Port, 1979; Wardrip-Fruin, 1982; Wolf, 1978). The cue for this distinction is less reliable when the preceding vowel is unstressed (Edwards, 1981) and would no doubt be substantially affected by the many different kinds of prosodic variables (e.g., phrase-final lengthening – Summers, 1987; Edwards, Beckman, & Fletcher, 1991) which affect the duration of speech sounds.

Various cues associated with the *closure*, rather than with the burst/aspiration stage, have been shown to be important for the voiced/voiceless distinction in oral stops. One of the most important of these is *vocal fold vibration during the closure* of voiced stops (often referred to as "buzz" in perception experiments) which appears as low-frequency energy (Lisker, 1978). This is especially characteristic of voiced oral stops in intervocalic position (e.g., the distinction between "rabid"/"rapid"); in perception experiments, the presence of voicing during closure has been shown to be one of the primary cues that underlies the voiced/voiceless distinction (Lisker, 1978). Another cue is the *duration of the closure* of intervocalic stops, which is less for voiced than voiceless stops (Lisker, 1957; Port, 1976; Slis & Cohen, 1969), although Lisker (1978) has also shown that the closure durations of voiced and voiceless stops in pairs of the "rapid"/"rabid" kind overlap considerably, even in citation-form speech.

Other cues include the *fundamental frequency at voicing onset*, which has been shown to be greater for voiceless stops than voiced stops in highly controlled, citation-form conditions (Haggard, Summerfield, & Roberts, 1981; House & Fairbanks, 1953; Lehiste & Peterson, 1961a; Mohr, 1971; Löfqvist, 1975; Umeda, 1981) and the *amplitude of the burst*, which is greater for voiceless stops (in similarly controlled contexts) compared with their voiced counterparts (Slis & Cohen, 1969; Repp, 1979; Zue, 1976).

4.3 Nasal consonants

Nasal consonants can often be quite easily detected from spectrographic and acoustic information by the presence of a *nasal murmur* (corresponding to the closure phase of the oral tract during nasal consonant production) whose amplitude is usually less than that of neighbouring vowels. Additionally, when the complete constriction in the oral tract (bilabial for [m], alveolar for [n], velar for [ŋ]) is released, this can result in an abrupt acoustic discontinuity that is visible as a change in the shape of the waveform at the nasal-vowel boundary or as an abrupt spectral change in spectrograms. An example of the segmentation of the [m] and [n] segments in "more money" is shown in Figure 4.23.

As discussed in Chapter 3, the spectra of nasal murmurs are characterised by both *nasal formants* (labelled N1, N2...) that are due to the nasal-pharyngeal tube and *oral antiformants* which depend on the shape of the oral side-branching resonator. (Following Fant, 1960, and Ochiai, Fukimura, & Nakatani, 1957, *oral formants* due to the mouth cavity are also expected in the spectra of murmurs, and they may show some continuity with F2 of adjacent vowels: however, they are likely to be very low in amplitude, both because of the oral closure and because of their expected proximity in frequency to the oral antiformants). While the frequencies of nasal formants are unlikely to vary substantially with place of

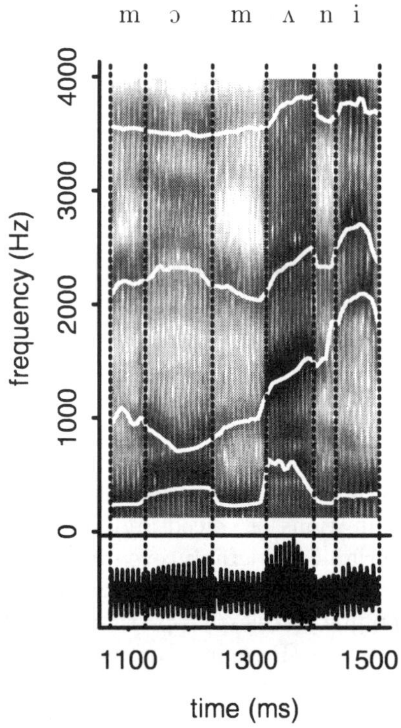

Figure 4.23: Spectrogram of "more money" showing automatically tracked formant frequencies. The low and intense first nasal formant is especially apparent in the two [m] segments. Note also the "abrupt" transition from the second [m] to the [ʌ] vowel and the much lower intensity of this [m] compared with that of the flanking vowels.

articulation (since the shape of the nasal-pharyngeal tube is expected to be similar for [m n ŋ]), from theoretical predictions of articulatory-acoustic models, it can be shown that the frequency of the first antiformant increases as the length of the oral tract decreases (see Chapter 3): the first antiformant is therefore predicted to be lowest in frequency for [m], intermediate for [n], and highest for [ŋ] (there are very few studies of higher antiformants in nasal consonants, but see Fujimura, 1962, for some comments on the second antiformant in [m] and [n]). Early studies of the acoustic structure of nasal consonants (e.g., Curtis, 1942; Delattre, 1954; Fujimura, 1962; Fujimura & Lindqvist, 1971; Hattori, Yamamoto, & Fujimura, 1958; House & Stevens, 1956; House, 1957; Nakata, 1959; Smith, 1951; Tarnóczy, 1948) have generally confirmed the following as acoustic characteristics of nasal consonants: nasal formants at approximately 700-800 Hz intervals, beginning with N1 at around 250-300 Hz; nasal formant bandwidths that are broader than the bandwidths of (oral) vowels; a first nasal formant of very high amplitude compared with that of higher formants; an antiformant in the 500-1000 Hz range for [m] and in the 1000-2000 Hz range for [n]. Many of these characteristics are also reported for nasal consonants in languages other than English (Recasens, 1983).

The locus theory of place of articulation should theoretically also be relevant to nasal (as well as oral) stops. In a synthesis and labelling experiment by Liberman et al. (1954), it was shown that place of articulation in nasal consonants could be cued by formant transitions in the vowel onglide that originated from different locus frequencies (see also Delattre, 1954; Larkey, Wald, & Strange, 1978). Since then, there have been various studies that have assessed the relative perceptual salience of murmur and formant transitions as cues for nasal place identification (e.g., Garcia, 1966, 1967; Nakata, 1959; Nord, 1976; Zee, 1981). In these experiments, new stimuli are created by cross-splicing murmurs and transitions with conflicting places of articulation (e.g., murmurs in [nʌ] syllables are replaced with those in [mʌ] syllables). The results from some of these studies show that place of articulation cues are often guided by transitions in these cross-spliced stimuli (e.g., Malécot, 1956; Recasens, 1983), although Kurowski and Blumstein (1984) show that murmurs and transitions are about equally important in cueing nasal stop place of articulation. In fact, the experiments of the last ten years, both in speech perception (Kurowski & Blumstein, 1984; Ohde, 1994; Repp, 1986, 1987; Repp & Svastikula, 1988) and in the acoustic analysis of natural speech data (Kurowski & Blumstein, 1987; Seitz, McCormick, Watson, & Bladon, 1990; Harrington, 1994) have suggested that the section of the waveform that straddles the murmur-transition boundary is where the crucial information is contained for place of articulation distinctions.

With regard to acoustic parameters, there is general agreement from recent studies that it is not viable to obtain a reliable estimate of the antiformant frequencies directly (Qi & Fox, 1992) even though this should provide direct information about the place of articulation differences. Instead, the assumption is made that the combination of nasal formants and antiformants causes gross differences in the shape of spectrum associated with different places of articulation. In Kurowski and Blumstein (1987), it is proposed that, since alveolars are expected to have an antiformant in the 1 to 2.5 kHz range, the change in acoustic energy from the murmur to the following vowel should be greater for alveolars in this frequency region (specifically in the Bark 11 to 14 or 1265 to 2310 Hz region).

In Qi and Fox (1992), a perceptually weighted linear predictive coding scheme is used to compute smoothed spectra for bilabial and alveolar nasal consonants. Their smoothed spectra show clear differences between bilabial and alveolar nasals in the 1 to 3 kHz range, and the spectra also show that the second computed peak (which cannot be directly identified with a particular formant) is higher for alveolars than bilabials. Some related data from Australian English is shown in Figure 4.24, which shows spectra of syllable-initial [m] and [n] segments taken from continuous speech utterances that have been smoothed using cepstral processing (see Chapter 6 for a discussion of cepstral

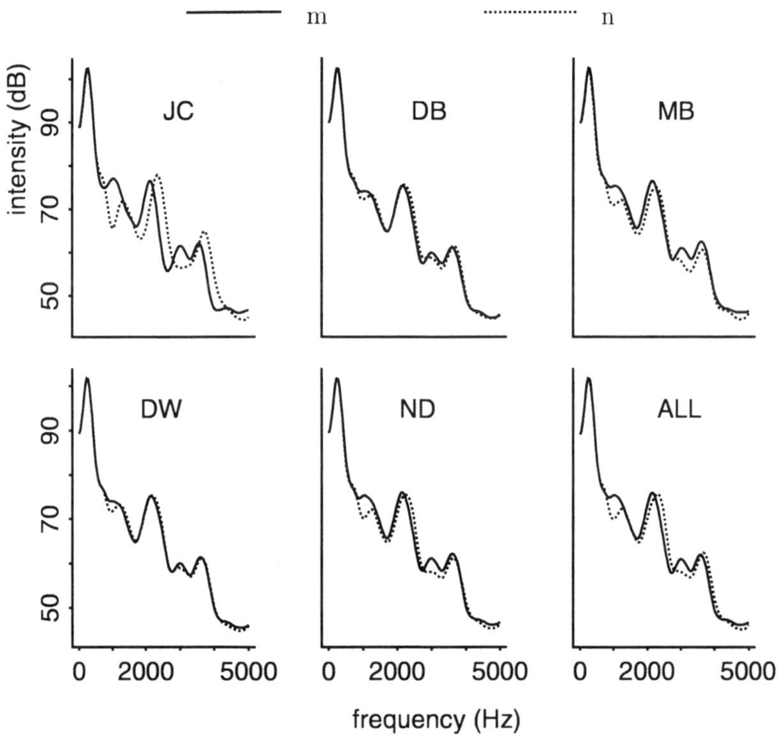

Figure 4.24: Averaged nasal spectra for the five male Australian English talkers in Harrington (1994). The spectra were calculated from a window centred 10 ms before the transition from the nasal consonant to the following vowel. Note the similar pattern of differences for all talkers between [m] and [n] in the 0.5-2 kHz range.

smoothing). The spectra were averaged for each label-type and each talker separately and were derived from a window that was centred 10 ms before the murmur-vowel onset boundary and that overlapped predominantly with the offset of the murmur (see Harrington, 1994, for further details). The averaged spectra in Figure 4.24 show differences between the bilabial and alveolar places of articulation that are similar in at least three ways to those of Qi and Fox (1992): first, there appears to be no difference between [m] and [n] in the low frequency region of the first peak; second, there is a greater dip for [n] in the 1 to 1.5 kHz range; finally, the peak for [n] at about 2 to 3 kHz is somewhat higher in frequency than that of [m] for all speakers.

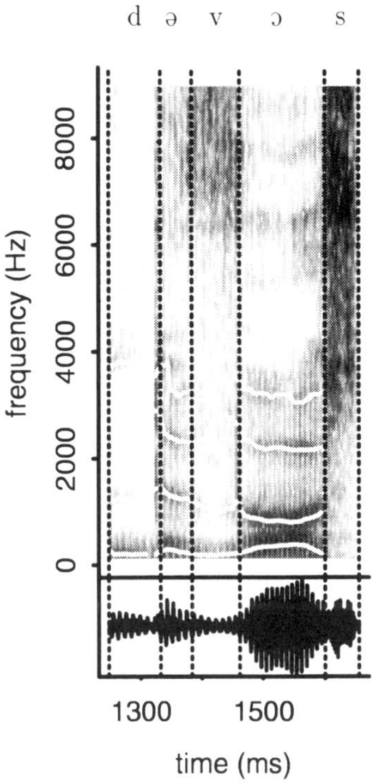

Figure 4.25: Spectrogram of "divorce": note the simultaneous periodicity below 500 Hz and noise above 6000 Hz for the [v].

4.4 Fricatives

Fricatives can usually be identified from spectrograms by aperiodic energy in a mid-high frequency range that extends throughout their production: Figure 4.1 on page 58, a spectrogram of "is this seesaw safe", shows typical acoustic characteristics of [s] in which there is a concentration of aperiodic energy above 2-3 kHz and some continuity in the fricative-noise with the second and third formants of the following vowel (see in particular the emerging formant-like structure for F2 and F3 in the [s] segments of "saw" and "safe"). In phonologically voiced fricatives, there may additionally be evidence of periodicity that occurs simultaneously with frication, although as some studies have shown (e.g., Haggard, 1978; Stevens, Blumstein, Glicksman, Burton, & Kurowski, 1992), the periodicity need not necessarily persist throughout the entire voiced fricative. Examples of simultaneous periodicity and frication are shown for the [v] of "divorce" in Figure 4.25.

The fricatives that will be considered include the sibilants [s ʃ z ʒ] and the

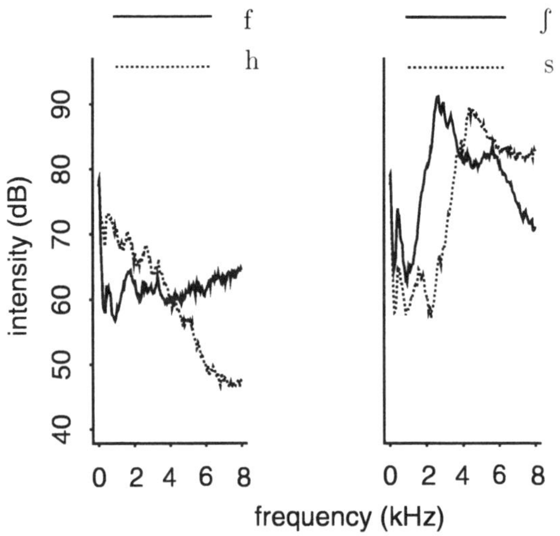

Figure 4.26: Averaged spectral slices calculated from a window centred at the midpoint of the fricatives [h f ʃ s]; see text for further details.

nonsibilants [f v θ ð h] that occur in English; some cues are also considered for separating the affricates [ʧ ʤ] from the fricatives [ʃ ʒ] at the same place of articulation.

4.4.1 Place of articulation distinctions

Many acoustic cues have been suggested for the place of articulation distinction within the fricative class including ones based on spectral shape (Hughes & Halle, 1956; Heinz & Stevens, 1961; Jassem, 1965; Lacerda, 1982; Manrique & Massone, 1981; Soli, 1981; Strevens, 1960; Yeni-Komshian & Soli, 1981), the amplitude level (Behrens & Blumstein, 1988; Gurlekian, 1981; Heinz & Stevens, 1961; Strevens, 1960), frication duration (Hughes & Halle, 1956; Jongman, 1989) and the emerging formant transitions into the following vowel (Mann & Repp, 1980; Yeni-Komshian & Soli, 1981).

Figure 4.26 shows spectra that have been averaged across tokens produced by five male talkers of Australian English in citation-form /CVd/ syllables. There were approximately 144 tokens per talker (716 in total) and approximately 180 tokens for each of the four fricative types [f s ʃ h] that were analysed. The spectra were calculated from the central 25.6 milliseconds of each fricative token.

The averaged spectra show some of the principal differences that have been noted in many of the previous studies cited above. First, both [s] and [ʃ] have a large concentration of energy in the central frequency range of the 0-10 kHz spectrum, but the energy concentration is at a higher frequency for [s] (just under 5000 Hz) compared with that of [ʃ] (just under 3000 Hz). Second, [f], in contrast to both [s] and [ʃ] lacks any significant energy concentrations: following the terminology in the feature system of Jakobson, Fant, and Halle (1952), [f] can be said to be characterised instead by a more *diffuse* spectrum. Third, the averaged spectrum for [h] shows spectral characteristics that are typical for vowels: peaks (corresponding to averaged formants) in the 0-4000 Hz region and a spectrum that falls off with increasing frequency. The reason for these vowel-like features is that the shape of the supralaryngeal vocal tract is generally very similar during [h] to the vowel that follows it (Lehiste, 1964). Finally, the average amplitude level of the spectrum is much less for both [f] and [h] (left panel of Figure 4.26) than for either [s] or [ʃ] (right panel of Figure 4.26). Although [θ] is not shown in the display, it has acoustic characteristics which are very similar to those of [f].

A significant energy concentration in a frequency location and spectral diffuseness are sometimes parameterised as *spectral moments* (Forrest, Weismer, Milenkovic, & Dougall, 1988; Nittrouer, 1995). The first spectral moment, or *spectral centre of gravity*, is a reflection of the frequency region in which most energy is concentrated: the first spectral moment for [s] should therefore be near 5000 Hz and somewhat lower than this value for [ʃ]. The second spectral moment characterises how diffuse, or spread, the energy in the spectrum is: therefore, on the assumption that [f] is the most diffuse of the fricatives under consideration, it should have a higher value on this parameter than either [s] or [ʃ]. Further details of spectral moments are given both in Nittrouer (1995) and also in Chapter 6 of this book.

Figure 4.27 shows the distribution of the four fricative types on these spectral moment parameters for five male talkers. The spectral moments were calculated from spectra in the 0-8000 Hz range. The ellipses in all cases are at just over two standard deviations from the mean and include 95% or more of the tokens from each fricative type. At least for this citation-form data, the ellipses show an effective separation on these two parameters for these four fricative types. The displays show that [s] has a higher spectral centre of gravity than [ʃ], which is in accordance with the earlier analysis; [h] has the lowest value on this parameter because of the vowel-like dominance of the lower part of the spectrum. With regard to the second spectral moment, [f] generally has higher values than [s] or [ʃ] for all talkers.

It was mentioned earlier that the amplitude level (obtained as e.g., the root-mean-square energy) can be used for place of articulation distinctions, and this has been shown to apply in particular to the separation of sibilants from nonsibilants (Ladefoged, 1971; Strevens, 1960). Figure 4.28 shows the distribution of the same fricative tokens on a parameter of RMS energy. These distributions, in which normal curves have been fitted to the data, show that [h] and [f] have a much lower RMS level than either [s] or [ʃ] (see also the averaged spectra in Figure 4.26): therefore, some of the fricatives in the [ʃ h f] set, which are confused on the spectral moment parameters, may well be separated when the RMS level is taken into account.

The analysis has so far focused on acoustic cues taken from a single spectral slice in the noise section of the fricative. However, there is evidence that the formants of abutting vowels may provide contributory information to the place

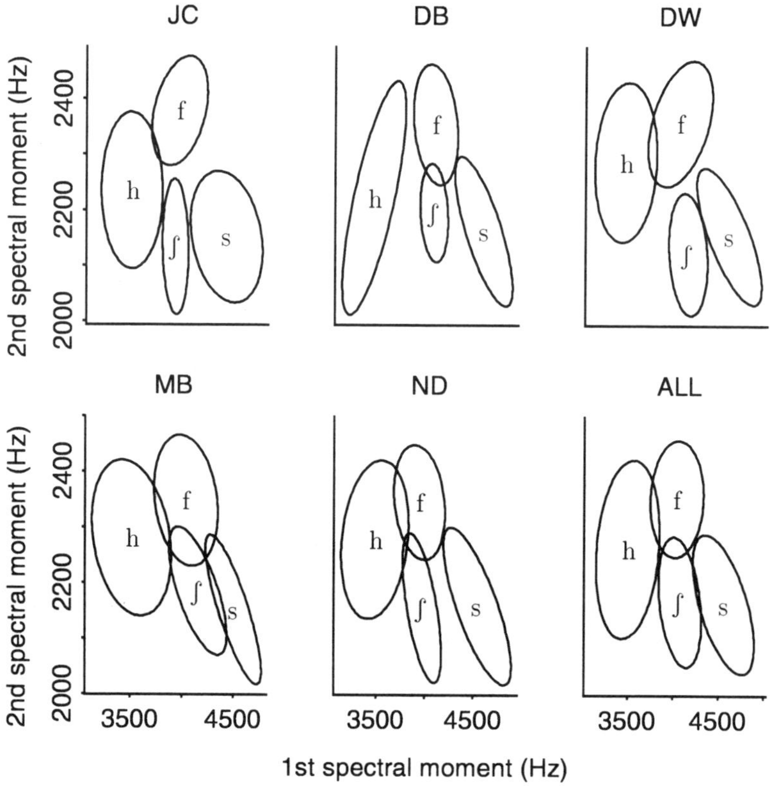

Figure 4.27: Distribution of the same fricative tokens used to calculate the averaged spectra in Figure 4.26 on the first two spectral moments for five male talkers.

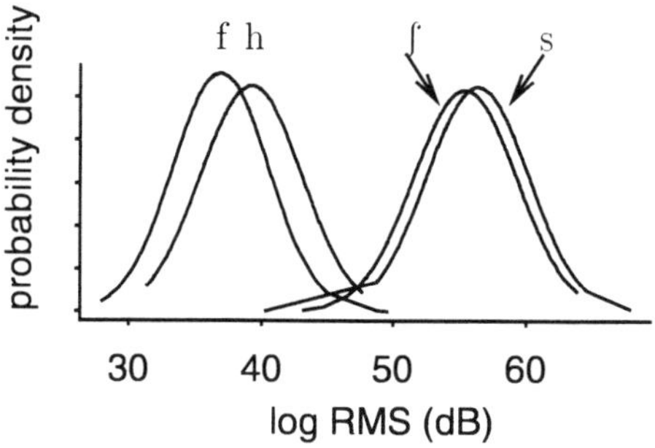

Figure 4.28: Fitted normal curves showing the distribution of the fricatives on the parameter dB RMS.

of articulation distinction in fricatives. Formant transitions have been considered to be important for the distinction between [f] and [θ] (as well as their voiced counterparts [v] and [ð]), which are insufficiently separated based on the spectral characteristics of the fricative noise alone. A perception study by Harris (1958) showed that, whereas listener identification of [s] and [ʃ] in various cross-spliced stimuli depended on spectral information in the noise, formant transitions were primary for the distinction between [f] and [θ]. A synthesis and labelling experiment by Delattre, Liberman, and Cooper (1955b) also pointed to the importance of F2 and F3 transitions for the distinction between [v] and [ð]; however, LaRiviere, Winitz, and Herriman (1975) show that when vowel transitions are removed in fricative-vowel syllables, listeners nevertheless correctly identify the transitionless noise as labiodental or dental in certain /fV/ and /θV/ syllables. More recently, Stevens (1985) has shown that the relative change in the amplitude of certain higher formants from the noise to the vowel is important for the distinction of [s] from [θ] and of [s] from [ʃ] (see also Hedrick & Ohde, 1994, for compatible evidence).

From a slightly different perspective, there are various perception experiments that have shown both that fricative noise is influenced by the formant structure of the following vowel (e.g., Yeni-Komshian & Soli, 1981) and also that the vowel can affect perceptual identification of the fricative (Mann & Repp, 1980). Specifically, Mann and Repp (1980) show that the same fricative token taken from a continuum synthesised from /s/ to /ʃ/ is more likely to be identified as /s/ in the context of a following rounded vowel like /u/ than in the context of an unrounded vowel like /a/. The reason for this is that a lowering of the spectral centre of gravity of noise is a cue both for /ʃ/ (as opposed to /s/) and for sounds produced with rounded lips (the spectral centre of gravity is lowered because lip-rounding increases the overall vocal tract length — see Chapter 3). Therefore, when listeners hear a token from an /s–ʃ/ continuum before a rounded vowel, they attribute a certain amount of the spectral centre of gravity lowering to the effects of the following rounded vowel, which they factor out (Fowler, 1984) i.e., they bias their responses towards /s/. This interesting experiment demonstrates that the vowel certainly can influence the perception of place of articulation in a preceding synthetic fricative, although this need not necessarily imply that a knowledge of the vowel is essential for separating natural [s] from [ʃ] in acoustic data, as the ellipses in Figure 4.27 would seem to suggest.

4.4.2 Affricate/fricative distinction

Various acoustic cues have been suggested for distinguishing the affricate [ʧ] from the fricative [ʃ], which are also likely to be relevant to the separation of the fricated stage of stop releases from fricatives at the same place of articulation (e.g., the frication following the closure of [t] from [s]). One of the most important of these is the *rate at which the amplitude of the frication increases* from segment onset: in an affricate, the peak energy level of the frication is reached soon after the release of the stop closure, whereas in the fricative, the build-up of acoustic energy is slower and it takes comparatively longer for the maximum amplitude level to be reached. The cue on which this distinction is based has been called *rise-time* following one of the first detailed studies on the affricate/fricative distinction by Gerstman (1957). Since then, Howell and Rosen (1983) have shown that the rise time is greater for [ʃ] than [ʧ] in isolated and continuous speech, where rise-time was defined as the time interval from the onset of frication to the maximum amplitude of frication (see also Cutting & Rosner, 1974, 1976). More recently, Weigelt, Sadoff, and Miller (1993) have applied a metric based on rise time to the distinction between both affricates and the frication stages of stops on the one hand and fricatives on the other. Their metric is based on calculating overall energy in the signal (expressed as dB RMS) and then differencing this energy parameter: in general, stops and affricates have higher values on this parameter than fricatives, which is compatible with the earlier analyses of rise time.

Other suggested cues for the fricative/affricate distinction are focused on the relative durations of the closure and frication noise (in distinguishing e.g., "why choose" from "white shoes"); relevant perception experiments include those of Repp, Liberman, Eccardt, and Pesetsky (1978) and Dorman, Raphael, and Isenberg (1980).

4.4.3 Voiced/voiceless distinction

Many of the acoustic cues that are relevant to the distinction between voiced and voiceless stops are also applicable to the separation of phonologically voiced fricatives from their voiceless counterparts. These include, for example, the *duration of the fricative noise*, which is less for voiced fricatives (Abbs & Minifie, 1969; Cole & Cooper, 1975; Massaro & Cohen, 1976), the *amplitude of the noise component*, which is less for voiced than voiceless fricatives (Soli, 1981) and the *fundamental frequency at vowel onset*, which has been shown to be lower following voiced compared with voiceless fricatives (Massaro & Cohen, 1976). In syllable-final position, *vowel duration* is greater before voiced fricatives than before their voiceless counterparts (Denes, 1955; Derr & Massaro, 1980; Flege & Hillenbrand, 1986; Soli, 1982); this cue is similarly applicable to the voiced/voiceless distinction of oral stops discussed earlier. In a recent detailed study of voiced and voiceless fricatives intervocalically and in clusters, Stevens et al. (1992) have confirmed the greater duration of the frication noise of voiceless fricatives and have also demonstrated the greater extent of the F1 transition in the vowel onglide when the fricative is voiced: this is once again similar in nature to the F1-transition cues, which are relevant to the voiced/voiceless distinction in oral stops.

Although voiced and voiceless fricatives are often presumed to be distinguishable based on the presence (voiced) or absence (voiceless) of vocal fold vibration (and therefore periodicity) during the fricative noise, acoustic analyses in fact show that combined periodicity and fricative-noise is not a necessary feature of voiced fricatives in English. Both Haggard (1978) and Stevens et al. (1992) show that sustained glottal vibration throughout the fricative noise is most likely to occur when fricatives are in intervocalic position; however, according to Stevens et al. (1992), while glottal vibration need not be sustained throughout the fricative, it always occurs in the VC and CV transitions (in VCV stimuli, where C is a phonologically voiced fricative) and the greater extent of voicing during the transitions is one of the most significant acoustic and perceptual cues for their separation from their voiceless cognates.

4.5 Approximants

Approximants, vowels and diphthongs belong to the class of *oral sonorants*, which are characterised acoustically by periodicity and F1-F3 in the 0-4000 Hz spectral range. In English, the approximants include the liquids /l/ ("leaf") and /r/ ("red") and the glides /w/ ("we") and /j/ ("you"). Once again, there can be many different phonetic realisations of the approximant set that can have a marked effect on the resulting acoustic structure; there is also considerable allophonic variability between accents of English (see Gimson & Cruttenden, 1994 for further details). For example, in most English accents, the approximants are realised with varying degrees of voicelessness following voiceless consonants, and in particular following voiceless stops (e.g., "please", "quite", "cute", "pry"); and many accents of English have two distinct kinds of /l/, a clear [l] which tends to occur in syllable-initial positions before vowels, diphthongs and /j/ and a dark (velarised) [ɫ] in other contexts – however, some accents (e.g., Irish English) have only a clear [l] and in Australian English there is much less distinction between the two kinds of /l/ compared with British English RP (Wells, 1982). There is also a good deal of variation in the realisation of /r/ which is a post-alveolar approximant in some accents (e.g., RP and Australian English), but has many other allophonic realisations including an alveolar flap (particularly in certain Scots accents) and (less commonly) a uvular approximant (Gimson & Cruttenden, 1994). Since these allophonic variations can have a substantial effect on the acoustic signal, it is important to be aware of them when classifying approximants from acoustic speech data.

The central difficulty in analysing approximants acoustically is their segmentation from other sonorants, in particular from unstressed vowels. Another problem is that, while their low amplitude can be used as a major cue for their distinction from other sonorants, it is also a source of confusion with other sonorant-like low-amplitude sounds that can include not only unstressed vowels but also nonsibilant fricatives (particularly if they are realised as approximants) such as /v/, /ð/, and /h/ (Weinstein, McCandless, Mondshein, & Zue, 1975; Espy-Wilson, 1994).

Figure 4.29 shows a spectrogram of the utterance "he will allow a rare lie" in which the segmentation of approximants is likely to be quite difficult. Most of the approximants are distinguishable from the abutting vowels by a decrease in energy in the F2-F4 region, although it is difficult to justify on acoustic grounds where precisely segmentation boundaries should be placed (and some might argue that, because an approximant-vowel sequence has no clearly defined discontinuity, it should not be segmented at all). Most approximants also have a

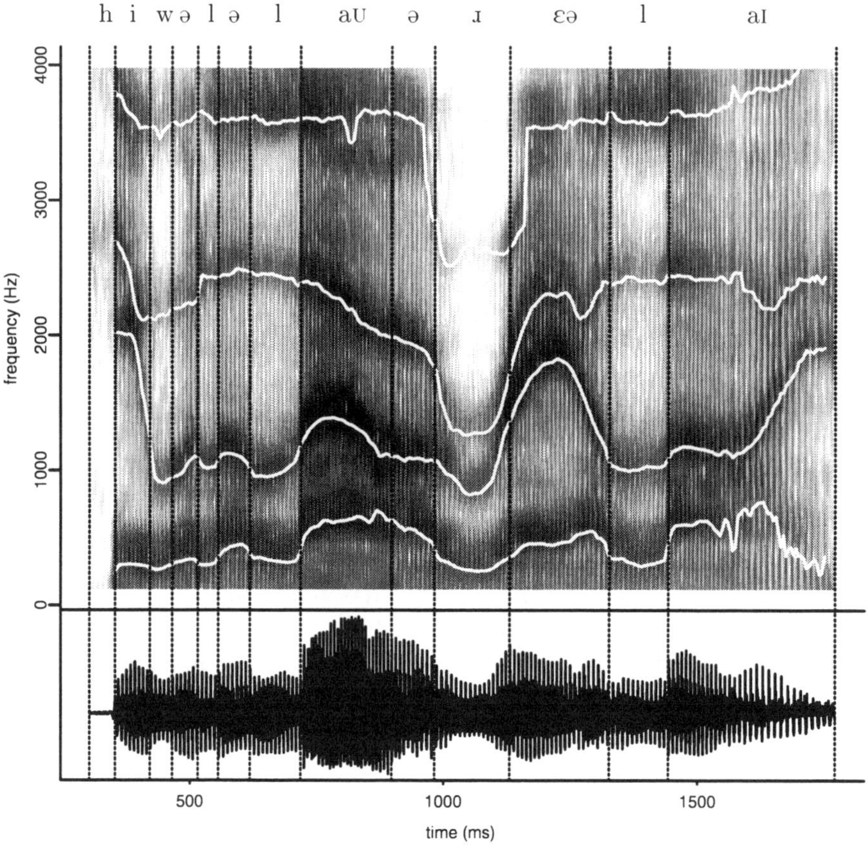

Figure 4.29: Spectrogram of "He will allow a rare lie". Note the decrease in amplitude for the approximants as shown by a decrease in the darkness of the striations on the spectrogram.

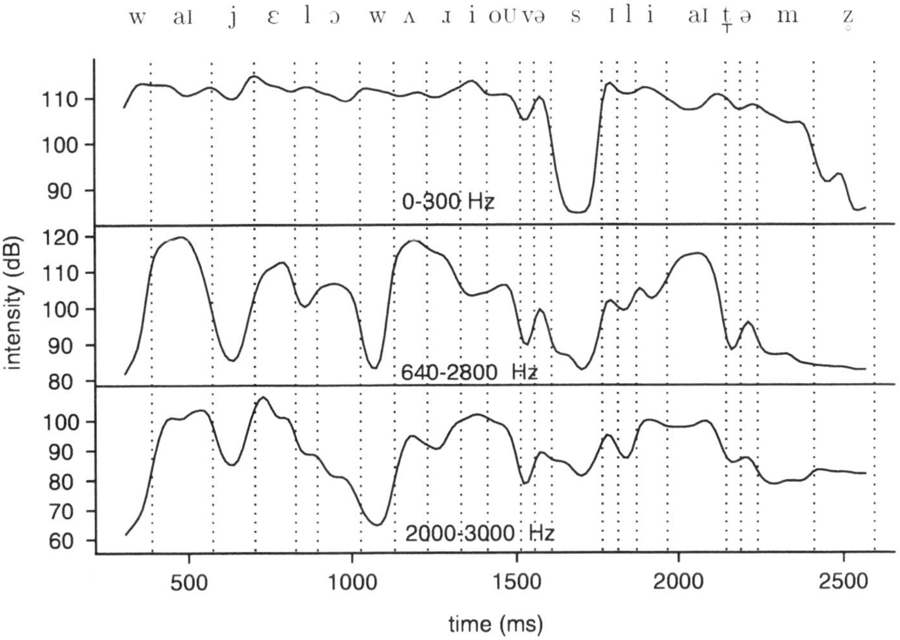

Figure 4.30: Summed energy values in three frequency bands for the utterance "why yell or worry over silly items". The top parameter (0-300 Hz) is used to separate sonorants from fricatives; the bottom two are used for the vowel/approximant distinction (approximants should be characterised by energy dips on these parameters).

distinctive formant pattern (Lehiste, 1964) that allows a further differentiation from other oral sonorants – these issues are considered in further detail below.

4.5.1 Energy dips

As suggested above, energy dips in selected frequency bands can often be used to segment approximants from other sonorants. In a recent analysis of approximants by Espy-Wilson (1992, 1994), some of the central parameters for detecting approximants included energy values in the 0-300 Hz, 680-2800 Hz, and 2000-3000 Hz bands. The first of these is used to separate sonorants in general from non-sonorants (sounds that are periodic have a concentration of energy in this frequency region); the basis for the second two is that approximants should show greater evidence of energy dips in these frequency ranges compared with most vowels.

Figure 4.30 and Figure 4.31 show summed energy values in these three bands for the utterances "why yell or worry over silly items" and "where were you while we were away" produced by an Australian English male talker. For the

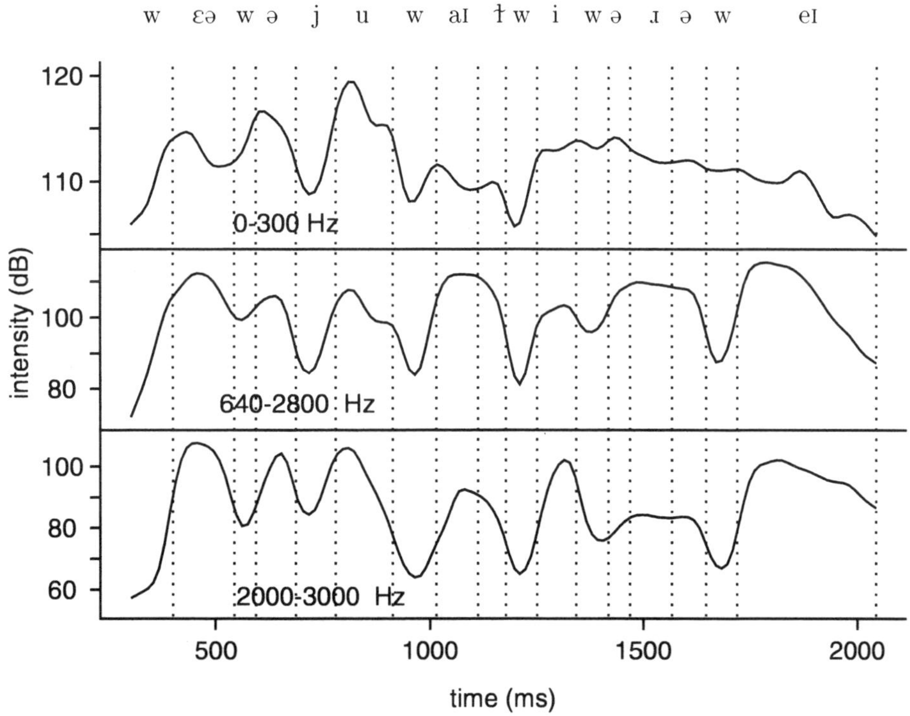

Figure 4.31: Summed energy values in three frequency bands for the utterance "Where were you while we were away?"

first of these utterances, the dip in the 0-300 Hz band can be clearly seen for the fricatives [s] and [z], but it is less apparent for the /t/ of "items" that was realised without complete closure (this is common before unstressed vowels in the database of Australian English from which this utterance was taken). There are troughs in the other two energy bands, although there are also dips for the [v] of "over", and the trough for [ɹ] (in band 2000-3000 Hz) is quite small. Focusing on these same bands (640-2800 Hz and 2000-3000 Hz) in the other utterance (Figure 4.31), there are clear dips for all approximants except for [ɹ] and for the [ɫ] of "while": the first of these difficulties can be resolved by considering formant frequencies; the second of these is problematic and occurs because of the heavily velarised [ɫ] of "while" which makes it acoustically almost indistinguishable from the following [w] (as Lehiste, 1964, and others have shown, both [w] and velarised [ɫ] have a very similar formant structure).

4.5.2 Formant frequencies

Some of the approximants have distinctive formant frequency values that may provide a further basis for their separation from other sonorants.

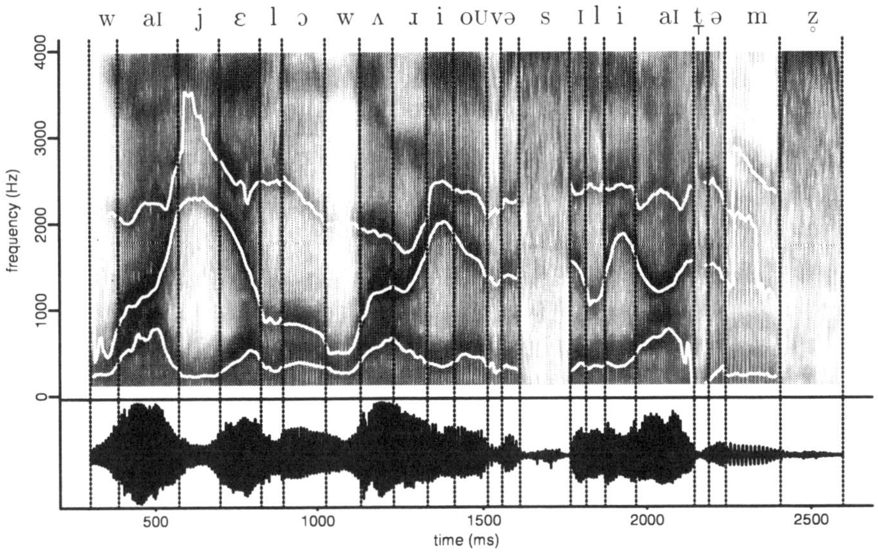

Figure 4.32: Wideband spectrogram showing the first three automatically tracked formant frequencies for the same utterance in Figure 4.30.

The glides [j] and [w] have a very similar formant structure to the corresponding close vowels [i] and [ʊ] respectively.[2] Thus F1 for [j] is usually low in the 200-300 Hz range while F2 is high at around 2000 Hz (Lehiste, 1964): these acoustic attributes give [j] a distinctive formant structure which differentiates it from other approximants and most vowels as Figure 4.32 and Figure 4.33 show. The glide [w] also has a distinctive formant pattern, but in this case F1 and F2 are low in frequency and close together: studies report F1 and F2 in the 300-400 Hz and 600-800 Hz ranges respectively (Dalston, 1975; Lehiste, 1964; Mack & Blumstein, 1983). A characteristic feature of [w] is that F2 should dip to a low value close to F1: this is once again evident in Figure 4.32 and Figure 4.33.

A well-known acoustic attribute of the liquid [ɹ] is its low third formant frequency value which is lower than F3 of all other sonorants. Typical values for F3 of [ɹ] in male talkers are 1300-1800 Hz (Dalston, 1975; Lehiste, 1964; Nolan, 1983): the [ɹ] segments of "were" and "worry" in Figure 4.32 and Figure 4.33 have the lowest F3 values.

It is more difficult to quantify the formant structure of [l] because of the large degree of contextual variability discussed earlier. At the very least, the two main allophones of /l/ (clear and dark) must be treated differently in an acoustic analysis. There is some agreement that clear [l] has F1 in the 250-400 Hz range (Bladon & Al-Bamerni, 1976; Dalston, 1975; Lehiste, 1964; Nolan, 1983; O'Connor, Gerstman, Liberman, Delattre, & Cooper, 1957). The same studies, and in particular those of Lehiste (1964) and Nolan (1983) show that F2 of [l] is influenced considerably by the following vowel: it is as high as 1500-1600

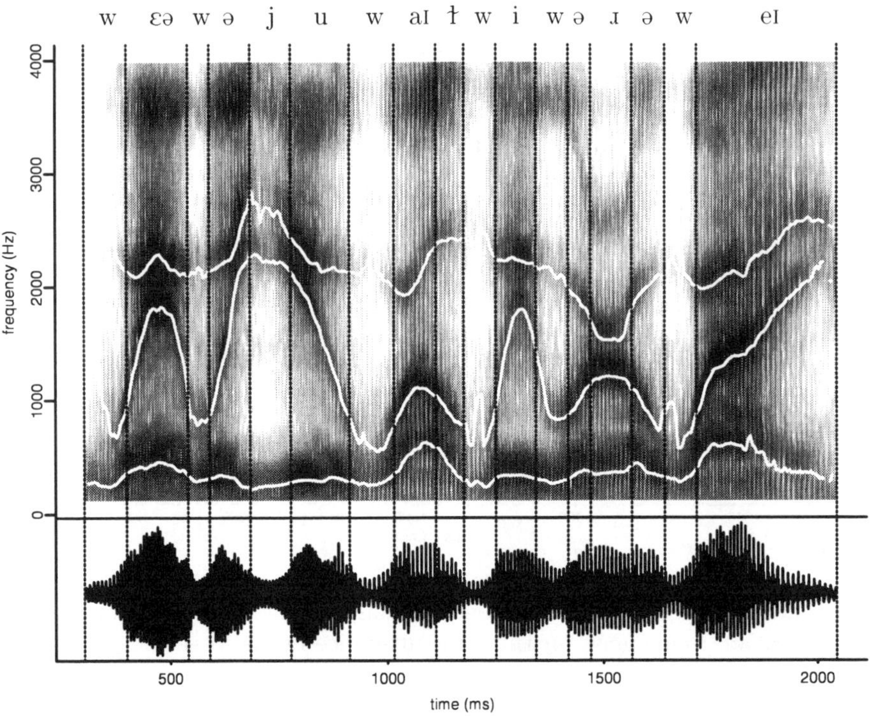

Figure 4.33: Wideband spectrogram showing the first three automatically tracked formant frequencies for the same utterance in Figure 4.31.

Hz when [l] occurs before vowels with a high F2 (i.e. [i]) but less than a 1000 Hz for [l] before back vowels with a low F2 target. The identification of F3 in [l] is problematic due to the presence of an antiformant (caused by the shunting effect of the mouth cavity behind the tongue blade), which tends to occur very close to F3 (Fant, 1960; Nolan, 1983): consequently, F3 is sometimes nearly cancelled by the antiformant, which can often cause F4 to be misinterpreted as the third formant frequency.

There is some suggestion that the formants from [l] into the following vowel may have a very short transition (Dalston, 1975; Polka & Strange, 1985), which, on spectrograms, can give the impression that the formants "jump abruptly" from the offset of [l] to the following vowel target. This attribute was recognised by O'Connor et al. (1957) for F1 in the synthesis study referred to earlier and a metric for identifying [l] based on this rapid spectral change is suggested in Espy-Wilson (1994).

It is generally agreed that F2 of dark [ɫ] is considerably lower than for clear [l] (this is because a dark, or velarised, [ɫ] has a back vowel quality that gives it a formant structure not dissimilar to [ʊ] i.e. with a low F2). Typical values of

F2 for velarised [ɫ] are in the 600-900 Hz range (Lehiste, 1964; Dalston, 1975) while F1 is reported to be in a similar, or slightly higher range than F1 of clear [l] in these same studies. Dark [ɫ] is much less prone to the coarticulatory influences of neighbouring vowels than clear [l] (or, as Bladon & Al-Bamerni, 1976, would say in their detailed acoustic study, [ɫ] has a greater *coarticulation resistance* than [l]). An acoustic consequence of these coarticulatory differences is that the F2 variation of dark [ɫ] due to context is much less than that of clear [l] (Lehiste, 1964; Nolan, 1983).

4.6 Prosody and juncture

4.6.1 Stress

In this section we present an overview of lexical-stress and accent in English and then consider some possible acoustic correlates of stress differences. The analysis is based to a large extent on a theory of metrical stress contrasts as discussed in Liberman and Prince (1977), Hayes (1984), Selkirk (1984), Nespor and Vogel (1986), and current models of intonational phonology that have been developed in various studies (e.g., Bruce, 1977; Pierrehumbert, 1980; Beckman & Pierrehumbert, 1986; Pierrehumbert & Beckman, 1988; Pierrehumbert & Hirschberg, 1990; Beckman, 1996; Ladd, 1996).

All languages have various mechanisms for highlighting some parts of spoken utterances. If we listen to any utterance in English, we will recognise that some syllables sound more prominent than others and, at another level, that some selected words or phrases are particularly salient. Stress is a general term that covers all manner of different kinds of contrasts in salience or prominence in spoken language.

It is important from the outset to recognise that there are at least two quite different kinds of "stress". We begin by considering the first of these, *lexical-stress* or *word-stress* which is concerned with the prominence relationships between the *syllables of a word.* English is a *quantity sensitive* language and builds its stress patterns from alternating *heavy* and *light* syllables (Vanderslice & Ladefoged, 1972; Beckman, 1986): this is the lowest level of contrast that is relevant for building lexical-stress patterns in English. From a phonological point of view, a syllable can be defined as heavy if it contains either a tense vowel or diphthong (e.g., "see", "say", "talk") or if it dominates a lax vowel followed by an obligatory syllable-final consonant (Beckman, 1996). Examples of this second type occur in monosyllables such as "put", "pat", "pet" and include the first syllable of "pattern" or the second syllable of "America".

Focusing now on the representation of lexical-stress patterns from the word level downwards, many models of lexical-stress include a level of the *foot* or *stress-foot* which obligatorily contains at least one heavy syllable followed optionally by any number of light syllables. When a word has only one stress-foot (i.e. there is only one heavy syllable in the word), then that is the syllable which in a more traditional treatment of lexical-stress would be designated at the *syllable with primary lexical stress.* A by-product of the constraint that words must

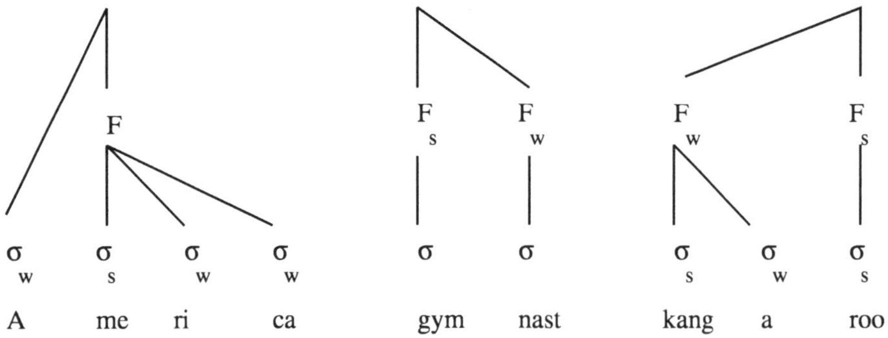

Figure 4.34: Lexical-stress relationships for three words. In "America", the contrast is between a strong, or heavy, second syllable and the other weak, or light syllables. (The /r/ phoneme is presumed to be ambisyllabic i.e. to be simultaneously in the coda of the second syllable – which makes it heavy – and in the onset of the third). In "gymnast" both syllables are heavy and the contrast is at the level of the foot. In "kangaroo", the contrast is both at the syllable and foot levels.

minimally have one stress-foot (and that a stress-foot must dominate a heavy syllable) is that there can be no monosyllabic words which end in lax vowels in English (e.g., no words such as /pɛ/ or /tæ/). The heavy syllable in a foot is sometimes known as the *head syllable* of the stress-foot.

There are of course also many polysyllabic words that have two stress-feet i.e. two heavy syllables. Examples of these would include words such as "Adelaide", "gymnast", and "kangaroo". Some bipedal words have the primary stress on the syllable of the first foot ("Adelaide", "gymnast") while others ("kangaroo") have the primary stress on the second foot when these words are said in isolation.

From the above discussion we can see that stress is concerned with prominence *relationships* (e.g., a particular syllable is stronger or more prominent than another) and that these relationships can occur at two different levels in a word-internal *hierarchical structure* consisting of feet and syllables (Figure 4.34).

The relational and hierarchical properties of stress also carry over into the second kind of "stress" known as *accent* or equivalently *sentence-level stress*. However, whereas lexical-stress is the pattern of prominence relationships between the syllables of a word, accent defines the prominence relationships between the *words in a phrase or utterance*. In the same way that we hear the second syllable of "America" as more prominent than any of the other syllables, the word "Bridge" is more prominent than any of the other words in B's reply to A's question:

A: Where did you leave the boat?
B: Under the Harbour Bridge.

Whereas in the following reply, "Harbour" is the most prominent word:

A: What bridge did you leave the boat under?
B: Under the Harbour Bridge.

This immediately points to another difference between lexical-stress and accent: leaving aside questions of stress-shift to be discussed below, the lexical-stress pattern of English words is usually fixed ("America" is in almost all conceivable contexts produced with primary stress on the second syllable), whereas the accentual pattern is variable and depends, among other things, on which parts of the utterance are *brought into focus* (in the second example, "Bridge" is deaccented because the information is "understood" or, to use terminology from Halliday, 1967, is *given*). As discussed in further detail in Section 4.6.4, when a word is accented, a *pitch-accent* is associated to the syllable that has primary lexical-stress. The result of this association is a change in pitch that often results in a clearly visible pitch peak or pitch trough in a record of the fundamental frequency of the utterance.

There is another level at which accentual prominence distinctions are made. By definition, the last accented word of any phrase is always the *nuclear accented word*, and it is usually perceived to be more prominent than any other accented words. We can therefore establish a second prominence contrast between the nuclear accented "Bridge" and accented "Under" in the first (default) example.

The accentual relationships are often sketched in a *grid*. The following shows the distribution of the accented and nuclear accented words for the first (default) version of "Under the Harbour Bridge". Words with no x marks are unaccented:

	Under	the	Harbour	Bridge
accent	x			x
nuclear accent				x

Having established the qualitative differences between lexical-stress and accent, we must now give some consideration to how they are related. Following Bolinger (1958) and the more recent expositions in e.g., Beckman and Edwards (1994) and Beckman (1996), an accent can dock onto any heavy syllable; when a word has more than one heavy syllable ("kangaroo") the default is for the accent to be attached to the heavy syllable in the strongest foot i.e. the syllable with primary lexical-stress. Therefore, when "Under" is accented, it is actually the primary stressed syllable "Un" that receives an additional prominence. Similarly, if "Adelaide" were accented, the first syllable (which has primary lexical-stress) receives an additional prominence due to accentuation. The combined lexical and accent prominence pattern for a neutral, declarative production of "Adelaide is in South Australia" is likely to be

	Ade	laide	is in	South	Australia
heavy	x	x		x	x
accent	x				x
nuclear accent					x

From this grid representation, a prominence pattern emerges for an entire utterance. Thus the second syllable of nuclear accented "Australia" is the most prominent syllable (this syllable is also often called the *tonic syllable*) and the first syllable of "Adelaide" the next most prominent syllable.

There is considerable evidence from many different kinds of studies that prominent syllables carry the most salient information for understanding an utterance. For example, various studies of lexical statistics (e.g., Aull & Zue, 1985; Huttenlocher & Zue, 1984) have shown that lexically stressed syllables are much more important for word distinctions than unstressed ones: these studies show that the confusion between lexical items increases dramatically if phonemes of stressed syllables are collapsed into a "broad class" representation, but far less so if the same is done to the phonemes of unstressed syllables. As far as accent is concerned, various studies show that one of its main functions is to highlight part of an utterance as important for the listener (Bolinger, 1972, 1975) and to signal new or important information (Nooteboom & Kruyt, 1987), and there are studies which show that accent helps to direct listeners' attention to points of information focus (Cutler & Foss, 1979).

4.6.2 The acoustic basis of stress distinctions

While on the one hand it would seem almost intuitive that rising and falling x-marks in the grid should have some identifiable correlate either in the production of speech or the speech waveform, in fact there is no evidence of any such relationship. There are many reasons why this is so. Perhaps the main one is that stress is a highly abstract concept. Although a segment like [i] is also an abstraction from the detailed production of speech, we can nevertheless relate it to quasi-predictable movements of tongue-dorsum raising and its acoustic consequences. The most concrete statement that we can make about stress on the other hand is that it sets apart, or demarcates, some syllables relative to others. Additionally, as we have seen, different sets of stress relationships are defined at multiple hierarchical levels and the cues that signal a contrast at one level (e.g., lexical-stress) may be different from those at another (e.g., nuclear accented as opposed to unaccented vowels). As discussed in Beckman and Edwards (1994), researchers investigating the acoustic and articulatory correlates of stress have not always distinguished carefully between the different kinds of stress and this has undoubtedly contributed to the conflicting results that have emerged from numerous experimental investigations.

If we begin by considering the lowest level in the hierarchy of stress contrasts i.e. the distinction between heavy and light syllables, then, since most light syllables have a nucleus that is /ə/ or that can reduce to a schwa-like vowel, the acoustic distinction can be based to a large extent on *vowel quality*: that is, heavy syllables have a more peripheral vowel quality and so their formants should occur closer to the edges of the formant (F1/F2) plane than those of light syllables. Additionally syllables with /ə/ nuclei or nuclei reducible to /ə/ are short in duration and low in intensity; and since light syllables can never be accented, they cannot be produced with the marked pitch change due to the association of a pitch-accent.

At the next level of the hierarchy we can consider the acoustic basis of the distinction between heavy syllables that do, and do not have, primary stress as in the contrast between minimal pairs of the "permit" kind in which the primary stress is on the first syllable when the word is a noun ("a *per*mit") , but the second syllable when it is a verb ("to per*mit*"). There are, however, various methodological difficulties with such an investigation. Foremost is that when these words are said in isolation, they are also necessarily overlaid with the stress contrasts at the higher levels i.e. the accent and nuclear accent levels. For example, a single-word production of the verb could only ever be:

	per	mit
heavy	x	x
accent		x
nuclear accent		x

or, if the first syllable were produced with a /ə/:

	per	mit
heavy		x
accent		x
nuclear accent		x

So the actual contrast which is being investigated in numerous studies that have dealt with these noun/verb contrasts (see Lehiste, 1970, for a review) is probably the above representation for the verb as opposed to the following for the noun:

	per	mit
heavy	x	
accent	x	
nuclear accent	x	

It is not surprising therefore, that pitch change has been reported as one of the main acoustic cues for the distinction of such noun/verb pairs – but this is almost certainly because there is a pitch-accent on the second syllable of the (accented) verb and the first syllable of the (accented) noun. A more exacting test of the acoustic differences that are due to such lexical-stress contrasts would have to be based on comparisons of the noun/verb pairs when the word occurs in an *unaccented* position. Interestingly, Huss (1978) found that when the minimal noun/verb pair "*im*port"/"im*port*" occurs in an unaccented position (i.e. no pitch-accent), the contrast between the pair is not perceptible (but see Sluijter & Heuven, 1996 for recent evidence that similar minimal-pair words are acoustically distinguishable in Dutch even when they are in a non-nuclear position).

At the level of contrast between *accented* (including nuclear accented) and *unaccented* vowels, a pitch change is undoubtedly one of the most significant cues for this distinction because, by definition, accented vowels are marked by a pitch-accent. Additionally, the accented/unaccented word pairs are often distinguished by durational and intensity differences. Thus in Harrington, Fletcher and Beckman (in press) the vowel of nuclear accented "Babber" was shown to have a greater intensity and duration than its unaccented (deaccented) counterpart in a dialogue such as

A: Can I speak to Dr Babber please?
["Babber" is nuclear accented]
B: Do you want Dr Anna Babber or Dr Clara Babber?
["Babber" is deaccented in both cases, and the nuclear accents fall on "Anna" and "Clara"]

There is also evidence from studies such as these (see also Engstrand, 1988; de Jong, 1995) that accented vowels are often phonetically more peripheral in the vowel space: for example, in the study by Harrington et al. (in press), accented [i] was found to have a higher F2 value at its target making it acoustically more distinct from the other vowels of the language.

4.6.3 Rhythm

A characteristic feature of word-stress in English and Germanic languages is that there is a very general alternation between heavy and light syllables. In English, words which have two abutting heavy syllables such as "gymnast" , "umpire" and "marsupial" (first two syllables are heavy) are much less common than the words with stress patterns in "apart", "pattern", and "rhododendron" in which heavy syllables are followed or preceded by light syllables. On the other hand, French and other Romance languages exhibit no such alternation, and there are no light syllables in the same sense as in English that tend to reduce to a central vowel. The idea that there is a quasi-alternating pattern of heavy and light (or what are often called "stressed" and "unstressed") syllables is evident in the sentence "Adelaide is in South Australia". In this case, there is a near alternation of an h-l (heavy-light) sequence:

Adelaide is in South Australia
h l h l l h l hl

The tendency for "stressed" syllables to be separated by "unstressed" ones in English also operates at the next hierarchical level of stress contrasts in some polysyllabic words that have two stress feet (Liberman & Prince, 1977; Hayes, 1984; Shattuck-Hufnagel, Ostendorf, & Ross, 1994; see also Kingdon, 1958; Bolinger, 1965). For example, an isolated production of the bipedal word *Japanese* has primary stress on the third *nese* syllable, but in the context of *Japanese food*, the primary stress tends to shift to the first syllable (Bolinger, 1965). There are many other examples: in all cases, words that are candidates for this kind of stress-shift must have two or more stress-feet and cannot have primary stress on the first foot in a citation-form production. Thus stress-shift is possible in "Chinese" and "kangaroo" because these have two feet and primary stress on the second foot. But "America" is not a candidate for this kind of stress-shift because it only has one foot; and neither is "gymnast" because although it has two stress-feet, the primary stress is on the first foot, not the second. As shown in an analysis of a large corpus in Shattuck-Hufnagel et al. (1994), at least one of the explanations of stress-shift is in terms of *pitch-accent clash* (another explanation, which we will not discuss in detail here, is that talkers may generally prefer an early pitch-accent placement to mark the beginning of a phrase). Consider that in producing "Japanese food", a talker is likely to accent both words and place the nuclear accent on "food". Since "Japanese" has a default primary stress on the third syllable this would result in a prominence pattern as follows:

	Japanese		food
heavy	x	x	x
accent		x	x
nuclear accent			x

But if we view the perception of prominence as largely the result of how a given syllable stands out in relation to other syllables (and in particular to those that immediately precede or follow it), then the perception of "food" as nuclear accented is likely to be diminished by the presence of the immediately preceding accented syllable. Talkers therefore tend to reorganise the rhythmic pattern of the first word to accent the first syllable rather than the second:

	Japanese		food
heavy	x	x	x
accent	x		x
nuclear accent			x

which allows a more salient contrast between the third syllable of "Japanese" and nuclear-accented "food".

This above account simply defines the "rhythm" of English as the tendency for stress contrasts to alternate at various levels of the stress hierarchy. Some phoneticians and phonologists have also made a claim that rhythm implies a *timing* constraint. This had been founded on earlier claims by e.g., Pike (1945) and Abercrombie (1964, 1967) (see also Halliday, 1980) that English is a stress-timed language with a tendency to make "feet" nearly isochronous (i.e. of approximately the same perceived duration). It is important to recognise that the definition of the foot in this isochronous sense is quite different from the one given earlier (the stress-foot defined earlier is strictly word-internal). Thus the Abercrombian foot would have intervals such as

|Ade|laide is in |South Au|stralia,

where | marks a foot boundary; and a claim is made that the perceived duration of each of these four feet is approximately the same. There is, however, no real evidence that talkers do make these intervals isochronous or even nearly isochronous (Crystal & House, 1990), nor is there any indication from studies of large corpora that the isochronous foot, as defined above, plays any significant role in how units of speech are timed relatively to each other (see e.g., Wightman, Shattuck-Hufnagel, Ostendorf, & Price, 1992 who reinterpret the timing evidence for the foot as presented in Lehiste, 1977 and Scott, 1982 in terms of preboundary lengthening).

4.6.4 Intonation and boundary tones

We present some details in this section of the model of intonational phonology for English that is particularly concerned with the relationship between an abstract linguistic-prosodic specification and the changing pitch contour. This system is also closely related to many aspects of autosegmental and metrical phonology (see Goldsmith, 1990, for a review) and has led to a practical system for the prosodic transcription of English known as the "tones and break indices" or ToBI system (Beckman & Ayers, 1994; Pitrelli, Beckman, & Hirschberg, 1994; see the reference to Beckman & Ayers, 1994, in this book for an email and WWW address for the ToBI materials).

In current models of intonational phonology, the production of speech is modelled as the alignment of a *tune* with the words of an utterance, or the *text*. This alignment has two main consequences: first, as described in Section 4.6.1, some words are accented; second, the words of the utterance are divided into *phrases*. The division of an utterance into phrases is hierarchical so that an utterance always consists of one or more intonational phrases and an intonational phrase always includes one or more intermediate phrases (Figure 4.35). The basis for the intonational phrase in Pierrehumbert (1980) is discussed in some detail in Ladd (1986): it is approximately synonymous with Halliday's (1967) tone group. The intermediate phrase has been proposed by Beckman and Pierrehumbert (1986) based on an analysis of prosody in English and Japanese: in all cases, the last accented word in an intermediate phrase is the nuclear accented word. Furthermore, since every intermediate phrase must contain minimally one accented word, citation-form words produced in isolation are necessarily also nuclear accented, as discussed earlier.

With regard to the tones, a *phrase tone* is associated with each intermediate phrase and a *boundary tone* with each intonational phrase in each utterance. The tones are in both cases one of two types high (H) or low (L) and there is a convention that phrase and boundary tones are appended with a hyphen and percentage sign respectively.

As far as the tones of accented words are concerned, we have already mentioned that a pitch-accent is associated to the vowel of each primary stressed syllable. Once again the number of tones is effectively reduced to two, low (L*) or high (H*). The most common type of pitch-accent, the H* tone target, is typically realised as a pitch peak close to the primary stressed vowel; an L* tone target is realised as a pitch trough. There are bitonal variations on these basic accent types that we will not deal with in any detail here: for instance an L+H* target means that that there is a more pronounced pitch rise to the H* tone target at the primary stressed vowel. An L*+H can produce a very similarly

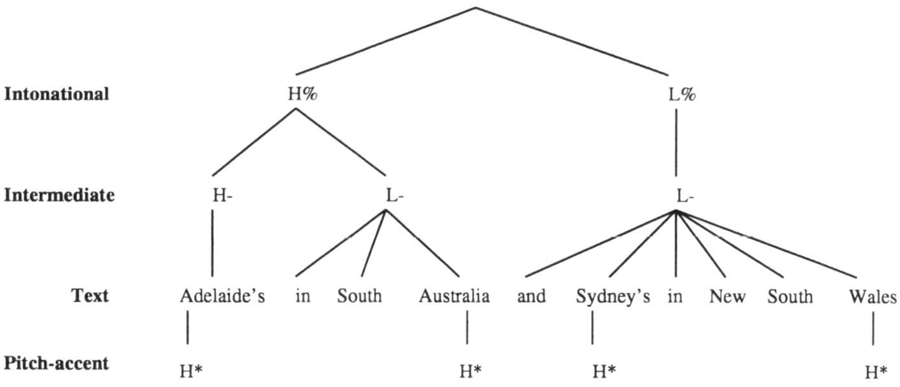

Figure 4.35: Hierarchical tree structure showing the alignment of a tune with the text. The f_0 contour for the utterance up to the first intonational boundary is shown in Figure 4.37.

shaped pitch contour on the accented word, but in this case it is as if the contour is displaced later in time: that is, in an L*+H pitch-accent, the *trough* is associated with the primary stressed vowel and it is followed by a marked pitch rise. Very clear examples of the high-toned pitch-accents (in fact, bitonal L+H*) are shown for the words "reduced" and "one" in Figure 4.38 (all the examples in this section are included on the CD-ROM in this book). Unfortunately, there is not always such a clear pitch peak in H* syllables when the preceding consonant is voiceless. This is because a voiceless consonant causes the pitch to start higher and to fall close to the vowel onset. The high falling pitch at the vowel onset can either obscure completely the pitch peak associated with an H* tone or else combine to produce a pitch inflection, rather a pitch plateau (these effects, due to what has been called *intrinsic pitch*, are well-documented and dealt with in further detail in e.g., House & Fairbanks, 1953; Lehiste & Peterson, 1961a; Lehiste, 1970; Mohr, 1971; Kohler, 1985; Löfqvist, 1975; Silverman, 1984; Umeda, 1981).

Perhaps the most distinctive part of the pitch contour is the interval between the tone target of the nuclear accented word and the edge of the phrase: the shape of the pitch contour in this interval is largely determined by the combination of the phrase and boundary tones. For example, in a neutral declarative reading of "Adelaide is in South Australia" (with nuclear accent on the last word), this interval extends from the [eɪ] of "Australia" to the end of the utterance. In such a neutral declarative production, the phrase and boundary tones are both low. The combination of these low tones with the H* pitch accent results in a characteristically sharply falling pitch. The pitch contour for this utterance (produced by a female talker of New Zealand English) is shown in Figure 4.36. The corresponding structure is:

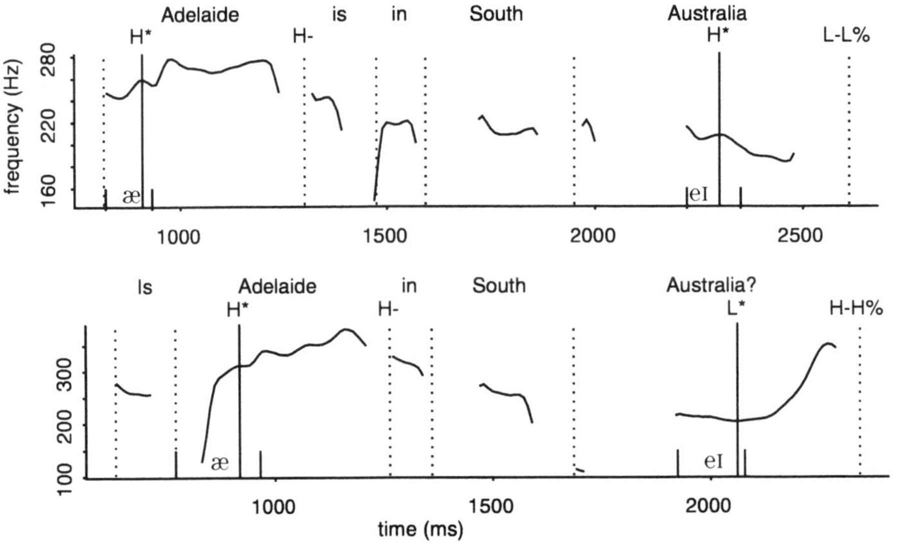

Figure 4.36: f_0 contour for a neutral declarative (*top panel*) and a "yes-no" question (*bottom panel*).

[[Adelaide] [is in South Australia]]
H- L- L%
H* H*

There are therefore two nuclear accented words (two intermediate phrases). For the present, our concern is with the falling pitch produced by the H*L-L% combination from [eɪ] of "Australia" to the end of the utterance.

In so-called "yes-no" questions on the other hand, we find that the pitch rises over this same interval. This is the result of an L* pitch-accent on the nuclear accented word in combination with high phrase and boundary tones. This is shown for the yes-no question in the bottom panel of Figure 4.36. Structurally, this corresponds to

[[Is Adelaide] [in South Australia?]]
H- H- H%
H* L*

In both these utterances considered so far, the phrase and boundary tones are either both low or both high. But it is possible for the phrase and boundary tones to have different values. In the top panel in Figure 4.37, the phrase-boundary tone combination is L-H%, which is typical in *continuation rise* utterances. The resulting pitch contour between the tone target of the nuclear

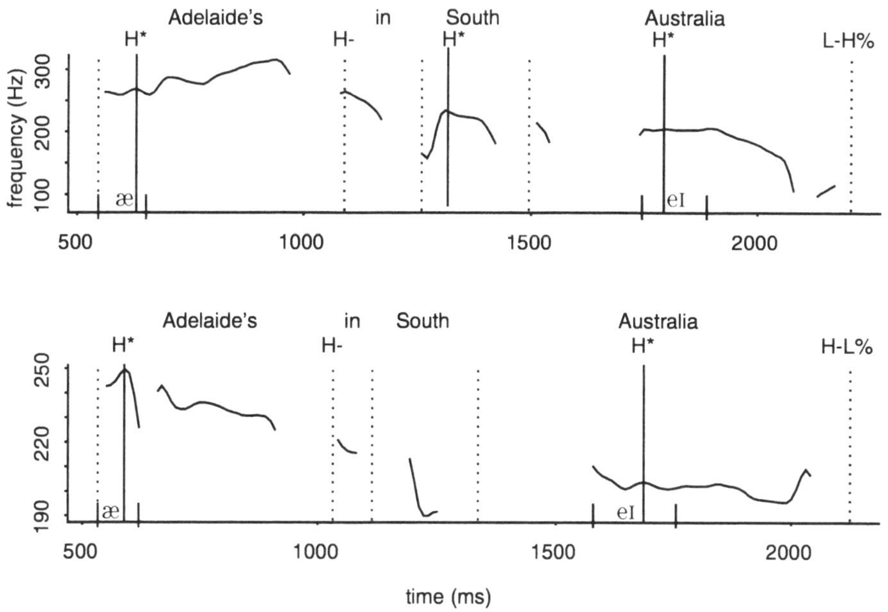

Figure 4.37: f_0 contour in L-H% (*top panel*) and H-L% (*bottom panel*) type phrases.

accented word and the edge of the phrase is a fall-rise: it falls due to the change from the H* pitch-accent to the L- phrase tone and then rises from L- to the H% boundary tone. This type of tune is common in the first phrase of sentences such as

[[When I get there]], I'll go and visit John

L- H%

H*

The fall-rising pitch contour is not very clear but nevertheless present in the top panel of Figure 4.37: the fall occurs from the right edge of [eɪ] to where there is a break in the pitch contour; the rise is then from where the pitch is again visible to the edge of the phrase. A more dramatic fall-rising contour can be seen on the nuclear accented word "one" in Figure 4.38.

The final logically possible phrase-boundary tone combination, which is perhaps the least common of all the four, is an H- phrase tone in an L% boundary tone which often results in a perceived level, or slightly falling pitch. In this case, the pitch stays more or less level due to the H* pitch-accent followed by a H- phrase tone and then falls slightly due to the L% boundary tone. (The fall is not as dramatic as in an L-L% type phrase). This kind of contour can occur in a somewhat bored, disinterested, and perhaps slightly impatient recital of lists.

For example, the utterance in the bottom panel of Figure 4.37 might be the first in a long recital of capitals of states in which the information seems so obvious to the talker that it is hardly worth the effort of saying.

In all the examples so far, the nuclear accent is the last word in the phrase and so the changes in pitch associated with the phrase-boundary tone combinations have been confined to relatively small stretches of the utterance. But this need not be so, and this emphasises the *phonological* nature of intonation: the same abstract specification of a tune can be realised as a multitude of different pitch shapes. Consider for example, the production of

[[Sydney isn't in South Australia!]]
L- H%
|
L+H*

in which the nuclear accented word, "Sydney" is four words away from the end of the phrase. In this case, the entire fall-rise contour that is squeezed into the extent of the single syllable "one" in "reduced their gods to one" is now stretched out from the [ɪ] of "Sydney" to the edge of the phrase (Figure 4.38): thus the pitch stays low over these intervening words (due to the L- phrase tone) and then rises on the last syllable (due to the H%) boundary tone. This contour is typical when a talker contradicts someone and is also mildly surprised by the listener's ignorance.

This is one sense in which intonation is phonological. Another is that intonation is simultaneously linear and hierarchical. The *linear* nature of intonation has been evident from the fact that sequences such as H*L-H% successively influence the pitch contour as it unfolds in time: first the H* causes a pitch peak, then the L- produces a trough, and so on. We have schematically shown the *hierarchical* nature of intonation by the bracketing of the utterances (e.g., intermediate phrases are within, or are dominated by, intonational phrases). The fact this hierarchical structure also has an influence on the pitch contour is most readily apparent in the examples of "Adelaide" in the top and bottom panels of Figure 4.37. In both cases, this word is nuclear-accented with the same tonal specification:

[Adelaide]
H-
|
H*

but in the top panel the pitch rises somewhat over the rest of this word, whereas in the bottom panel it is level or slightly falling. What could account for these differences? A probable explanation is that in the first utterance, the H- intermediate phrase is within a superordinate H% intonational-phrase: therefore, the pitch contour from the [æ] of "Adelaide" to the end of that word (edge of the phrase) takes on some of the characteristics we observed for an H-H% contour

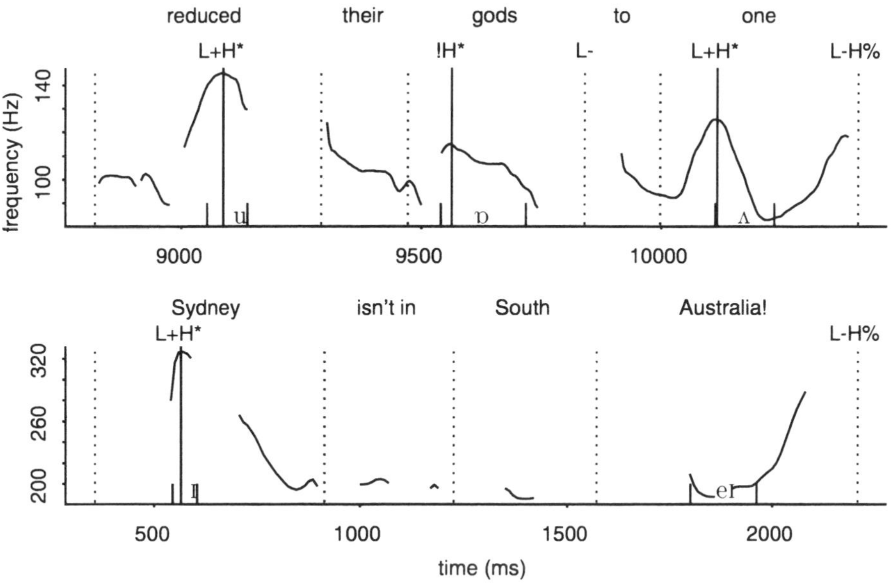

Figure 4.38: f_0 contour for L-H% phrases when the nuclear accent falls on the last (*top panel*) and first (*bottom panel*) word of the phrase.

i.e. a rising pitch. In the bottom panel, on the other hand, the H- intermediate phrase is dominated by a L% intonational phrase. Here, then, the pitch contour between [æ] and the end of the word has all the characteristics of an H-L% type phrase i.e. a level or slightly falling pitch.

The above is a very brief overview of intonational phonology. Above all, we have not considered declination effects in which the pitch contour has been shown to drift downwards over the course of an utterance. Nor have we considered downstepped pitch-accents which are often triggered by a preceding bitonal pitch-accent in the same phrase. There are also many other possible uses of low-toned pitch-accents that are certainly not just restricted to yes-no questions. These issues and those discussed above are dealt with comprehensively in the ToBI training materials referred to earlier and more recently in Ladd (1996).

Cues to phrase-boundaries

The tonal specification defined by pitch-accents phrase and boundary tones is a major part of the prosodic specification of an utterance. Another is how the utterance is divided into phrases. A phrase implies that there is a prosodic coherence between words and a also a perceptible break, prosodic discontinuity, or juncture between words that span a phrase boundary. In current models of intonation and prosody, the degree of perceived disjuncture at a boundary is dependent on the *boundary strength*. There is some evidence from various studies (Ladd & Campbell, 1991; Price, Ostendorf, Shattuck-Hufnagel, & Fong, 1991; Wightman et al., 1992) of at least a four-level hierarchy of boundary strengths. Within this system, intermediate phrase boundaries are perceived to be stronger than the normal (nonphrasal) boundaries between words and intonational boundaries are stronger than intermediate boundaries. Studies have shown that trained and naive listeners can perceive boundary strength reliably (Pijper & Sanderman, 1994) and that incorporating at least five levels of boundary strength improves the naturalness of synthetic speech quality over systems using fewer levels (Sanderman & Collier, 1996). Acoustic analyses of large corpora have shown that there are cues that differentiate nonphrase final word boundaries, intermediate boundaries, and intonational boundaries (Price et al., 1991; Wightman et al., 1992).

One of the main cues to a phrase boundary is phrase-final lengthening (Lehiste, 1973; Klatt, 1975; Beckman, Edwards, & Fletcher, 1991) in which the duration of the syllable rhyme (the nucleus plus any post-vocalic consonants) that precedes the boundary is lengthened (and greater lengthenings are associated with boundaries of greater strength). For example, the "aide" of "Adelaide" is lengthened because of the presence of the following H- intermediate phrase boundary in all the productions discussed earlier. In studies of large corpora, the durational differences associated with boundary strengths only emerge after the duration of the phonetic segments have been normalised (Lehiste, 1979; Campbell, 1990; Wightman et al., 1992) by e.g., subtracting the mean duration for the segment in a large corpus and dividing by its standard deviation; tempo (rate) variations also have to be factored out (Wightman et al., 1992).

Declination reset is also common at phrase boundaries. Specifically, declination implies that there is a continual downwards drift of fundamental frequency over the course of an utterance, but at major prosodic boundaries, the downwards trend is interrupted and the fundamental frequency is raised above the declination "trend line". Furthermore, the greater the discontinuity implied by the boundary, the greater the extent of reset as Cooper and others have shown (Cooper & Paccia-Cooper, 1980; Cooper & Sorensen, 1981; Ladd, 1988; see also O'Shaughnessy & Allen, 1983, and Thorsen, 1985, 1986).

Recent studies emphasise that, while there is a close correspondence between syntactic and prosodic boundaries (e.g., Collier & 't Hart, 1975), they do not necessarily coincide (Nespor & Vogel, 1986). The extent to which prosodic boundaries correspond to syntactic boundaries is the major concern of a study by Price et al. (1991). Their study shows that there is a good correspondence between major prosodic boundaries and major syntactic boundaries but that 'prosodic boundaries need not correlate perfectly with syntactic boundaries'. One of the most reliable associations between prosodic and syntactic boundaries was for clauses i.e. for syntactic phrases containing both a subject and predicate, which were in most cases marked by the phrase-final lengthening and pause insertion of the kind described above.

4.6.5 Word boundaries

Although listeners can usually hear the sequence of words intended by the talker, separating words in an acoustic signal of continuous speech is remarkably difficult. Since there is nothing in the acoustic waveform that corresponds directly to the white spaces that separates one word from the next in written form, the problem of just how listeners manage to perceive the succession of words has been an intensively studied field of investigation in recent years. Faced with the difficulty of finding word boundaries from the acoustic signal directly (the "bottom-up" approach), some researchers (e.g., McClelland & Elman, 1986) have proposed that word boundaries are found by matching the lexicon against phonemic or sub-phonemic units derived from the acoustic signal. This is the approach taken by Marslen-Wilson (1978, 1986, 1987) in earlier versions of the cohort model. In this model, words can be eliminated based on considerations of the cohort size that corresponds to the first few phonemic units obtained from the speech signal. For example, in the sequence /trɛspəsɪŋɪzfəbɪdn/, we already know by the fourth phoneme /p/ that the first word must be "trespass" since there are no other words that begin with this phonemic sequence: therefore a word-boundary can be hypothesised before /ɪ/ and, since /ɪŋ/ is not a word and since /ɪŋɪ/ is a very unlikely word-onset, the word "trespassing" can be hypothesised. Notice that one of the advantages of matching the lexicon to the phonemic or sub-phonemic string is that it may not be necessary to analyse all parts of the acoustic signal into phonetic or phonemic units (e.g., for the previous example, an acoustic analysis of the /əs/ part of the signal is redundant since the sequence is predictable from the word-initial sequence /trɛsp/).

As Christophe, Dupoux, Bertoncini, and Mehler (1994) have recently noted, one of the problems with a model in which the lexicon is an integral part of the word-segmentation strategy is that it cannot explain how infants, who do not yet have an internalised mental lexicon, can segment the acoustic signal into words. In their study, they show that infants as young as three weeks can perceive word boundaries (they used a sucking paradigm to show that they can hear the difference between the same phonemic string that does and does not span a word-boundary): this would of course suggest that there must be some direct information in the acoustic speech signal itself which can be used for word-segmentation.

In some models of lexical-access, prosodic cues are presumed to provide such direct, pre-lexical information for word-segmentation (e.g., Grosjean & Gee, 1987; Cutler & Norris, 1988). Cutler's model of lexical-access for English (see also Fear, Cutler, & Butterfield, 1995) exploits the fact that over 90% of words in English begin with "strong" syllables (Cutler & Carter, 1987) that are approximately equivalent to the "heavy" syllables that are the basis of lexical-stress contrasts defined earlier. Based on these lexical statistics, Cutler proposes that strong syllables should be identified in the speech signal. These should be easier to detect than "weak" (light) syllables, both because they are more salient, and because they are less prone to phonetic variability. A lexical search is begun at each strong syllable that is found. There are also experiments to show that when listeners misperceive a word boundary, the most likely type of error is the incorrect insertion of a boundary before a strong syllable (Butterfield & Cutler, 1988; see also Cutler & Butterfield, 1990 and Cutler & Butterfield, 1991 for compatible evidence in an analysis of word boundary perception in clear speech).

Other researchers have shown that there are *allophonic distinctions based on considerations of syllable-structure* (e.g., Church, 1983, 1987; Quené, 1993) that may be used to help resolve the word segmentation problem. Before considering these direct acoustic cues to word segmentation in further detail, we review briefly the kinds of word-boundary ambiguity that can arise in English.

The first kind occurs because different words can have the same phonemic form. This includes homophones of the "read"/"red" kind; words that have the same phonemic composition but a different stress pattern (e.g., "*con*trast"/-"con*trast*"; "*per*mit"/"per*mit*"); and words that become phonemically equivalent as a result of word-reduction processes of continuous or fast speech (e.g., "support" which can become homophonous with "sport"; "parade"/"prayed" which can both be /preɪd/ etc.). The second kind of ambiguity arises when a sequence of words has the same phonemic form as either another word, or another sequence of words. For example, the following pairs could all map onto the same phonemic form in continuous speech: "conform"/"can form"; "waiter"/"way to"; "propagate"/"prop a gate"; "attire"/"a tyre"; "nitrate"/"night rate". This second set can be further grouped depending on whether the pairs have a similar or a different syllable-structure.

At this point we encounter the well-known difficulty of the basis on which a phonemic string should be syllabified. But if we assume a form of the *maximum onset principle* (Hoard, 1971; Kahn, 1976; Pulgrum, 1970; Selkirk, 1982), in which as many word-internal phonemes as possible are syllabified with a following vowel providing the resulting sequence is phonotactically legal (thus "constrain" is syllabified as /kən.streɪn/ because there are no words in English that begin with /nstr/), then "attire"/"a tyre" have a similar syllable structure in which the /t/ is syllable-initial in both cases, whereas the syllable structure for "nitrate"/"night rate" is different (the /t/ is syllable-initial according to the maximum onset principle for "nitrate" but syllable-final for "night rate").

Word sequences that have the same phonemic composition but a different syllable-structure are the easiest to disambiguate for the reason that when a phoneme is in different syllable-positions, the resulting allophones are sometimes different in quality. There is certainly a good deal of evidence from perception experiments to show that listeners can distinguish between these kinds of ambiguous word string (e.g., Gårding, 1967; Hoard, 1966; Lehiste, 1960; Nakatani & Dukes, 1977; Quené, 1985). For example, Quené (1985) shows that over 80% of Dutch phrases differing only in syllable structure (e.g., /mu.rok/ vs. /mur.ok/) are correctly identified by listeners, while Hoard (1966) reports a score of 88% for listeners on a similar experiment in English. A nonexhaustive summary of the allophonic variation due to syllable-structure differences that may be useful for word-boundary disambiguation is summarised in Table 4.2. These allophonic differences are specific to English and do not necessarily carry

Allophonic differences	**Author**	**Examples**
aspiration of syllable initial stops	Lehiste (1960)	grey tie/great eye Nye trait/nitrate keeps ticking/keep sticking
devoicing of sonorant consonants	various	ice-cream/I scream
glottal reinforcement of syllable-final stop	various	might rain/my train
ratio of closure duration to duration of entire stop	Boucher (1988)	free Danny/freed Annie
intervocalic flapping	various	at ease/a tease win terrain/winter rain
glottal stop preceding word-inital vowels	Gårding (1967) Hoard (1966) Lehiste (1960)	an aim/a name an iceman/a nice man
velarisation of word final /l/	Bladon and Al-Bamerni (1976)	fee label/feel able
shortening of consonants in clusters	Lehiste (1960) Christie (1977)	plum pie/plump eye help us nail/help a snail

Table 4.2: Some of the allophonic cues that can be used for word boundary disambiguation in word strings of the same phonemic composition, but differing in the location of a medial word boundary.

over into other languages (Barry, 1983; Quené, 1985).

When competing word strings have the same syllable-structure, their acoustic distinction is more difficult and must be based on quite subtle durational (rather than quality) differences that can be neutralised by the effects of continuous speech. Various durational cues have been suggested. An experiment by Nakatani and Schaffer (1978) was based on reiterant speech in which a talker produced a natural phrase or sentence using [ma] syllables: for example, "tasty food" and "bold design" are produced as /mama#ma/ and /ma#mama/ in which all aspects of the prosodic structure are, as far as possible, retained. The results of their perceptual and acoustic experiment showed that the duration of the first syllable (longer in ma#mama) was the principal durational cue that could be used to distinguish between the phrases. Polysyllabic shortening (Lehiste, 1970; Lindblom & Rapp, 1973; Lyberg, 1977; O'Shaughnessy, 1981), in which the duration of a syllable decreases as the number of syllables in the word increases (i.e. the duration of "speed" progressively decreases in "speed", "speedy", "speediness") was proposed by Jones (1956) as the prime cue for distinguishing between "waiter" and "way to" ([weɪ] is shorter in "waiter"); however, beyond citation-form speech, this cue is likely to be of limited value since Harris and Umeda (1974) have shown that polysyllabic shortening is less likely to occur in continuous speech.

More recently, Quené (1993) has proposed two durational cues for distinguishing between phonemically identical strings that differ in terms of the location of the medial word boundary (see also Quené, 1992). These are the *duration of the initial consonant* (longer in V#CV than in VC#V strings, where # is the location of the word boundary) and the *energy rise-time* (the duration from the vowel onset to the time at which the amplitude reaches 90% of its maximum value). In this experiment, which used Dutch word-string pairs that were similar in form to "no notion"/"known ocean" pairs in Nakatani and Dukes (1977), word-stimuli were presented to listeners after shortening or lengthening the medial consonant and vowel rise time durations. In addition, the phrases had been read such that either the first or second syllable was accented in order to determine the contribution of accent location to word segmentation. Quené (1993) found that the two durational cues were of primary importance for word boundary identification and that accent only contributed to word segmentation to the extent that it enhanced them. However, as Quené (1993) notes, this does not necessarily run counter to the proposals in Cutler and Norris (1988) because their word segmentation strategy is based at the level of the heavy/light (strong/weak), rather than the accented/unaccented distinction.

Notes

1. It is not immediately obvious that the line $y = x$, or in this case formant onset = formant target, bisects the regression line at the locus frequency. However, the fact that they do bisect can be demonstrated algebraically. The theory that underlies the locus equations is that the extension of the straight line passing through vowels with different formant onset (f_O) and different formant target (f_T) frequencies also passes through a locus frequency (k_L) that is a constant for all vowels when the preceding stop has a constant place of articulation (Figure 4.39). If the times at which f_O and f_T occur are t_O and t_T, respectively (t_O and t_T are also variables), the following two (straight-line) equations can be derived, assuming that the locus frequency occurs at time zero

$$f_O = mt_O + k_L \tag{4.6}$$
$$f_T = mt_T + k_L, \tag{4.7}$$

where m is the gradient. Rearranging Equations 4.6 and 4.7 in terms of m

$$m = (f_O - k_L)/t_O \tag{4.8}$$
$$m = (f_T - k_L)/t_T \tag{4.9}$$

from which it follows that

$$(f_T - k_L)/t_T = (f_O - k_L)/t_O \tag{4.10}$$

Cross-multiplying:

$$t_O(f_T - k_L) = t_T(f_O - k_L)$$
$$t_O f_T - k_L t_O = t_T f_O - t_T k_L$$

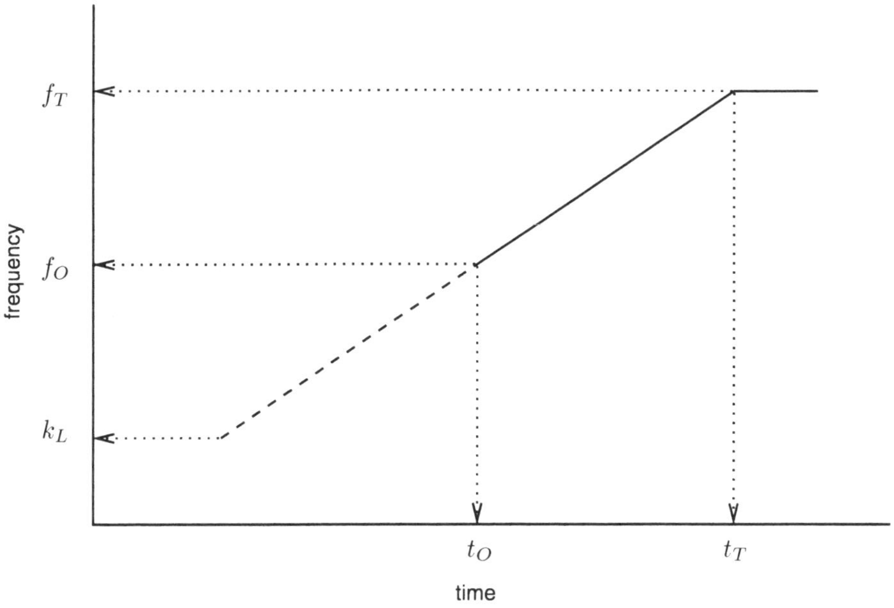

Figure 4.39: Schematic formant transition to the vowel target. t_O: onset time of formant transition; t_T: onset time of vowel target; k_L: the locus frequency; f_O: the frequency at transition onset; f_T: the frequency of the vowel target.

Rearranging:

$$\begin{aligned} t_O f_T &= t_T f_O - t_T k_L + k_L t_O \\ &= t_T f_O + k_L(t_O - t_T) \end{aligned}$$

$$\begin{aligned} f_T &= (t_T/t_O)f_O + k_L(t_O - t_T)/t_O \\ &= (t_T/t_O)f_O + k_L - k_L(t_T/t_O) \\ &= (t_T/t_O)f_O + k_L(1 - t_T/t_O) \end{aligned}$$

or denoting (t_T/t_O) by α

$$f_T = \alpha f_O + k_L(1 - \alpha). \tag{4.11}$$

Equation 4.11 represents the line in the f_O/f_T plane that passes through vowels with formant onset frequency f_O and formant target f_T (f_O and f_T are variables), which converge to the same constant locus frequency k_L (the line represented by Equation 4.11 is analogous to the regression lines in Figure 4.19). In order to determine the point at which Equation 4.11 bisects Equation 4.12,

$$f_O = f_T. \tag{4.12}$$

Equations 4.11 and 4.12 are solved as simultaneous equations, first for f_O

$$\begin{aligned} f_O &= \alpha f_O + k_L(1 - \alpha) \\ f_O - \alpha f_O &= k_L(1 - \alpha) \\ f_O(1 - \alpha) &= k_L(1 - \alpha) \\ f_O &= k_L. \end{aligned}$$

When $f_O = k_L$, $f_T = k_L$ by Equations 4.11 or 4.12. Therefore, Equations 4.11 and 4.12 bisect at (k_L, k_L), the locus frequency.

2. This is true of Australian English - in other accents in which [u] is a back vowel, there may be greater similarity between [w] and [u].

5
TIME-DOMAIN ANALYSIS OF DIGITAL SPEECH SIGNALS

The advances that have been made in digital electronics and computing since the 1960s have had an increasing impact on speech research and nowadays computers are used in almost every kind of experimental speech investigation. There are many advantages of using a computer in speech analysis. Among the most important are that there can be no degradation in the quality of the speech signal once it has been digitally encoded: it can be copied over and over again without any loss of information at all. On the other hand, an analogue tape-recorded speech signal suffers deterioration both with the passage of time and especially if it is repeatedly copied. Another advantage that the computer offers is easy retrieval of information. These days it is possible to store several hours of speech data on a computer disk. Moreover, if the data has been appropriately indexed, or labelled, then speech signals of a particular sound in context (such as all /n/ sounds that precede high vowels), or produced by a particular talker-group, can be very easily extracted and analysed (this should be compared with having to edit out appropriate sections of hours' worth of tape-recorded speech). The software that is included with this book includes a search engine of just this kind.

There are many other advantages to using a computer for speech analysis. A waveform can be displayed and analysed to an accuracy of well under a millisecond and there are techniques for analysing a speech signal for the efficient calculation of energy, pitch, and formants and other frequency information that can only be applied to digital speech signals.

All of these advantages must be seen in the context of the increasing importance of speech technology research which is concerned with the development of human-machine communication systems. Machines for converting text into intelligible speech have existed for some time, and they have many applications in telecommunications and as aids for the speech handicapped. Although speech recognition technology is somewhat less advanced, there are now machines commercially available that can recognise speech based on a limited vocabulary of words and a constrained grammatical system and that require a certain amount of talker training. The automatic recognition of a talker's voice represents an-

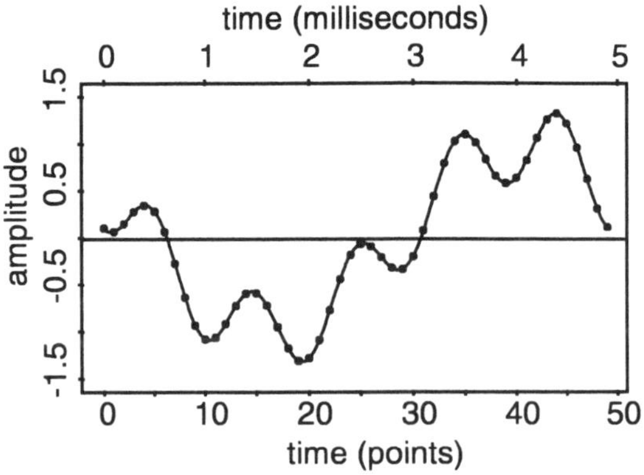

Figure 5.1: The dots represent the digital equivalent of the analogue signal (solid line) at a sampling frequency of 10000 Hz.

other growing branch of speech technology which has numerous applications in areas such as voice access to secure systems and the forensic analysis of speech. The field of digital coding, which is concerned with techniques for compressing the speech signal, has become increasingly important to the telecommunications industry.

5.1 Sampling and quantisation

We will discuss in this section some very basic principles of sampling and quantisation which are prerequisites to storing any time-varying signal on a computer.

The process by which a signal is converted to a digital representation is known as *analogue-to-digital* or *A/D conversion* (*D/A conversion* is the reverse process). In A/D conversion, a continuously changing time-signal is converted into a series of discrete values. This process is similar in principle to making a motion picture in which continuously changing events are represented as a series of static picture frames that are spaced at equal intervals of time. In both cases, the viewer (of the film) or the listener (of the digitised speech signal) perceives a continuous event if the interval between successive frames or digital values is sufficiently small.

The fixed time interval between digital values, or *samples*, is known as the *sample period* and will be represented by T. The sample period depends on the *sampling frequency* (abbreviated as f_s), which defines the number of data points that are assigned to the continuous speech signal per second. The sample period in seconds is the reciprocal of the sampling frequency in Hz: in Figure 5.1, the sampling frequency is 10000 Hz and so $T = 1/f_s = 1/10000$ second or 0.1 ms.

The number of points in any digital time signal depends on the sampling frequency and the duration of the analogue signal that has been digitised. The total number of points N in a digital speech signal of duration t seconds digitised at a sampling frequency f_s Hz is given by

$$N = tf_s \ points. \tag{5.1}$$

Rearranging Equation 5.1, the duration can be calculated from the number of points and the sampling frequency as follows:

$$t = N/f_s \ seconds. \tag{5.2}$$

It would seem that if we represent a continuously varying signal in a digital form, there must nevertheless be some loss of information. Or to put this in a slightly different way: does it matter that in digitising a signal we have cut out pieces of the signal between the points? It is here that Nyquist's (1928) theorem is important. Nyquist showed that a bandlimited continuous signal — that is one that contains only a certain range of frequencies — can be *exactly* reconstructed from a corresponding digital signal providing that the sampling frequency is chosen to be at least twice the highest frequency contained in the continuous signal. For example, if the highest frequency component in a continuous signal is 100 Hz, then the signal must be digitised with a sampling frequency of at least 2 x 100 = 200 Hz in order for the continuous signal to be identically reconstructed when the digital signal in passed back through D/A conversion.

If a continuous signal is digitised with a sampling frequency of less than twice its highest frequency, a phenomenon called *aliasing* occurs. An example of aliasing is illustrated in Figure 5.2 in which analogue sinusoids are digitised at less than (top) and more than (bottom) twice their frequencies, respectively. In the top panel, the analogue sinusoid has a frequency of 15 Hz, while the bottom analogue signal has a frequency of 5 Hz. Both sinusoids are now digitised with a sampling frequency of 10 Hz. In the bottom panel, the frequency of the analogue signal is faithfully reconstructed in digital form because the sampling frequency (10 Hz) is twice the frequency of the 5 Hz analogue sinusoid: both the analogue and the digital signals have 5 repetitions per second i.e. a frequency of 5 Hz. In the top panel, however, the frequencies of the analogue and digital signals are divergent: they have frequencies of 15 Hz and 5 Hz, respectively. This means that at a sampling frequency of 10 Hz, the 15 Hz and 5 Hz analogue signals are indistinguishable from each other. We can also say that when a 15 Hz continuous sinusoid is digitised at 10 Hz, it is *aliased* onto the 5 Hz sinusoid.

In the case of speech, most of the frequencies of interest lie below 10000 Hz, and in fact the large majority of speech sounds are distinguishable from acoustic information in the 0–5000 Hz range. Listeners can understand speech signals from an even more compressed bandwidth as shown by our ability to communicate over the telephone, which is usually bandlimited between about 300 Hz and 3400 Hz. If want to preserve the frequency content of a speech signal below 10000 Hz, then a sampling frequency of at least 20000 Hz is mandatory. But additionally, we have to use a *lowpass filter* that discards as much of the information as possible above 10000 Hz because otherwise the higher frequency information would be aliased onto the frequency range that we want to represent

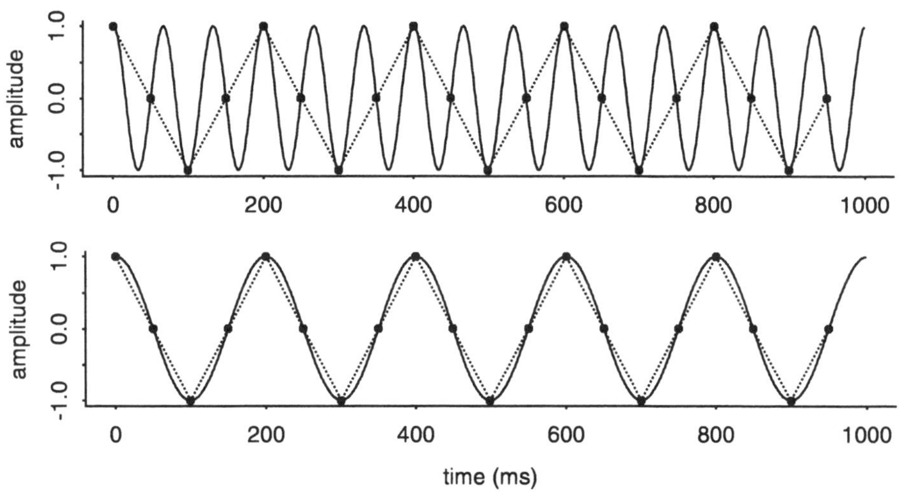

Figure 5.2: The analogue signals are represented by solid lines. At a sampling frequency of 10 Hz, the 15 Hz (*top*) and 5 Hz (*bottom*) analogue sinusoids are aliased onto the same 5 Hz digital sinusoid.

faithfully in digital form. Therefore, the first stage in digitising a speech signal must be to apply a lowpass filter to the continuous speech signal, which filters out as much information as possible from the signal above a certain frequency; and second, the filtered signal must be sampled at twice, or greater than twice, that frequency. The lowpass filter which rejects higher frequency information from the analogue signal is sometimes called a *presampling filter*.

When a speech signal is stored on a computer, the continuously changing amplitude values must also be converted into discrete digital values. This process is known as *quantisation*. The level of quantisation is expressed in *bits* or *binary digits*: n-bit quantisation means that the amplitude range is represented as 2^n equally spaced amplitude levels. Figure 5.3 and Figure 5.4 shed further light on how quantisation is achieved. Figure 5.3 is a tree-diagram showing the expansion into 3-bit quantisation: with each additional bit of quantisation, the number of discrete amplitude levels is doubled resulting in 8 (2^3) amplitude levels in this case. Since we have to deal with positive and negative amplitude values in speech analysis, binary values with a leading zero are sometimes used for positive values, while those with a leading 1 are used for negative values. This process is further illustrated in Figure 5.4 which shows the relationship between binary numbers and negative and positive values for three-bit quantisation of a signal.

Quantisation once again results in a degree of degradation of the analogue signal because continuous amplitude values are converted into one of a number of discrete amplitude steps: the more discrete steps there are, the more faithfully the signal will be digitally encoded. This raises the question of what

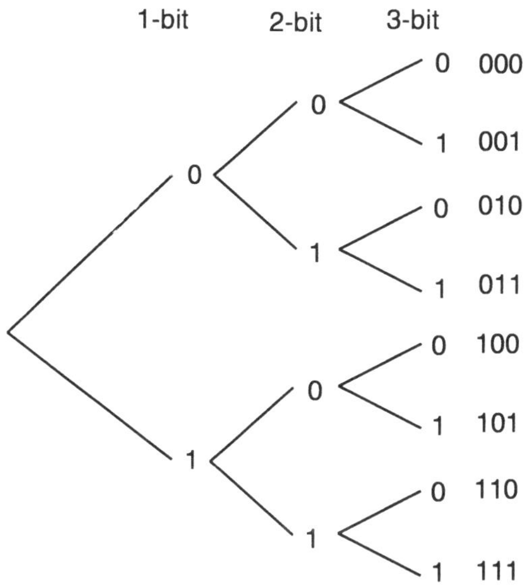

Figure 5.3: The relationship between binary values and three-bit quantisation.

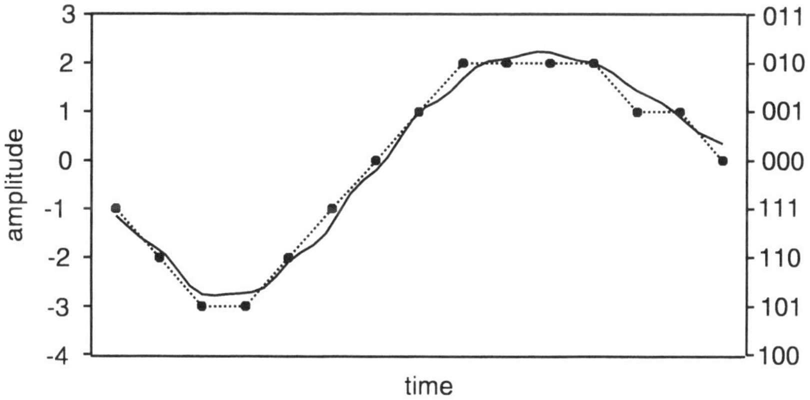

Figure 5.4: The relationship between binary numbers and negative and positive quantised values in three-bit quantisation of a continuous signal.

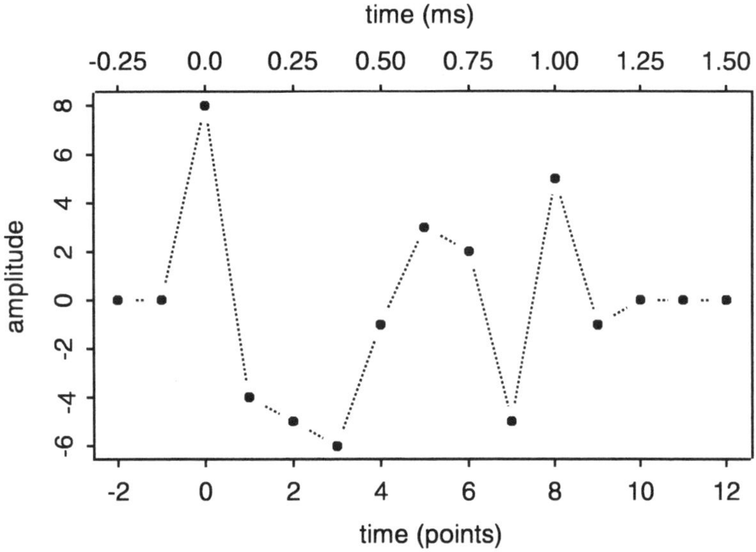

Figure 5.5: A digital signal of length $N = 10$ points digitised with a sampling frequency of 8000 Hz.

level of quantisation is needed for speech analysis. It can be shown that each additional quantisation bit accounts for approximately a 6 dB improvement in the dynamic range of the analogue-to-digital converter (see Owens, 1993, p. 25). Since the dynamic range for speech is in the order of 40–60 dB, 9-bit quantisation or greater is essential. Usually, 12-bit quantisation is preferred for the analysis of high quality speech: following from the earlier discussion, this level of quantisation encodes signals as $2^{12} = 4096$ discrete amplitude levels ranging from -2048 to +2047.

5.2 Definition of a digital signal

The result of digitisation and quantisation is a digital signal which is defined only for discrete time and amplitude values. Since we will be applying various mathematical operations on digital signals at various stages in this book, we must first be clear about how we want to represent the signal in numeric terms. First, we will use a bold notation — $\boldsymbol{x}$, $\boldsymbol{y}$ etc. — when referring to digital signals their entirety. A vector notation will be used to refer to all the values of the signal:

$$\boldsymbol{x} = [8, -4, -5, -6, -1, 3, 2, -5, 5, -1]$$

means that the signal $\boldsymbol{x}$ has the above 10 amplitude values which occur at discrete intervals of time: these are shown in Figure 5.5.

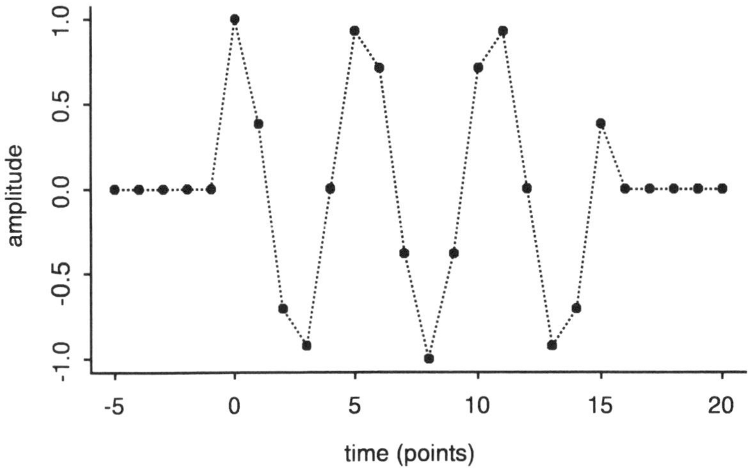

Figure 5.6: The digital sinusoid defined by Equation 5.3.

Second, we will often need to refer to digital values within an individual signal. The common practice is to define a signal as having N data points that occur at time $t = 0, 1, 2, 3...N-1$ points. The specific digital values of $\boldsymbol{x}$ are then indexed relative to the time (in points) at which they occur. For example, $x[0] = 8$ literally means: at time $t = 0$ data points, $\boldsymbol{x}$ has an amplitude value of 8. Since signals are typically zero-indexed (they are usually defined to start at time t=0), and since we have defined the total number of data points in the signal to be N, the expression for the last data in the signal must be $x[N-1]$.

Once we have this notation in place, we can conveniently use an equation to define some signals. Consider, for example, the digital sinusoid in Figure 5.6. One way to refer to define it is $\boldsymbol{x} = [0, 0, 0, 0, 0, 1, 0.38, \text{-}0.71 \ldots]$. But a more compact way, which also shows that the signal is a sinusoid that is defined to be zero outside a certain range of values is

$$\begin{aligned} x[n] &= \cos(6\pi n/N) \quad 0 \leq n \leq N-1 \\ &= 0 \qquad\qquad\quad\ \ \text{elsewhere} \end{aligned} \tag{5.3}$$

This digital sinusoid in Figure 5.6 is of length $N = 16$; so the data point at $n = 5$ would be given by $\cos(6\pi \times 5/16) = \cos(1.875\pi)$. When n is less than zero or greater than $N-1$ ($N-1 = 15$ in this case), the signal is zero-values as Figure 5.6 shows.

Lastly, time defined in terms of the number of points can be easily converted into an equivalent time values in seconds (or milliseconds) if the *sample period* is known: for a sample period T, the time at which a data point $x[n]$ occurs is nT seconds.

5.3 Simple operations on signals

At various stages in this book, we will perform numerical operations on signals. We will therefore briefly extend some of the notation that is needed to represent such operations.

An example of a simple arithmetic operation is

$$\boldsymbol{y} = 4\boldsymbol{x} + 3 \tag{5.4}$$

This means that a new signal $\boldsymbol{y}$ is created by multiplying all the digital values or *elements* of $\boldsymbol{x}$ by 4 and then adding 3 (to all elements).

The following operation creates a new signal from two signals $\boldsymbol{a}$ and $\boldsymbol{b}$:

$$\boldsymbol{y} = \boldsymbol{ab}. \tag{5.5}$$

In this case, a new signal $\boldsymbol{y}$ is created by multiplying successive elements of $\boldsymbol{a}$ and $\boldsymbol{b}$. For example, if $\boldsymbol{a}$ and $\boldsymbol{b}$ are defined as two signals that are zero-valued outside the range $0 \le n \le N-1$

$$\begin{aligned} a[n] &= \quad [4, 5, 9] \\ b[n] &= \quad [3, 0, -2] \end{aligned}$$

then by Equation 5.5, $y[0] = 12$, $y[1] = 0$, $y[2] = -18$ or

$$y[n] = [12, 0, -18].$$

It will sometimes be necessary to sum all the data points (resulting in a single value). The notation for this is $\sum$. Thus, for $\boldsymbol{y}$ as defined in the last example, $\sum \boldsymbol{y} = 12 + 0 - 18 = -6$. The expression

$$\sum \boldsymbol{ab} \tag{5.6}$$

multiplies the signals $\boldsymbol{a}$ and $\boldsymbol{b}$ element by element and then sums the product. Thus for the previous example

$$\sum \boldsymbol{ab} = \sum \boldsymbol{y} = -6.$$

The $\sum$ notation can be used to sum the elements over a range of n. For example, the summation of the first five data points of a signal $\boldsymbol{x}$ would be written as

$$\sum_{n=0}^{4} x[n]. \tag{5.7}$$

A very important operation in both digital signal processing and speech analysis is *time-shifting*. In this case, the sample data points are displaced in time but their amplitude values are unchanged. The notation for delaying a time signal $\boldsymbol{x}$ by p points (p is a positive integer) is $x[n-p]$; when $\boldsymbol{x}$ is advanced by p points, the notation is $x[n+p]$.

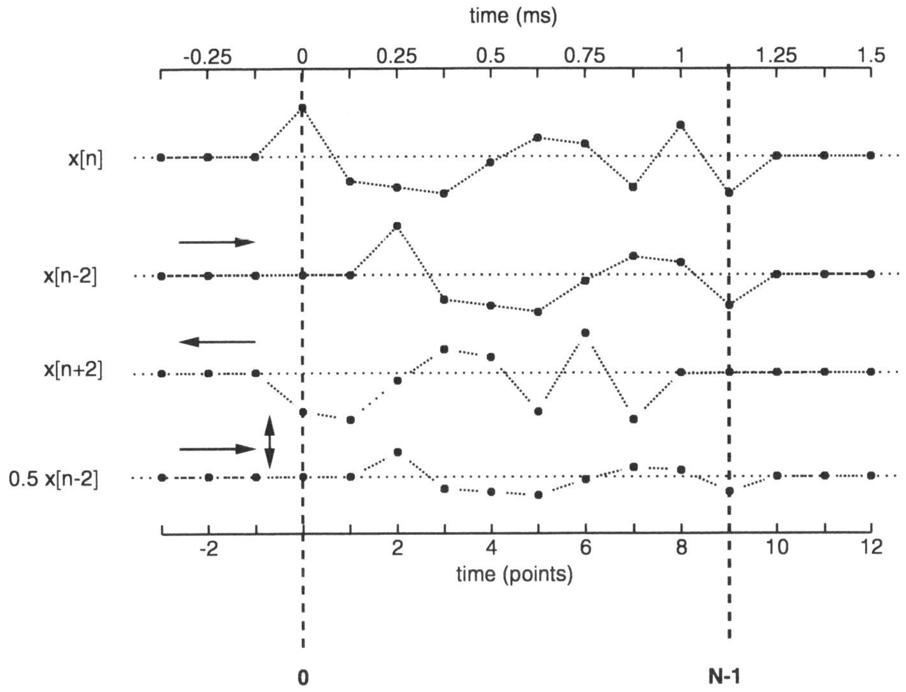

Figure 5.7: A signal $\boldsymbol{x}$ which is delayed by two points (*second row*), advanced by two points (*third row*), and delayed by two points and rescaled (*fourth row*). In all cases, the time-shifted signals are defined to be zero outside the range $0 \leq n \leq N-1$.

The second row of Figure 5.7 shows $\boldsymbol{x}$ delayed by two time points: the signal is therefore a copy of itself, but it starts two time points later. A closer inspection of the delayed signal in fact shows that the last two values of $\boldsymbol{x}$ have been cut off (or set to zero): this is because we have assumed that n is zero outside the range $0 \leq n \leq N-1$ of the original signal.

A less common type of time shift in speech processing is shown in the third row of Figure 5.7 in which $\boldsymbol{x}$ is advanced by two time points. Once again,we have assumed that the shifted signal is zero-valued outside the limits of the original signal. Many types of operation involve weighted or scaled shifting in which the signal is not only displaced in time, but is also weighted by some value. In the fourth row of Figure 5.7, the values of $\boldsymbol{x}$ are halved, and the signal is delayed by two time points: this corresponds to $0.5x[n-2]$.

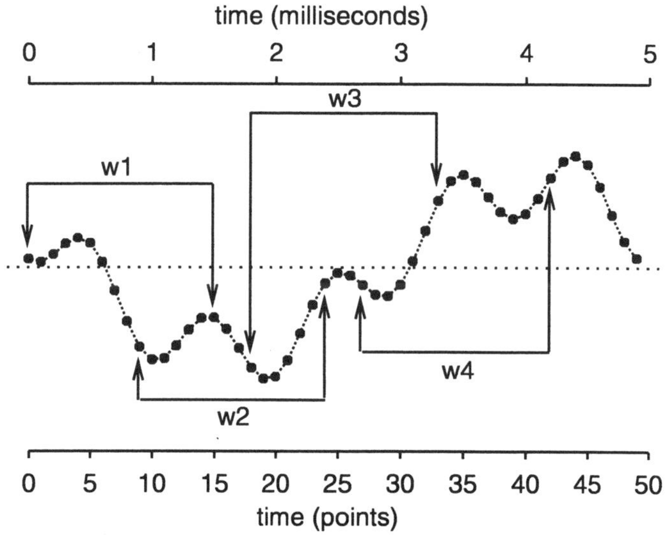

Figure 5.8: A signal divided into 4 overlapping windows of length $N = 16$ and step size $M = 9$.

5.4 Windowing signals

In speech analysis, we often need to process the speech signal is sections rather than in its entirety. We already alluded to this in discussing the frequency analysis of speech data in Chapter 2: a spectrogram of speech data is not produced by Fourier analysing the entire utterance but is derived instead by dividing up the data into a number of chunks of equal duration and Fourier transforming each of these. In digital signal processing, each of these chunks, or sections, is referred to as a *window* of data.

There are four main parameters associated with a window. The first parameter is the *starting frame*, which defines where the window starts relative to the signal: if we wish to identify a section from the beginning of the speech signal, the starting frame is 0. The second parameter is the *length of the window* (denoted as N), and the third parameter is the *step size*, *frame shift*, or *shift* of the window (denoted as M). Figure 5.8 shows the division of a signal using a window of length $N = 16$, of shift $M = 9$, and with a starting frame $n = 0$.

The fourth parameter defines the *window type*. When the window shown in Figure 5.8 is applied to the data, none of the values of the original signal is changed: this type of window is known as a *rectangular window*. Some windows progressively attenuate the signal values towards the window's edges: two of these are *Hamming* and *Hann* windows (Hamming, 1989). The application of a Hamming window to some speech data is shown in the right panel of Figure 5.9: notice that the values towards the centre of the window are left unchanged, while the values towards the left and right edges of the window are close to

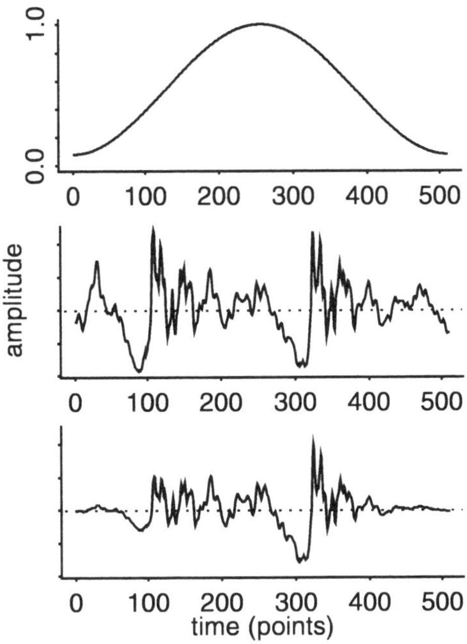

Figure 5.9: *Top*: A 512-points Hamming window. *Middle*: a section of a speech waveform. *Bottom*: the results of applying a Hamming window to the waveform in the middle panel. Note the main differences between the original and Hamming-windowed waveform are at the edges of the signal.

zero.

When a window (of any kind) is applied to speech data, the speech signal is *multiplied by* the window in the sense intended by Equation 5.5. We can consider this in further detail for the rectangular window, whose equation is defined as

$$\begin{aligned} w[n] &= 1 \quad (0 \leq n \leq N-1) \\ &= 0 \quad \text{elsewhere.} \end{aligned} \tag{5.8}$$

In other words, a rectangular window is 1 over the whole of its length. Consequently, when a rectangular window is multiplied with part of a signal, the values are unchanged. A Hann window is defined as

$$\begin{aligned} w[n] &= 0.5 - 0.5\cos(2\pi n/(N-1)) \quad (0 \leq n \leq N-1) \\ &= 0 \quad \text{elsewhere,} \end{aligned} \tag{5.9}$$

and the equation for the Hamming window is

$$\begin{aligned} w[n] &= 0.54 - 0.46\cos(2\pi n/(N-1)) \quad (0 \leq n \leq N-1) \\ &= 0 \quad \text{elsewhere.} \end{aligned} \tag{5.10}$$

We can now see why these windows have the largest effect at the edges. The first value of the Hann window is

$$\begin{aligned} w[0] &= 0.5 - 0.5\cos(2\pi 0/(N-1)) \\ &= 0.5 - 0.5\cos(0) \\ &= 0.5 - 0.5 \\ &= 0. \end{aligned}$$

So when part of a signal is multiplied with a Hann window, the value at the left edge will be zero. On the other hand, data points at the centre of a Hann window are close to one. For example, for a 512-point Hann window, the middle value is near $n = 256$, so

$$\begin{aligned} w[256] &= 0.5 - 0.5\cos(2\pi 256/(512-1)) \\ &\approx 0.5 - 0.5\cos(2\pi/2) \\ &\approx 0.5 - (0.5 \times -1) \\ &\approx 0.5 + 0.5 \\ &\approx 1. \end{aligned}$$

As we shall see in later chapters, it is often necessary to apply a Hamming or Hann window before carrying out in various kinds of frequency analysis.

5.5 Some common time-domain parameters

Once the speech signal has been digitised, we can process or parameterise it to provide us with a clearer indication of the salient differences between the major speech sound categories. The parameters that are applied directly to the speech signal are often called *time-domain parameters*; in a subsequent chapter we consider *frequency-domain parameters* that operate on the frequency representation of the speech signal.

5.5.1 RMS

One of the simplest kinds of time-domain operations is the calculation of the signal's root-mean-square (RMS) amplitude, which can be used to give an indication of the signal's loudness. RMS is obtained by squaring all the values, then averaging them, and finally taking the square root of the average, producing a single value for the signal that it is applied to. The purpose of squaring the values is to convert all negative values into positive values, since otherwise the values would tend to cancel each other out (when summed) resulting in an amplitude measure that would be close to zero for most kinds of speech waveforms. As an example, the RMS of a signal $\boldsymbol{x} = [4, -1, 0, 8]$ is

$$\begin{aligned} RMS(\boldsymbol{x}) &= \sqrt{(4^2 + (-1)^2 + 0^2 + 8^2)/4} \\ &= \sqrt{(16 + 1 + 0 + 64)/4} \\ &= \sqrt{20.25} \\ &= 4.25. \end{aligned}$$

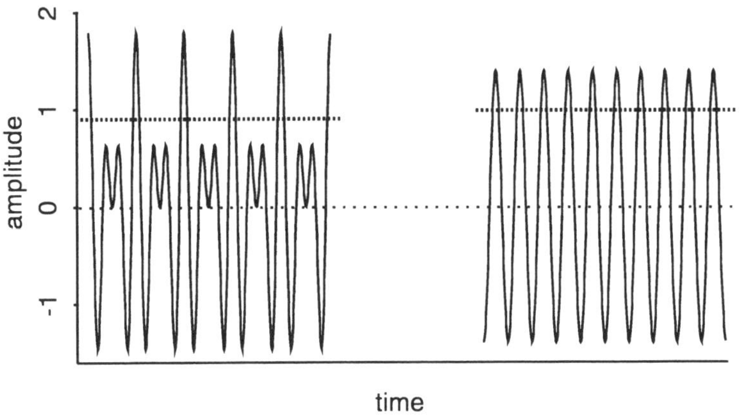

Figure 5.10: The RMS values of the two periodic signals are shown as the dotted horizontal lines in each case.

The RMS is compared for two periodic signals in Figure 5.10: although the first has a higher peak amplitude, the RMS amplitude of the second is slightly greater.

In calculating RMS amplitude for speech data, we often wish to know how RMS changes from the beginning to the end of an utterance or word: this requires windowing the data and then applying an RMS calculation to each window. Figure 5.11 shows RMS calculations on the word *stamp* using different window lengths. The RMS signal is smoother using longer windows because with high values of N a large number of data points are being averaged (in all cases the frame-shift $M = N/2$). The shortest window shown is less than one pitch period, and so the RMS envelope for $N = 100$ shows small cycle-to-cycle variations during the vowel due to the periodic nature of the waveform.

There are no definitive rules for how long the window should be, although one measure that is sometimes used is to make the window roughly the same length as the duration for which the vocal tract is assumed to be "stationary" i.e. changes minimally with time. If we assume the vocal tract is stationary for roughly 10–20 ms, the window length would need to be approximately 200–400 points for a sampling frequency of 20 kHz based on this criterion.

5.5.2 Zero-crossing rate

Another time-domain parameter that can be easily applied to digital speech signals is the *zero-crossing rate* (ZCR). A zero-crossing occurs whenever the signal crosses the x-axis (where the amplitude is zero). The number of zero-crossings bears a direct relationship to a sinusoids's frequency: specifically, a sinusoid of frequency n Hz has $2n$ zero-crossings per second. Consequently, the frequency of a sinusoid in Hz can be estimated by counting the number of zero-

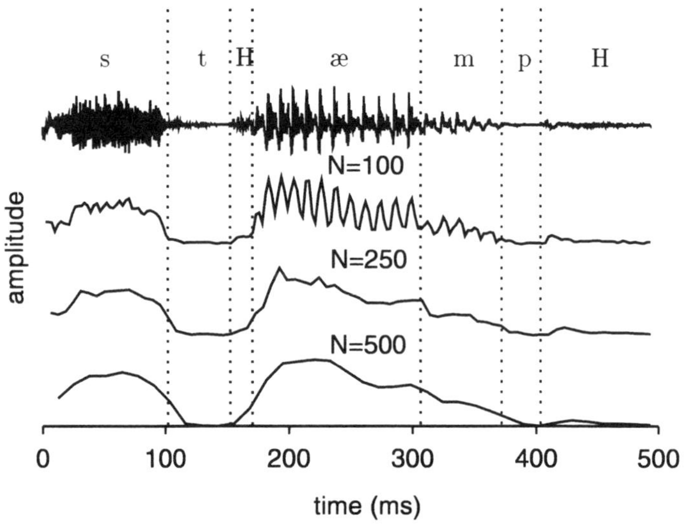

Figure 5.11: RMS of the word "stamp" calculated with windows of different lengths. "H" denotes the stop burst.

crossings per second and dividing by 2. Where Z is the number of times that the signal crosses the x-axis, the zero-crossing rate in Hz is defined as

$$ZCR = \frac{Zf_s}{2N} \quad \text{Hz}, \tag{5.11}$$

where f_s is the sampling frequency in Hz and N the number of points in the window. For example, in Figure 5.12, $Z = 8$, $N = 500$ and $f_s = 20000$ Hz, so

$$\begin{aligned} ZCR &= (8 \times 20000)/(500 \times 2) \quad \text{Hz} \\ &= 160 \quad \text{Hz} \\ &= f_0 \text{ (the sinusoid's frequency in Hz).} \end{aligned}$$

For speech data, in which the energy in the signal is distributed across a much broader range of frequencies, the zero-crossing rate is fairly well correlated with the frequency at which there is a major energy concentration. In vowels and sonorants, the zero-crossing rate tends to follow the first formant frequency because this is usually the part of the spectrum of greatest amplitude. For example, in Chapter 2, we saw that the first formant frequency for [æ] of "stamp" was at around 500 Hz; the zero-crossing rate calculated on the corresponding time-waveform (specifically on the waveform in Figure 2.3 of Chapter 2) is a little over this value at 635 Hz. On the other hand, since voiceless fricatives have most of their energy concentrated in the upper part of the spectrum, ZCR is considerably higher than for voiced sounds. In Chapter 2, we saw that the spectrum for the [s] of "stamp" had most of its energy concentrated

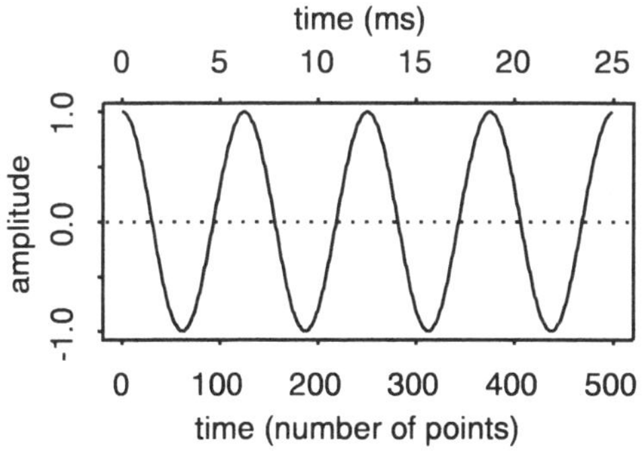

Figure 5.12: The zero crossing density is the number of times the signal crosses the x-axis (8 in this case).

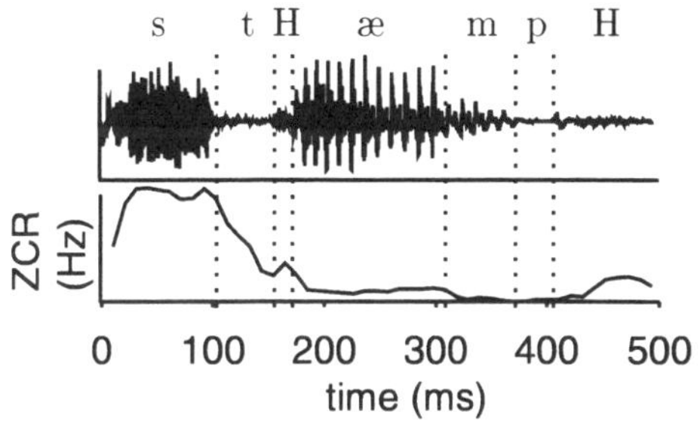

Figure 5.13: Zero-crossing rate for the word "stamp" calculated using a rectangular window of length 400 and frame-shift 200 points. "H" denotes the burst release of the two stops.

in the frequency range from 4.5-9 kHz (Figure 2.15): ZCR calculated for the corresponding time waveform evaluates to 5967 Hz.

This analysis suggests that ZCR can be used to distinguish some voiced from voiceless sounds. Figure 5.13 shows the zero-crossing rate from the onset to offset of *stamp* calculated using a rectangular window of length 400 and frame-shift 200 points. The Figure shows a high ZCR value throughout the voiceless sounds [s] and the fricated release of *stamp*, low values throughout the [æ] and values close to zero at the [m] and the closure of the [p] (closures are expected to have a low ZCR value because the signal deviates minimally from an amplitude of zero). It seems that ZCR effectively separates voiceless fricated sounds from voiced sonorants, but since these closures (which are of voiceless stops) have low ZCR values, ZCR by itself does not provide an effective distinction between voiced and voiceless sounds.

5.5.3 Short-time autocorrelation

The final parameter to be considered is the short-time autocorrelation function, which can be used to estimate the fundamental frequency of voiced speech. The autocorrelation function, as its name suggests, provides a measure of how correlated part of a signal is with another part of itself. Autocorrelation is always calculated at a *lag value* relative to the start of the signal at $n = 0$. For example, for a signal $\boldsymbol{x}$ of length 7 points, a calculation of the autocorrelation at lag 2 means that the points $x[0]$, $x[1] \ldots x[4]$ are compared respectively with the points $x[2]$, $x[3] \ldots x[6]$: the result of this comparison is a single value that varies between -1 and +1 corresponding to maximally uncorrelated and maximally correlated respectively. More generally, the autocorrelation of a signal $\boldsymbol{x}$ of length N points at lag k correlates the samples at $x[0]$, $x[1] \ldots x[N-k-1]$ with the samples at $x[k]$, $x[k+1] \ldots x[N-1]$.

The autocorrelation function can be used to estimate a speech signal's fundamental frequency. This is because, since the shapes of the waveforms of successive pitch periods are very similar to each other, the correlation between pitch periods in different parts of the signal will be high. Seen from another point of view, if we can determine the lag value at which the autocorrelation function attains maximum values, we can use this information to estimate the duration of a pitch period.

This principle is shown for a two-cycle sinusoid at three different lag values in Figure 5.14. In all cases, the original signal is cut up into two different signals $\boldsymbol{a}$ and $\boldsymbol{b}$ of the same length which are then time-aligned in the panel below. In very general terms, the greater the extent to which signal $\boldsymbol{b}$ can be overlaid point by point onto $\boldsymbol{a}$, the better the correlation between them. Comparing the three panels, it is clear that the correlation is perfect at lag 16, which is also the duration of the pitch-period (in points), poorest at lag 10, and intermediate at lag 4.

Usually when estimating the duration of a pitch period, autocorrelations are carried out for lags 0, 1, 2, $\ldots k$, where k is thought to include at least two pitch periods, and then graphed in the form shown in Figure 5.15. The actual calculation of the autocorrelation is obtained by multiplying the two part-signals point by point and then summing them: at lag k, the short-time autocorrelation, $R[k]$, is given by:

$$R[k] = \sum_{n=0}^{N-k-1} x[n]x[n+k]. \qquad (5.12)$$

Notice that when $k = 0$, $R[0] = \sum(x[n])^2$: expressed in words, the autocorrelation at lag 0 is a measure of the signal's power obtained by squaring all the values of $\boldsymbol{x}$ and then summing them. (Therefore, the RMS of a signal is equivalently given by $\sqrt{R[0]/N}$). The autocorrelation function of a signal is usually normalised by dividing by $R[0]$: since $\boldsymbol{R}$ has a maximum at $k = 0$, a division by $R[0]$ ensures that $\boldsymbol{R}$ always varies between ± 1. The left and right panels of Figure 5.16 show the autocorrelation calculations on two sections of the "stamp" signal taken from the central parts of the [s] and [æ]. The shapes

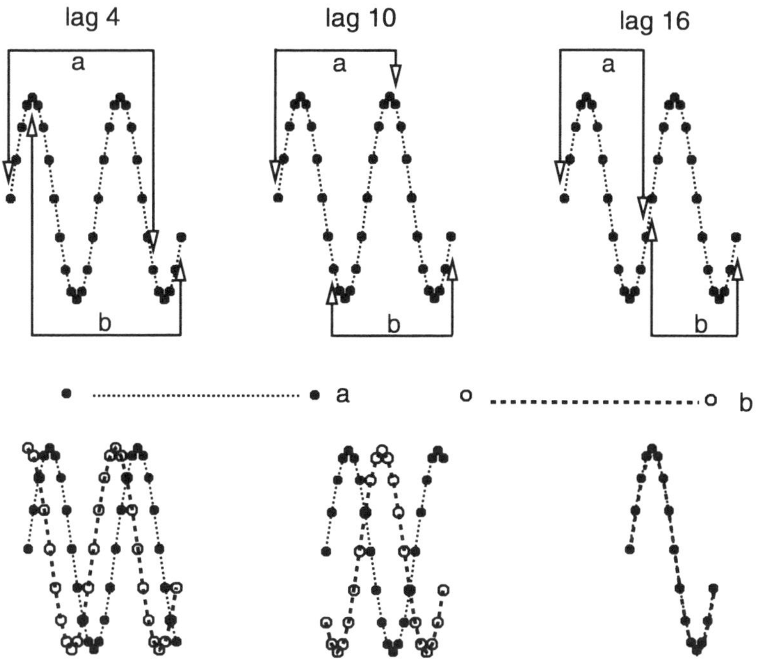

Figure 5.14: A representation of the initial stages in calculating the short-time autocorrelation of a signal at three different lag values. The top row shows a two-cycle sinusoid windowed at lag 4 (*left*), lag 10 (*middle*), and lag 16 (*right*). In the bottom row, the two windows are overlaid point by point. For lag 16, which is also the sinusoid's pitch-period duration, the two windows overlap exactly i.e. there is maximum autocorrelation at this lag value.

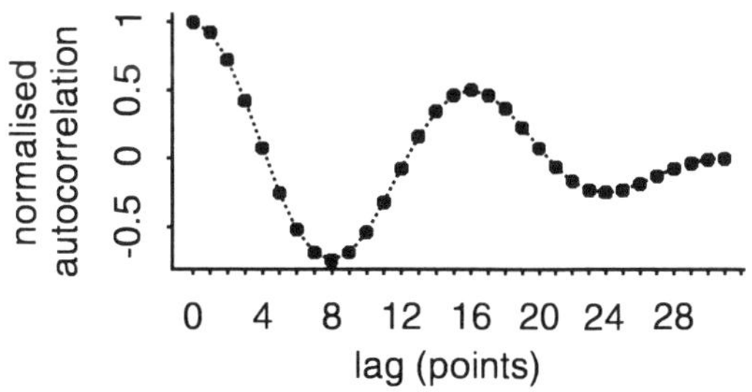

Figure 5.15: The autocorrelation function of the two-cycle sinusoid in Figure 5.14.

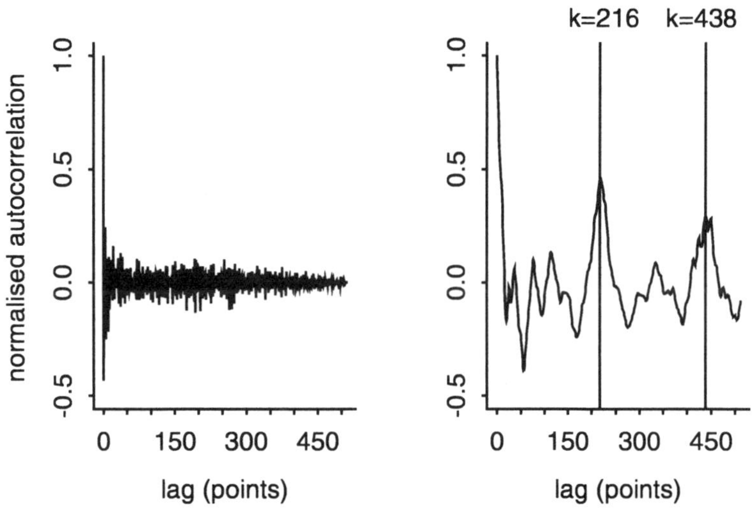

Figure 5.16: Autocorrelation functions of sections of [s] (left) and [æ] (right) signals.

of the autocorrelation functions are markedly different for these voiceless and voiced segments: for voiced [æ], there are distinct peaks at lags $k = 216$ and $k = 438$, whereas the autocorrelation function for [s] is less patterned and shows no evidence of any peaks.

These calculated peaks in the right panel of Figure 5.16 can be used to estimate the fundamental frequency of the signal as follows. At a sampling frequency of 20 kHz, the sample period $T = 0.05$ ms. Therefore, a lag at $k = 216$ corresponds to $216 \times 0.05 = 10.8$ ms. The next peak occurs at $k = 438 - 216 = 222$ points or 11.1 ms after the first peak. These are the estimated pitch period durations for two periods of [æ], and they agree quite well with the hand measured estimations of the second and third periods in Figure 2.3 of Chapter 2.

5.6 Convolution and time-domain filtering

In the acoustic theory of speech production in Chapter 3, we discussed how the production of speech can be modelled by passing a source through a filter. The source can be either a periodic waveform produced by vocal fold vibration, a waveform of noise caused by a turbulent airflow, or a mixture of the two, while the filter is defined by the shape of the vocal tract. In speech production, the vocal tract filter is imprinted upon the source signal and it is this "imprint" of the filter in the resulting speech waveform that is mostly responsible for conveying differences of phonetic quality.

The source-filter model of speech production is carried over into speech synthesis in which a source signal is passed through a filter to produce the resulting speech output. The source-filter model also underlies many models of speech recognition: in this case, we are starting with a speech waveform and attempting to reconstruct the source signal and filter characteristics that generated it.

In digital speech processing, the process of passing a source through a filter to obtain speech output is modelled by an operation known as *convolution*. The source is a periodic or aperiodic waveform, depending on the voicing status of the sound, while the filter is described in terms of *transfer-function coefficients* that can also be thought of as *weightings* on the source signal. The reason that the term *transfer-function* is used is because its coefficients transform a source signal into an output signal — in more general terms, they are responsible for leaving the imprint on the source signal on which the phonetic quality differences depend.

In order to apply techniques for either synthesising a waveform or decomposing a natural speech waveform into source and filter characteristics, we must first explain in more precise detail the process of convolving a source signal with transfer function coefficients. In the following sections, we will discuss this entirely in terms of the convolution of time-domain waveforms; in the next chapter, we will show how convolution can also be described as the multiplication of their corresponding frequency-domain representations.

The digital filter model that is outlined in this chapter and the next and then applied to digital speech synthesis and analysis rests on various assumptions. First, the filters are always assumed to be *linear* and *time-invariant* or LTI. The reason for this is partly that the relationships between time and frequency domain filtering are considerably simplified if the filtering operation is assumed to be LTI and the mathematics of LTI systems are very well understood. Also, to the extent that the production of speech can be modelled as a series of stationary processes – that is, we have to assume that the changing supralaryngeal vocal tract shape can be represented as a series of static shapes over a small time intervals of about 10-15 milliseconds – a digital LTI model can, within certain limitations, be successfully applied to digital speech synthesis and analysis. Indeed, the assumption that the vocal tract can be modelled as a cascade of LTI filters is at the very core of digital formant synthesis and analysis systems to be described in Chapters 7 and 8. Another assumption that must be made for the digital LTI filter model to be applicable to speech production is that the source signal in speech is *independent* of the vocal tract filter which also underlies much of the acoustic theory of speech production discussed in Chapter 3. Although there are many instances of complicated interactions between the source and filter (see, e.g., Klatt & Klatt, 1990), we can nevertheless make considerable progress in understanding the nature of speech production in many different ways – such as deriving nomograms from tube models of vowels and synthesising intelligible speech – by assuming a model in which the source and filter are independent.

Before describing in further detail the mechanism in digital terms by which a source is passed through a filter, we will make a few remarks about what characterises a filter as linear time invariant. A system is defined to be linear under two main conditions. First, there is the condition of *superposition.* If we pass a signal $\boldsymbol{x_1}$ through a filter and obtain an output $\boldsymbol{y_1}$ and then pass another signal $\boldsymbol{x_2}$ through the same filter to obtain a second output signal $\boldsymbol{y_2}$, and finally a third signal that is the sum of $\boldsymbol{x_1}$ and $\boldsymbol{x_2}$ to obtain an output $\boldsymbol{y_3}$, then if the filter is LTI, $\boldsymbol{y_3} = \boldsymbol{y_1} + \boldsymbol{y_2}$. Or, as Moore (1989) succinctly states: "the output of the [LTI] system in response to a number of independent inputs presented simultaneously should be equal to the sum of the outputs that would have been presented if each input were presented alone". Second, LTI systems have the property of *homogeneity*, which implies that if the input signal is scaled in magnitude by a certain factor, then the output will be scaled by the same factor. So if the system is LTI, and the input to the system is $k\boldsymbol{x_1}$, the output must be $k\boldsymbol{y_1}$. Third, LTI systems have the property of *time-invariance.* This means that the filtering operation is insensitive to time-shifting in the sense defined earlier. Therefore, if the output after filtering a signal $x[n]$ is $y[n]$, then the result of the same filtering operation on $x[n-p]$ — i.e. the original signal time-shifted by p units — must be $y[n-p]$. Another important property is that the *order* in which LTI filters is applied to an input signal has no significant effect on the output signal. So if a signal is to be passed through two separate LTI filters, $\boldsymbol{h_1}$ and $\boldsymbol{h_2}$, we could either pass it through $\boldsymbol{h_1}$ and then feed the output from that through the filter $\boldsymbol{h_2}$ or *vice-versa*: in both cases, the same output is obtained.

Lastly, one of the most important consequences of using LTI filters that allows many time-frequency correspondences to be straightforwardly defined in mathematical terms is that when a sinusoid is passed through an LTI filter, the filter only affects the sinusoid's amplitude and phase but not its frequency. So if the complex waveform in Figure 2.11 of Chapter 2 were passed through an LTI filter, or a cascade of LTI filters, the output would consist of a complex waveform at the same *frequencies* as those shown in Figure 2.11 but at different amplitudes and phases.

5.6.1 Convolution as weighted differencing

We will devote the following section to describing just two points that are fundamental to understanding a digital model of speech production. First, when a source, either periodic or aperiodic, is passed through the vocal tract filter, this process can be defined mathematically as the *convolution* of the source and the filter; and second, this convolution makes the output signal – that is, the speech waveform that we see in an acoustic record of speech – have the property that it can be decomposed into the *sum of scaled and delayed versions* of itself.

We will begin with a simple example of two signals, $\boldsymbol{x}$ and $\boldsymbol{h}$, that are to be convolved with each other. It will be convenient to think of these as representing the source and filter, respectively, in speech production, although their structure for a real speech signal would actually be far more complicated. We will also define a signal $\boldsymbol{y}$ that is the result of convolving $\boldsymbol{x}$ and $\boldsymbol{h}$: the signal

$\boldsymbol{y}$ is analogous to the resulting speech waveform. Mathematically, we can write

$$\boldsymbol{y} = \boldsymbol{x} * \boldsymbol{h}, \tag{5.13}$$

where $*$ denotes (linear) convolution. In speech synthesis, we would be starting with $\boldsymbol{x}$ and $\boldsymbol{h}$ and convolving these to construct $\boldsymbol{y}$; in speech analysis, we would begin with a speech waveform (output signal) $\boldsymbol{y}$ and attempt to estimate the source and filter that could have generated $\boldsymbol{y}$. We will develop further the interpretation of $\boldsymbol{h}$ as the impulse response of the (vocal tract) filter in the next chapter.

The filter $\boldsymbol{h}$ can often be broken down into *recursive* and *nonrecursive* filters that are convolved with the output and input signals respectively so that equation Equation 5.13 can be rewritten as two convolutions:

$$\boldsymbol{y} * \boldsymbol{a} = \boldsymbol{x} * \boldsymbol{b}. \tag{5.14}$$

In order to explain convolution as weighted differencing, we consider the nonrecursive part of the filter in Equation 5.14 by assuming that $\boldsymbol{a} = 1$ so that the output signal $\boldsymbol{y}$ is calculated from $\boldsymbol{x} * \boldsymbol{b}$. Figure 5.17 shows the principal process by which the signal $\boldsymbol{b}$ is convolved with the signal $\boldsymbol{x}$. As this figure shows, convolution involves sliding the data points of $\boldsymbol{b}$ in reverse order past $\boldsymbol{x}$: the four panels represent the successive stages as $\boldsymbol{b}$ advances past $\boldsymbol{x}$ one point at a time. In order to determine the output from convolution, the two signals must be multiplied data point by data point and then summed.

The temporal index of the output is given by the position of $b[0]$ relative to the time axis. For example, in the top right panel, $b[0]$ is positioned at $n = 1$; the sum of the product of the signals in this panel therefore gives the value of $y[1]$. In this case, there are two nonzero values:

$$\begin{aligned} y[1] &= x[1]b[0] + x[0]b[1] \\ &= 2 \times 0.5 + 4 \times 0.2 \\ &= 1 + 0.8 \\ &= 1.8. \end{aligned}$$

The process of convolution that has just been demonstrated Figure 5.17 is equivalent to a filter that *weights delayed values of an input signal.* The same sense of *delayed* is intended as defined in Section 5.3: a signal $\boldsymbol{x}$ that is delayed by p points starts p points later in time (and is denoted as $x[n-p]$ — see Figure 5.7). When $\boldsymbol{x}$ is convolved with $\boldsymbol{b}$, the convolution of the two signals is also given by

$$\begin{aligned} y[n] &= x[n]b[0] + x[n-1]b[1] + x[n-2]b[2] + \ldots + x[n-p]b[p] \quad (5.15) \\ &= x[n] * b[n]. \end{aligned}$$

From Equation 5.15, we can therefore write $\boldsymbol{y} = \boldsymbol{x} * \boldsymbol{b}$ in the present example as

$$y[n] = x[n]b[0] + x[n-1]b[1] \tag{5.16}$$

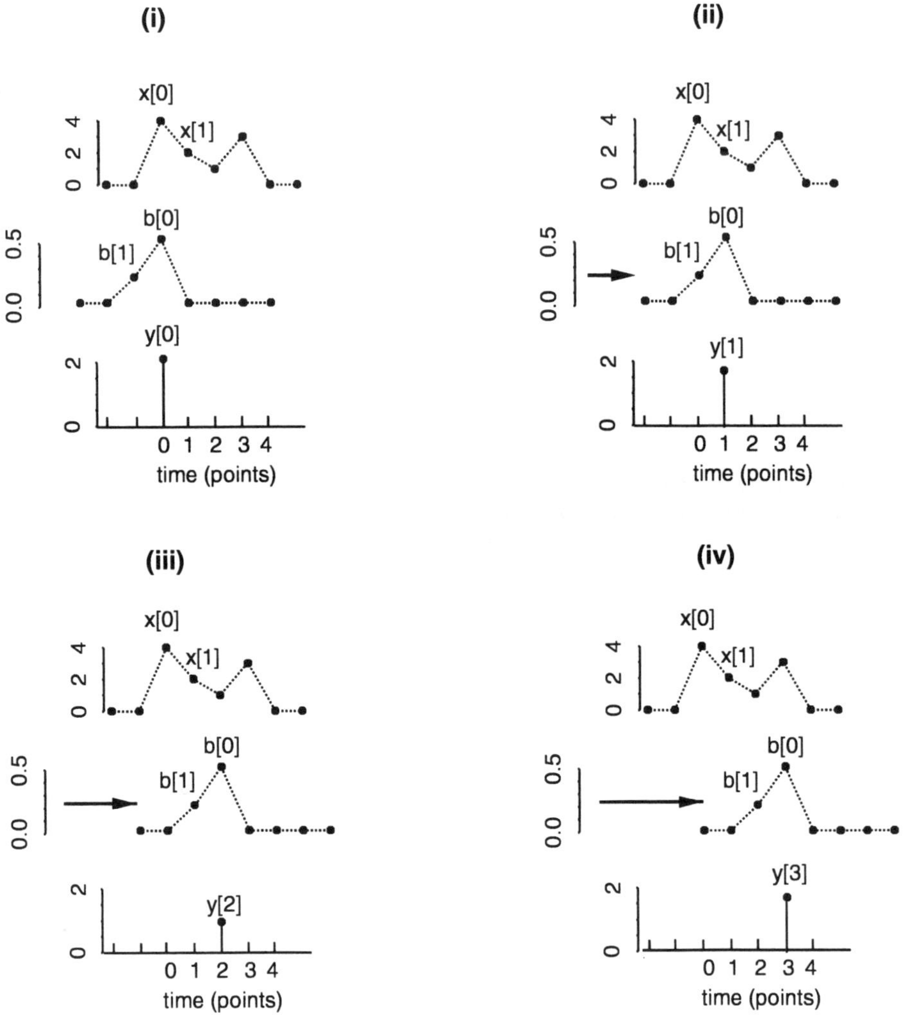

Figure 5.17: The process by which the filter $\boldsymbol{b}$ slides past the input signal in reverse order when the two are convolved with each other to produce the output signal $\boldsymbol{y}$. The four panels represents the calculation of $y[0] \ldots y[3]$. In all cases, $y[n]$ is obtained by lining up $\boldsymbol{b}$ in reverse temporal order with $\boldsymbol{x}$, multiplying the two signals point by point, and then summing the result.

This is evaluated for the current example as follows:

n	$=$	0	1	2	3	4
$x[n] \times b[0]$	$=$	4×0.5	2×0.5	1×0.5	3×0.5	0×0.5
$x[n-1] \times b[1]$	$= +$	0×0.2	4×0.2	2×0.2	1×0.2	3×0.2
$y[n]$	$=$	2	1.8	0.9	1.7	0.6

The example considered so far has been one in which the source signal $\boldsymbol{x}$ is weighted by (convolved with) a filter $\boldsymbol{b}$. This type of filter is known as a *non-recursive* or *finite-impulse-response* (FIR) filter because the output has a finite number of values (five in the present example). We are now going to consider how to derive $\boldsymbol{y}$ in a *recursive* filter i.e. when $\boldsymbol{b} = 1$ so that

$$\boldsymbol{y} * \boldsymbol{a} = \boldsymbol{x}. \tag{5.17}$$

This type of filter is also known as an *infinite-impulse-response* (IIR) filter because, in contrast to the nonrecursive filter that we have looked at so far, it generates an infinite number of (usually decaying) values. The reason for the name *recursive* is that the calculation of $y[n]$ depends on $y[n-1]$: seen from another point of view, once a sample value of $\boldsymbol{y}$ has been calculated, it is fed back into the weighting process to calculate the next sample value of $\boldsymbol{y}$ (so yet another name that is sometimes used is *feedback* filter).

Following from the demonstration above that convolution is the same as differencing delayed samples, Equation 5.17 can be rewritten as

$$a[0]y[n] + a[1]y[n-1] + a[2]y[n-2] + \ldots a[p]y[n-p] = x[n] \tag{5.18}$$

and rearranged as:

$$a[0]y[n] = x[n] - a[1]y[n-1] - a[2]y[n-2] - \ldots a[k]y[n-p]. \tag{5.19}$$

The first coefficient of the filter is simply a scale factor for the entire signal and is almost always assumed to be one. So the left-hand side of the above equation is just $y[n]$.

We will now consider an example using the same source and filter i.e.

$$\begin{aligned} \boldsymbol{x} &= 4 \quad 2 \quad 1 \quad 3 \\ \boldsymbol{a} &= 0.5 \quad 0.2. \end{aligned}$$

From Equation 5.19, the first value of the output, $y[0]$ is given by

$$y[0] = x[0] - 0.5y[-1] - 0.2y[-2]. \tag{5.20}$$

However, since all signals discussed in this book are zero-valued when n is outside the range $0 \leq n \leq N-1$, the last two terms in the right-hand side of Equation 5.19 are zero, and so $y[0] = x[0] = 4$. The next output value $y[1]$, which is dependent on the prior calculation of $y[0]$, is given by

$$\begin{aligned} y[1] &= x[1] - 0.5y[0] - 0.2y[-1] \\ &= 2 - (0.5 \times 4) - 0 \\ &= 0. \end{aligned} \tag{5.21}$$

The third output value is

$$\begin{aligned} y[2] &= x[2] - 0.5y[1] - 0.2y[0] \\ &= 1 - (0.5 \times 0) - (0.2 \times 4) \\ &= 0.2. \end{aligned} \tag{5.22}$$

Values of $\boldsymbol{y}$ for higher values of n are iteratively calculated in a similar way.

5.6.2 Relationship of filtering to speech production

As we shall see in later chapters, the weightings on the *output* are directly related to the formant frequencies while the weightings on the *input* (source) signal are related to antiformant frequencies.

Since vowels can be uniquely characterised by formant frequencies (because the assumption is made that there are no side-branching resonators), an appropriate digital model for a vowel could be

$$\boldsymbol{y} * \boldsymbol{a} = \boldsymbol{x} \quad (a[0] = 1), \tag{5.23}$$

in which the filter characteristics are defined by $\boldsymbol{a}$ and $\boldsymbol{x}$ is the source signal. As we shall see later, the length of $\boldsymbol{a}$ is directly dependent on the number of formants that are used to model the speech signal; more importantly, vowels of different phonetic quality have different coefficient values, i.e. different values of $\boldsymbol{a}$.

Since nasal (and fricative) consonants have both resonances and anti-resonances, a more appropriate model is one in which there are weightings on both the input and output signal:

$$\boldsymbol{y} * \boldsymbol{a} = \boldsymbol{b} * \boldsymbol{x} \quad (a[0] = 1), \tag{5.24}$$

which, following from the analysis above, could also be written as a difference equation using terms of $y[n-p]$ (weighted by $a[n]$) and $x[n-p]$ (weighted by $b[n]$).

These issues of digital formant synthesis and analysis are are explored in further detail in Chapters 7 and 8. Before this, we will need to recast the digital recursive and nonrecursive filters in the frequency-domain. We will do this by first developing the digital implementation of the Fourier transform, known as the *discrete Fourier transform*, and then another frequency-transform known as the *z-transform*. With the mechanism of the z-transform, we will be able to express a combination of recursive and nonrecursive filters in terms of the *impulse response* of the vocal tract filter that is convolved with a source in digital speech synthesis or estimated in an analysis technique such as linear predictive coding.

Further reading

Two useful books on signal processing that do not assume an engineering/signal processing background are Steiglitz (1996) and Rosen and Howell (1992). The first of these assumes a higher level of mathematics than the present book but can be understood with a basic grasp of calculus. This book develops many of the principles of digital signal processing from an analysis of wave equations that are also relevant to some of the material in Chapter 3. Rosen and Howell (1992) is based on a more elementary knowledge of mathematics than the present book and deals almost entirely with *analogue* filtering, although the principles that are described are usually just as relevant for an understanding of the digital speech processing techniques described here.

6 Frequency-Domain Analysis of Digital Speech Signals

In this chapter, we reconsider many of the frequency-domain concepts discussed in earlier chapters in terms of digital analysis. We begin by discussing some properties of digital sinusoids and the *discrete Fourier transform*, which is the principal operation for deriving spectra from digital time signals. Having established the procedure for obtaining digital spectra, we can consider how they can be further parameterised to highlight some of the salient spectral differences between speech sounds. One of the parameterisations, known as *cepstral analysis*, will provide the first technique for separating the acoustic source from the filter. In the final part of this Chapter (6.6), we will reanalyse time-domain convolution in terms of frequency-domain multiplication, thereby providing another kind of model for relating the output to the filter and the source.

6.1 Digital sinusoids

In Chapter 2, a sinusoid was defined as the height of a point above a horizontal line as it moves at constant speed around a circle. In the digital domain, this statement is also valid with the qualification that the point P moves in discrete jumps between equally spaced points around the circle and such that the time taken between jumps, which is the sample period T, is constant. The resulting waveform is known as an N-point sinusoid where N is the number of equally spaced points $(0, 1, 2 \ldots N-1)$ around the circle. Since the time taken between points is the sample period T, the duration of any N point sinusoid is always $(N-1)T$ seconds. A 16-point sinusoid (duration = $15T$ seconds) is shown in Figure 6.1.

6.1.1 Amplitude

There are direct parallels between the digital and analogue cases (discussed in Chapter 2): the amplitude of a digital sinusoid is varied by changing the radius of the circle.

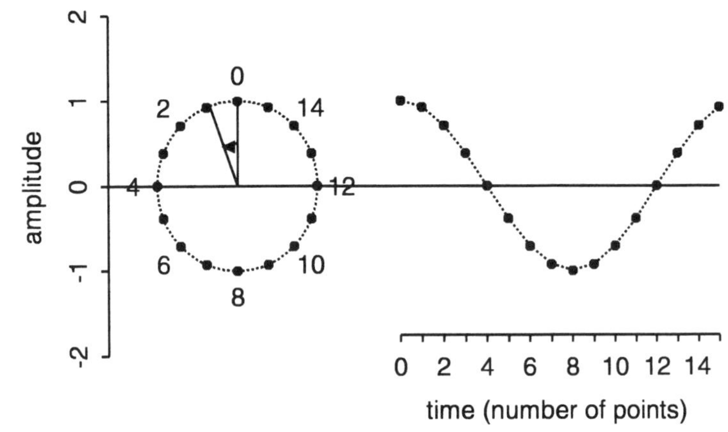

Figure 6.1: A 16-point digital cosine wave generated when a point P moves around the circle in discrete "jumps" (from one number to the next) starting at the top of the circle. The time taken between jumps, which is the sample period, T, is constant.

6.1.2 Phase

As discussed in Chapter 2, phase defines the position of the first point, $x[0]$, on the circle. If $x[0]$ is at the top of the circle, the phase is 0 radians. In the case of a digital sine wave (Figure 6.2), $x[0]$ starts a quarter of a cycle earlier (and so the phase is $-\pi/2$ radians).

6.1.3 Frequency

There are three separate definitions of frequency that are interrelated and that need to be distinguished.

In the first case, the frequency of a sinusoid can be defined by the *number of cycles*, or revolutions, made by a point P around the circle per N points. The frequency of the sinusoids in Figure 6.2 and Figure 6.1 is 1 cycle because a single pitch-period sinusoid is generated when P completes one revolution around the circle. In Figure 6.3, the N-point sinusoid is derived when P moves around the circle three times: although both sinusoids in Figure 6.3 and Figure 6.1 have the same duration (and are composed of the same number of points), the frequency of the sinusoid in Figure 6.3 is three times that of the one in Figure 6.1.

Second, the frequency of the sinusoid can be defined as the *number of cycles per second* which is measured in Hertz. In Figure 6.1 and Figure 6.2, one cycle (one complete revolution around the circle) is completed every NT seconds. Therefore, the frequency of the sinusoids is 1/NT Hz. More generally, the frequency of a sinusoid derived from a point P that rotates at k cycles every NT seconds is given by

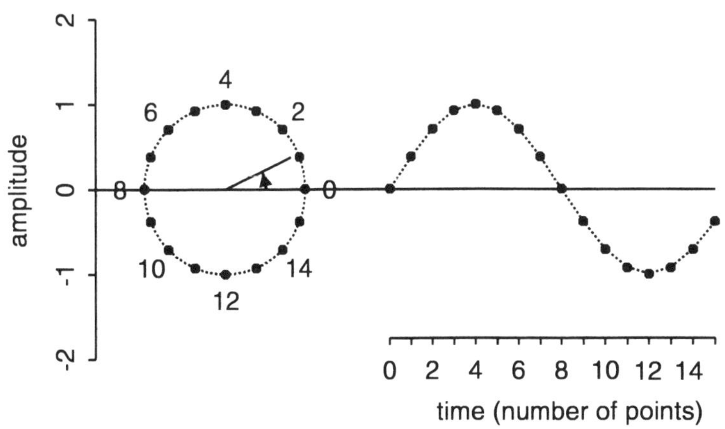

Figure 6.2: A 16-point digital sine wave. This sinusoid is defined to have a phase $-\pi/2$ radians, which implies it starts a quarter of a cycle earlier than the digital cosine wave in Figure 6.1.

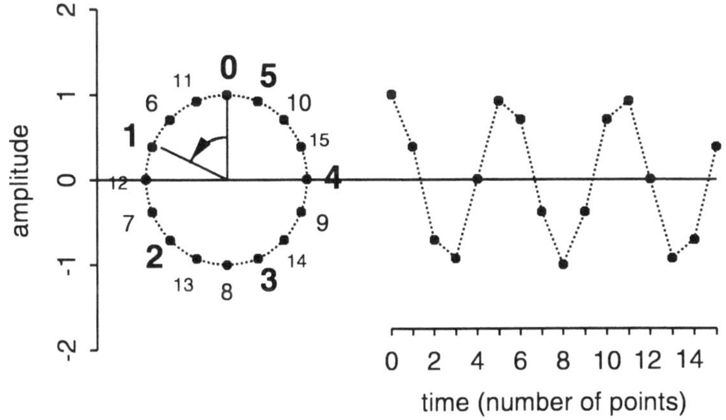

Figure 6.3: This sinusoid is generated when a point P moves around the circle three times in 16 equally spaced jumps: note the sample period T is the same as that of the one-cycle sinusoids of Figure 6.1 and Figure 6.2. The points in the largest font (from $x[0]$ to $x[5]$) are generated on the first cycle.

$$f_{Hz} = \frac{k}{NT} \tag{6.1}$$

where f_{Hz} is the frequency in Hertz. The third definition is in terms of the *radian frequency* ω, which is defined by the angle between any two data points on the circle:

$$\omega = 2\pi k/N \quad \text{radians per sample.} \tag{6.2}$$

For example, since the sinusoids in Figure 6.2 and Figure 6.1 are generated from one cycle, their radian frequency is

$$\begin{aligned} \omega &= 2 \times \pi \times 1/16 \\ &= \pi/8 \quad \text{radians per sample.} \end{aligned}$$

For the sinusoid in Figure 6.3, ω is

$$\begin{aligned} \omega &= 2 \times \pi \times 3/16 \\ &= 3\pi/8 \quad \text{radians per sample} \end{aligned}$$

i.e. three times that of the sinusoid in Figure 6.1. Notice that ω can vary between 0 and 2π radians/sample; although, as we shall see in the discussion of the Nyquist frequency below, from the point of view of the sinusoid's frequency in Hertz, it effectively varies between 0 and π radians/sample. Finally, a radian frequency can be converted into a frequency in Hertz (number of cycles per second) from the relationship:

$$f_{Hz} = \frac{\omega}{2\pi T} \quad \text{Hz} \tag{6.3}$$

6.1.4 Equation for a sinusoid

The equation for any digital sinusoid of amplitude A and phase ϕ can be succinctly stated in terms of the radian frequency:

$$x[n] = A\cos(\omega n + \phi) \quad 0 \leq n \leq N-1. \tag{6.4}$$

The equation for the sinusoid in Figure 6.2 is therefore

$$x[n] = \cos(\pi\frac{n}{8} - \frac{\pi}{2}) \quad 0 \leq n \leq 15.$$

Therefore, the amplitude of the fifth data point, $x[4]$, would be given by

$$\begin{aligned} x[4] &= \cos(8\frac{\pi}{16} - \frac{\pi}{2}) \\ &= \cos(0) \\ &= 1. \end{aligned}$$

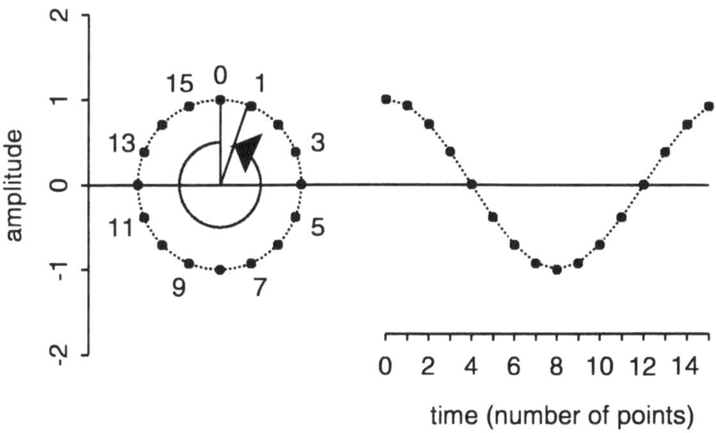

Figure 6.4: A 15-cycle, 16-point cosine wave (compare with the 1-cycle sinusoid in Figure 6.1).

6.1.5 The Nyquist frequency

In the discussion of digital signals in the preceding chapter, it was mentioned that only the frequency range from 0 Hz to the Nyquist frequency, or half the sampling frequency, can be accurately reconstructed in digital signals: any frequencies greater than the Nyquist frequency are aliased onto those of a lower frequency.

The effect of the Nyquist frequency can also be demonstrated in a different way: any sinusoid of k cycles and phase ϕ radians is aliased onto a sinusoid of $N - k$ cycles and phase $-\phi$ radians if k is greater than half the number of data points ($N/2$).

Consider as an example of this the 16-point sinusoid in Figure 6.4 in which $k = 15$ cycles (and the phase is 0 radians). In this case, k is greater than 8 (half the number of data points), and so it is aliased onto a sinusoid of $16 - 15 = 1$ cycle: compatibly, the sinusoids derived from the circles in Figure 6.4 and Figure 6.1 are identical. The reason for this is that the data points in both cases are at the same height (amplitude) on the circle but on *opposite sides*: compare, for example, the height of the 3rd data point in Figure 6.4 with the height of the 3rd data point in Figure 6.1.

The highest frequency sinusoid (in Hertz) that can be generated is one in which k, the number of cycles, is half the number of data points. Substituting $k = N/2$ in 6.2, this sinusoid has a radian frequency of $\omega = (2\pi N)/(2N) = \pi$ radians/sample. The same substitution in 6.1 shows that such a sinusoid has a frequency in Hertz of $N/2NT = 1/2T$ Hz i.e. half the sampling frequency (which is $1/T$ Hz).

6.2 The discrete Fourier transform

In Chapter 2, we saw that the purpose of a Fourier transform was to convert a waveform into a set of sinusoids such that, when the sinusoids are summed, the original waveform is reconstructed. The operation which converts a digital waveform into (digital) sinusoids is the *discrete Fourier transform* (or DFT). A variation on the DFT is the *fast Fourier transform* (or FFT). The FFT is simply a computationally more efficient version of the DFT under which certain assumptions are made about the number of points that are to be Fourier-transformed: but since the DFT and FFT produce identical results (but in a different way), we shall simply refer to this operation as the DFT, while recognising that, in practice, it is certainly preferable to carry out the transform using the FFT (there are many summaries of the computational advantage of the FFT — in the speech literature, see, e.g., Owens, 1993, or Witten, 1982).

It can be shown that any digital waveform of any length can be converted into a set of sinusoids by the DFT, which, when summed, produce the original waveform. In this case, a DFT produces the same number of sinusoids as there are points in the waveform that is transformed (so N sinusoids for a waveform of N points); and these occur at 0 to $N-1$ cycles (where cycles has the definition given earlier). Consider therefore the DFT of an 8-point waveform:

$$\boldsymbol{x} = [-8, -8, -4, 5, -2, 4, 7, 9]. \tag{6.5}$$

We know from the above discussion that the result of applying a DFT to the waveform in Equation 6.5 is 8 sinusoids at 0, 1, 2 ... 7 cycles. These are shown inside the rectangle in Figure 6.5. Various observations can be made about these sinusoids. First, the sinusoid at k = 0 cycles is always a straight line. This zero-cycle sinusoid is sometimes called the *d.c. offset*, and it is equal to the sum of the original time signal's values. Second, sinusoids at greater than $N/2 = 4$ cycles are identical to those at lower frequencies (compare, for example, the 6-cycle sinusoid with the 2-cycle sinusoid). This follows both from the earlier discussion of the Nyquist frequency and because higher-frequency sinusoids (for which ω is greater than π radians) generated by the DFT have a phase that is opposite in sign to their corresponding lower frequency sinusoids. Third, some of the sinusoids do not seem to be very sinusoidal, but this is either because they are very low in amplitude (such as the 3-cycle sinusoid) or because they are supported by a very small number of data points in these examples. The fact that they really *are* sinusoidal can be demonstrated by superimposing a continuous sinusoid at the same amplitude, frequency, and phase (Figure 6.6).

Fourier analysis can be reversed by summing all the sinusoids that are derived from the DFT. This process, known as *Fourier synthesis*, results in an exact reconstruction of the digital time signal to which the DFT was originally applied (Figure 6.5).

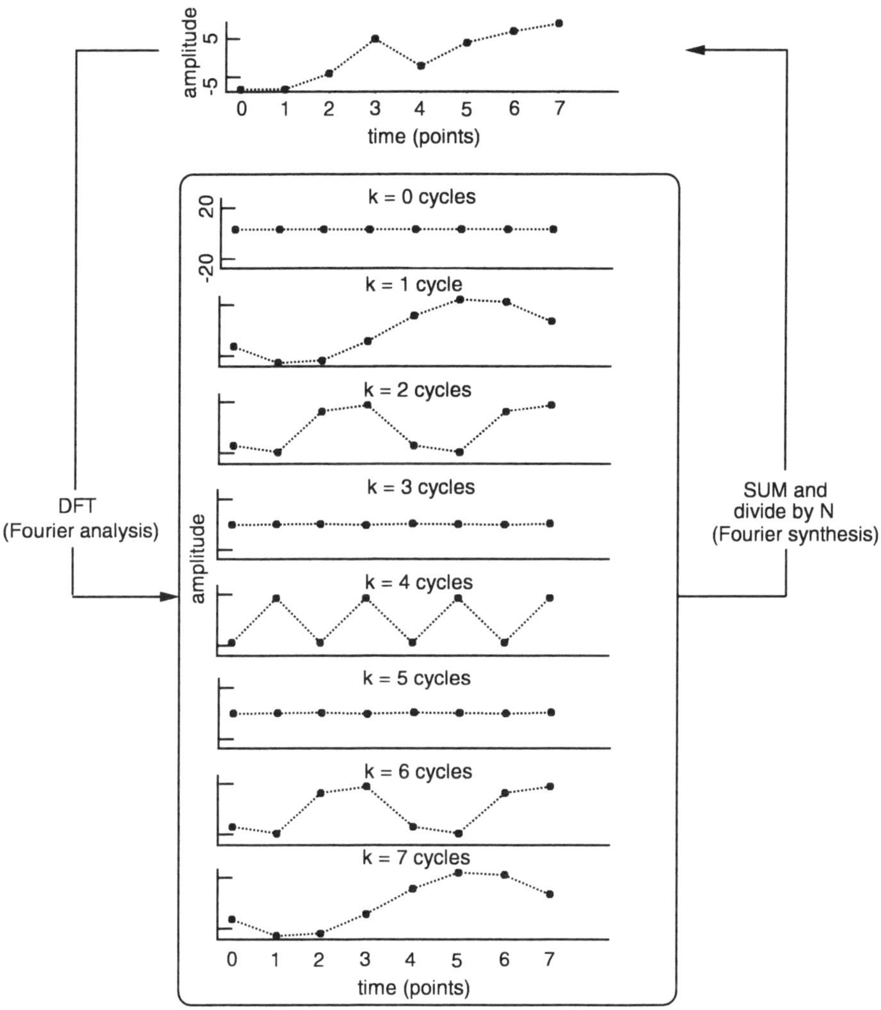

Figure 6.5: An 8-point time-signal (*top panel*) and its decomposition into sinusoids (inside the rectangle). This decomposition is known as *Fourier analysis* and is accomplished by the discrete Fourier transform. When this process is reversed (*Fourier synthesis*), the sinusoids are summed point by point (and the result is also divided by N, which is 8 in this case). Summation results in an exact reconstruction of the original digital time signal.

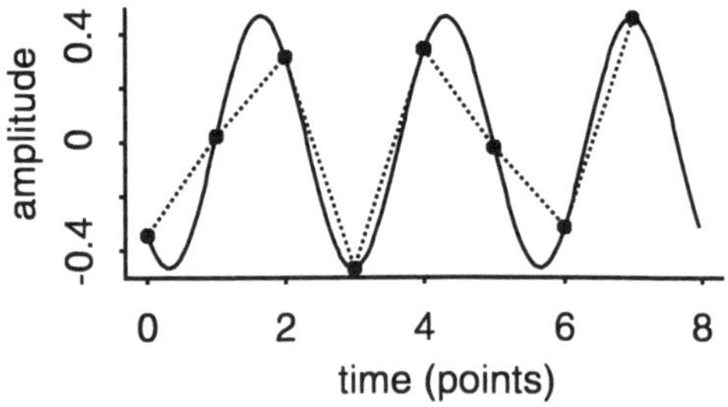

Figure 6.6: Superimposition of a continuous sinusoid on a digital sinusoid supported by a small number of data points. The digital sinusoid is the one at $k = 3$ cycles in the previous figure.

6.3 Spectra derived from the DFT

As discussed in Chapter 2, a spectrum is a display of the sinusoids' amplitude (y-axis) against frequency (x-axis). In digital spectra, the x-axis is typically a measure of the radian frequency or, if the sampling frequency is known, the frequency in Hertz. Furthermore, since the DFT of a speech signal always produces a "reflected" spectrum, in which the sinusoids at higher frequencies are simply copies of those at lower frequencies, it is also conventional to display only the frequencies up to the Nyquist frequency (up to π radians if the axis is radian frequency; or up to half the sampling rate if the x-axis is in Hz). As an example, and assuming a sampling frequency of 8000 Hz, the spectrum of the digital time signal from the preceding section is shown in Figure 6.7. Notice that there are five frequency components equally spaced between 0 Hz (0 radians) and the Nyquist frequency. The more general case is as follows. The spectrum resulting from a DFT applied to an N-point time signal consists of $(N/2) + 1$ unreflected values that are equally spaced on the frequency axis between 0 Hz (0 radians) and the Nyquist frequency. It is also common to display the y-axis in decibels rather than as amplitude of air pressure. Following from the discussion in Chapter 2, the conversion from amplitude of air pressure values (A) to decibels is as follows:

$$I = 20 \log_{10}(A). \tag{6.6}$$

6.4 Some points of procedure in applying a DFT

We now briefly discuss some further criteria that should be considered when calculating spectra of speech data using a discrete Fourier transform. While the points of detail expressed in the rest of this section are important, they can be omitted until a more thorough analysis using the DFT is required.

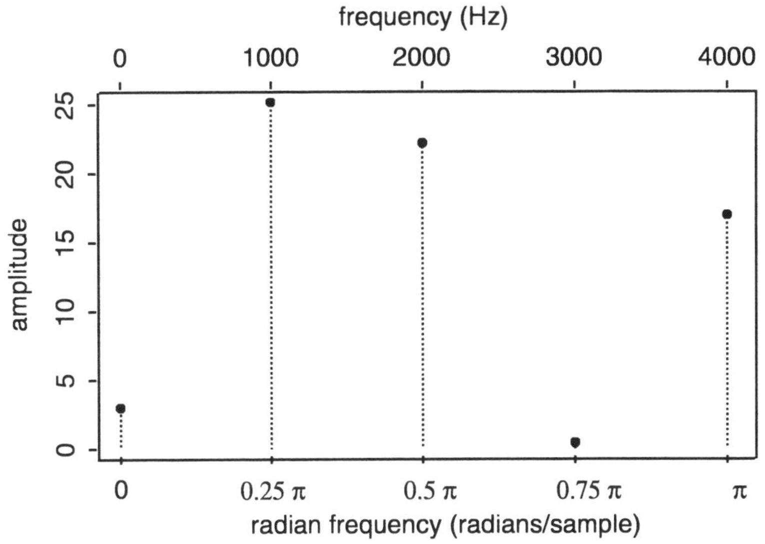

Figure 6.7: Amplitude spectrum of the waveform in Figure 6.5. The horizontal axis shows the radian frequency and the frequency in Hz assuming a sampling frequency of 8000 Hz.

6.4.1 Hamming and Hann windows

In carrying out a DFT, there is an implied windowing such that all the values are set at zero except those that are Fourier-transformed by the DFT. Essentially, this means that the signal is multiplied by a rectangular window that is unity for the signal's extent and zero elsewhere.

However, because the signal is multiplied by a rectangular window to isolate the values of interest, the DFT is effectively applied not only to the section of the signal but also to the rectangular window with which it is multiplied. Furthermore, since a DFT of a rectangular window results in high-frequency components that are due to the window's sharp edges (discontinuities), these are mixed in with the time signal that is of interest: specifically, the time signal's spectrum is contaminated with that of the rectangular window.

The adverse effects of the rectangular window in the spectrum can be considerably reduced by multiplying the time signal with a smooth window such as a Hamming or Hann window before applying a Fourier transform (see Chapter 5, for further details).

Since a smooth window distorts the original signal (by attenuating it at the edges), it may seem that the degradation is at least as severe as the effects that are introduced into the spectrum by the rectangular window. From the point of view of speech analysis, however, we are often interested in the spectrum of a central part of a speech sound (because the edges are likely to be increasingly influenced by coarticulatory effects of adjacent sounds), and so the fact that there is progressive attenuation towards the edges may in any case be desirable.

6.4.2 Length of the window

The length of the window is the number of points in the signal to which the DFT is applied: an N-point DFT means that the discrete Fourier transform is being applied to a time signal of length N points.

The length of the window in calculating a DFT from speech data is chosen according to two main criteria. First, as discussed earlier, the DFT is usually carried out using the fast Fourier transform algorithm, and it is a property of the FFT that the window length has to be power of 2: therefore, in most practical applications, the section of speech signal to be Fourier-transformed is of length 32, 64, 128, 256, 512, or 1024 points.

The second criterion is to do with the frequency resolution of the spectrum: as discussed in Chapter 2, time and frequency resolution are inversely proportional. This inverse relationship between time and frequency resolution also applies to digital signals, and it follows directly from one of the properties of the DFT discussed earlier: an N-point DFT results in $N/2 + 1$ points in the spectrum that are equally spaced between zero (Hz or radians) and the Nyquist frequency. Consequently, when N is large, the DFT produces a large number of spectral components and therefore a small frequency interval between them i.e. a high-frequency resolution.

We can consider an application of these two criteria in calculating the required window lengths to produce narrowband (45 Hz) and wideband (300 Hz) spectra for a sampling frequency of 20000 Hz. The interval between frequency components in the spectrum (f_{int}), the sampling frequency (f_s), and the number of points in the window (N) to which the DFT is applied are related by

$$f_{int} = f_s/N \tag{6.7}$$

or

$$N = f_s/f_{int} \tag{6.8}$$

From (6.8), we can see that the number of points in the window must be $20000/45 = 444$ points to produce a spectrum with a spacing between frequency components of 45 Hz. However, applying the constraint that the window lengths must be a power of 2, the choice is between $N = 256$ points ($f_{int} = 78.125$ Hz) or $N = 512$ points ($f_{int} = 39.0625$ Hz). For the wideband spectrum with $f_{int} = 300$ Hz, $N = 67$ points (by (6.8)); assuming N has to be a power of 2, then $N = 64$ results in $f_{int} = 312.5$ Hz at a sampling frequency of 20000 Hz.

Two spectra calculated from the same [ɜ] vowel (taken from Australian English "curd" produced by an adult male talker) but with different window lengths are shown in Figure 6.8: the spectrum in the top right panel, in which $N = 512$, clearly has more spectral detail than the one in the bottom left for which $N = 64$. Notice that there are so few spectral components in this second spectrum ($N/2 + 1 = 33$ components) that it appears to be "jagged". In fact, a smoother spectrum can be obtained even from a short window by a technique known as *padding the data window with zeros*: if the DFT is applied to

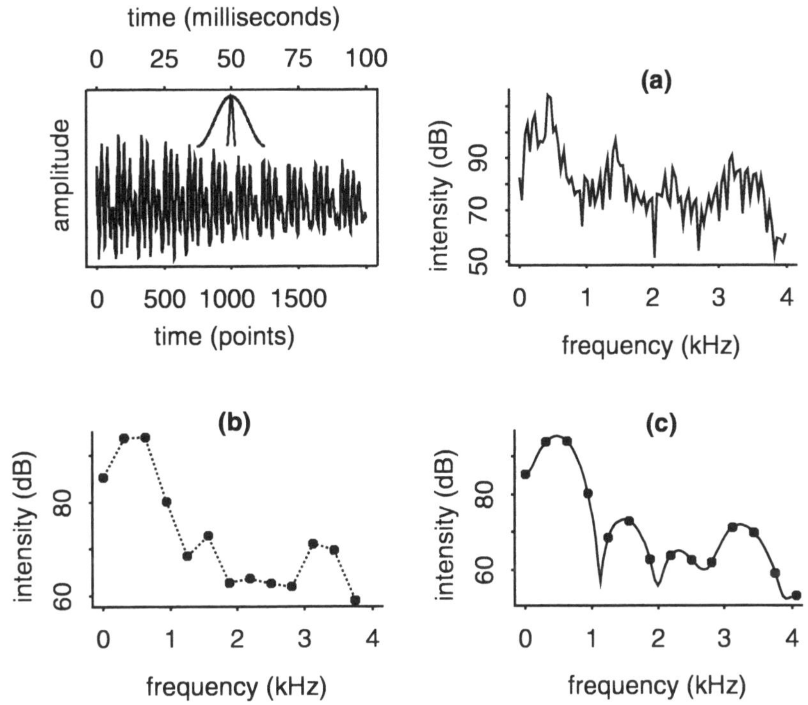

Figure 6.8: Three different types of spectra calculated from the Hamming-windowed sections of the waveform of [ɜ] in the top left panel. The spectra are derived from *(a)*: a 512-point DFT. *(b)*: a 64-point DFT. *(c)*: a 512-point DFT applied to 64 points of the waveform augmented by 444 zeros. This third spectrum also has the spectral points from (b) superimposed on it.

a 512-point signal that includes the original 64 points of the signal followed by (or preceded by — it makes no difference) $512 - 64 = 448$ zeros, the result is a much smoother spectrum, as shown in the bottom right panel of the same figure. Notice that the points of this smoother spectrum pass through those of the 64-point DFT.

The possibility of padding a window with zeros means that a spectrum can be produced with any frequency resolution: for example, an effective spacing of 45 Hz between frequency components could be produced by padding the window of length $N = 444$ points with 68 zeros to bring it up to a 512 point window. It is clear, however, that when a spectrum is padded with zeros, some frequency components are introduced into the spectrum that are not present without zero-padding; thus this technique must be used with caution (see Hamming, 1989, for a further discussion).

6.4.3 Amplitude normalisation of spectra

The simplest way of converting amplitude of air pressure into decibel values is according to the formula in (6.6). However, as discussed in Chapter 2, the decibel is a *relative* and not an absolute scale: consequently, the dB values should be normalised relative to a reference value. There are two main possibilities that can be considered: *self-normalisation* and *level-normalisation*.

In self-normalisation, one of the components of the spectrum is assigned an arbitrary dB value, and then all the other spectral values are scaled relatively to this reference value. The panels in the second and third rows of Figure 6.9 show two kinds of self-normalisation that are applied to two different signals of different mean amplitude (shown in the top panel). The spectra in the second panel are derived by self-normalisation relative to the spectral component of greatest amplitude (which happens to be 0 dB). In the third panel, self-normalisation is applied by fixing the 2000 Hz spectral component at 20 dB.

A problem with any kind of self-normalisation is that it cannot preserve in the spectrum any mean amplitude differences between different time signals. As Figure 6.9 shows, the spectra of the two time signals with very different average amplitudes have similar mean amplitudes in their spectra if self-normalisation is applied. One way of preserving these mean amplitude differences is to normalise relative to an arbitrary value which is independent of the spectrum. This can be done by fixing an arbitrary amplitude value (e.g., a low amplitude value such as 10^{-2}) at 0 dB and then rescaling all spectra in relation to this 0 dB value. Self-normalisation using such a reference value is shown in the fourth panel of Figure 6.9 in which the amplitude differences between the time signals are preserved in the spectra.

6.4.4 Preemphasis

In Chapter 2, we saw that the spectra of voiced sounds are characterised by a downward trend in which frequencies in the upper part of the spectrum are attenuated at a rate of approximately 6 dB per octave. The downward slope in the spectrum is caused by a combination of the glottal source spectrum, which slopes at -12dB per octave, and the radiation effect due to the lips which causes a spectral boost of $+6$dB per octave (producing a net trend of -6dB/octave). In the spectral analysis of voiced speech, the -6dB/octave trend is often compensated for by a preemphasis factor of $+6$dB/octave (thereby removing the downward trend of the voiced spectrum). One of the reasons that this is done is to boost the intensity of high frequencies that would otherwise be very low due to the downwards sloping spectrum. Since the -6dB/octave trend is produced only for voiced speech in which there is a glottal source, there should be no need to apply preemphasis to voiceless speech sounds, although in practice, the variable application of preemphasis only to voiced speech may be difficult.

The easiest way to simulate an approximate 6 dB/octave rise is to subtract a scaled and delayed version of the signal from itself. The delay is one time-point and the scale-factor, a is set close to, and just less than, 1 (usually at e.g.,

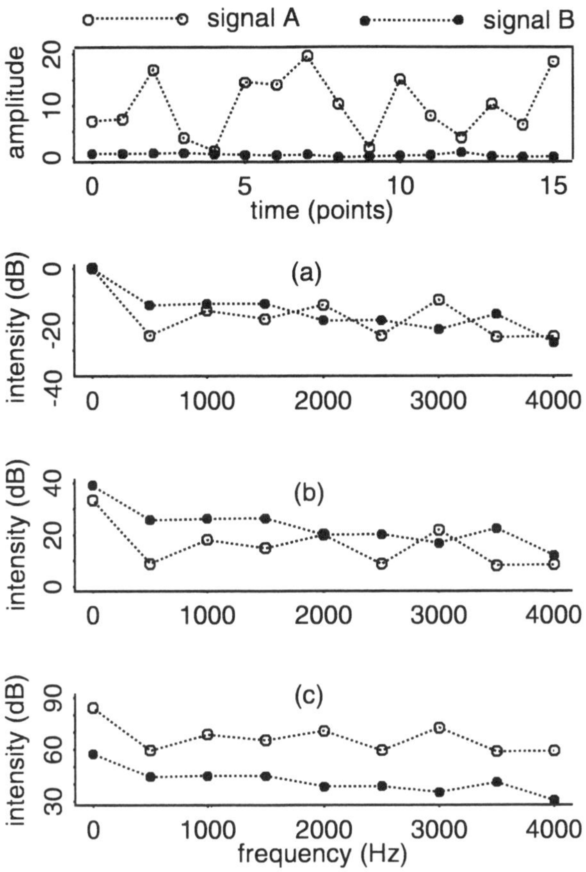

Figure 6.9: Three different kinds of amplitude normalisation in deriving the spectra of two signals (shown in the *top panel*) that have different mean amplitudes. *(a)*: self-normalisation relative to the spectral component of greatest amplitude. *(b)*: self-normalisation relative to 20 dB at 2000 Hz. *(c)*: level normalisation when 0 dB is fixed relative to an amplitude of 1/100 units. Only this third type of amplitude normalisation preserves the amplitude differences between the time signals in the spectra.

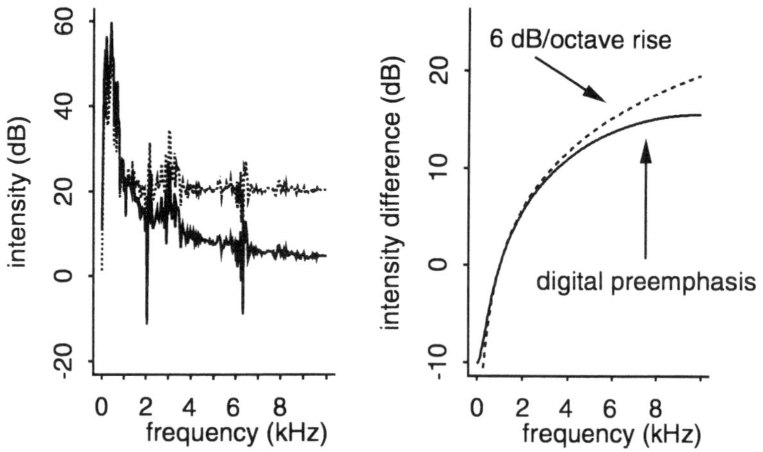

Figure 6.10: *Left*: the spectrum of a vowel with (dotted) and without (solid line) preemphasis. *Right*: the dB difference between the two spectra on the left, which is shown as the solid line in the right panel, follows an approximate 6 dB/octave rise.

0.95). Thus if we have a speech signal $\boldsymbol{x}$, an approximate 6 dB/octave rise can is obtained from[1]

$$x[n] - ax[n-1].$$

The spectrum of an [ɔ] vowel with, and without, digital preemphasis is shown in the left panel of Figure 6.10: in support of the above analysis, the dotted line spectrum (with digital preemphasis) has the same shape as the solid line one, but its amplitude values at higher frequencies are boosted. The amount by which the preemphasised and original spectra differ is shown in the right panel of the same figure: notice that digital preemphasis produces only an approximation to a true 6 dB/octave rise.

6.5 Spectral parameterisations

6.5.1 Relationship to time-domain parameters

The root-mean-square energy and the autocorrelation function discussed in the time-domain parameterisations of the previous chapter can also be calculated from spectral data. The RMS of a signal $\boldsymbol{x}$ is equivalently given by $1/\sqrt{N}$ times the root-mean-square of the amplitude components returned by a DFT.

Second, the short-time autocorrelation can be derived by applying an inverse discrete Fourier transform (IDFT) to the square of the amplitude values returned by the DFT. An IDFT, as its name suggests, is simply the reverse of a DFT and can be used to convert a spectrum back into a time signal. If we apply a DFT to a time signal, square the amplitude values, and then apply an IDFT

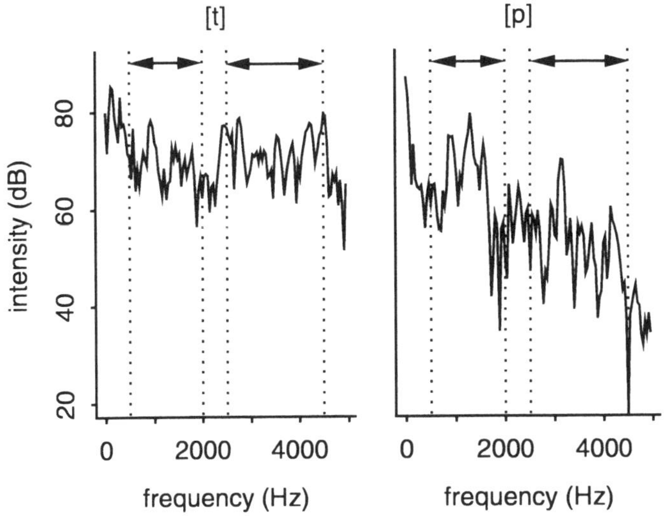

Figure 6.11: Spectra for [t] and [p] showing the frequency bands in which there is a major difference between the sounds.

to the squared amplitude values, the result is the short-term autocorrelation function of length $N/2$ representing lags $0 \ldots N/2$ of the original time signal.

6.5.2 Filter-bank analysis

One of the simplest forms of spectral data-reduction that is often used in speech analysis is the summation of amplitude values in one or more selected frequency bands. For example, a comparison between spectra of the [t] and [p] releases in the word "stamp" shows that there is considerably more energy in the upper frequency range in the spectrum of the alveolar release compared with that of the bilabial. This difference could be expressed by summing the amplitude values in two appropriate frequency bands (e.g., 500-2000 Hz; 2500-4500 Hz) to produce two values for each sound (Figure 6.11). It is important that any summation should not be done on decibels directly (because these are logarithms and summing logarithms is equivalent to multiplication) but on the prior stage of amplitude components produced by a DFT. If decibel values have already been calculated, they should be converted back to the (linear) amplitude of air pressure scale before averaging by rearranging (6.6) as

$$A = 10^{I/20}, \tag{6.9}$$

where A is the amplitude of air pressure and I the corresponding value in decibels. Thus the sum of 60 dB and 65 dB would be

$$A = 10^{60/20} + 10^{65/20}$$

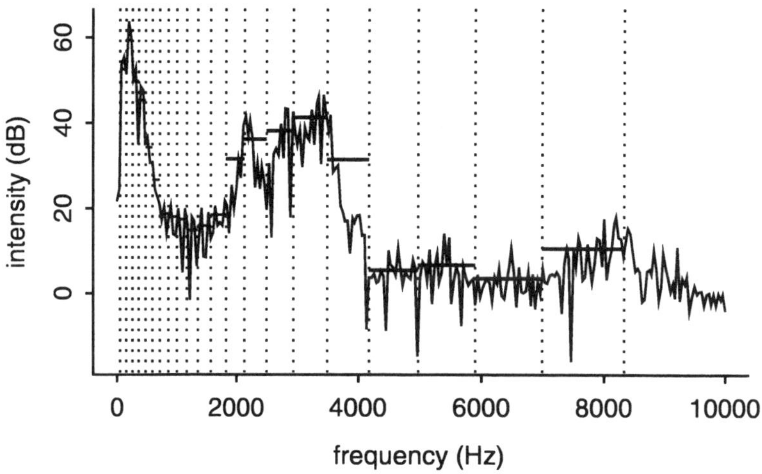

Figure 6.12: A spectrum of [i]. The dotted lines are at one Bark intervals. The horizontal lines are the average level within each band.

$$= 10^3 + 10^{3.25}$$
$$= 1000 + 1778$$
$$= 2778$$

converting to decibels:

$$I = 20 \times \log_{10}(2778) \quad \text{dB}$$
$$= 68.9 \quad \text{dB.}$$

Sometimes the spectral difference between two classes of sounds is expressed as the *ratio of energy* in two frequency bands. For the present example, this would be done by summing the amplitudes in the two bands for each of [t] and [p] (so two summations per sound) and then dividing one by the other to produce a dimensionless number. For the spectra in Figure 6.11, the ratio of energy in the upper to the lower band is 0.32 for [p] and 1.77 for [t].

A filter-bank analysis may be carried out using a perceptual scale such as the mel or Bark scales: Figure 6.12 shows the spectrum of an [i] vowel in which the spectral energy falling within 1 Bark intervals is averaged (the filter bands are centred at $1, 2, 3 \ldots 21$ Bark and are all 1 Bark wide), thereby reducing the spectrum to a set of 21 values.

A data-reduction of spectra to around twenty values that represent auditory spacings on the frequency scale can effectively highlight the salient differences between different classes of speech sounds; a technique for further reduction of this 20-dimensional Bark-space to a smaller number of dimensions is discussed in Chapter 9 of this book. Various computationally efficient techniques that make use of the fast Fourier transform (FFT) for a filter bank analysis are discussed in Rabiner and Juang (1993).

6.5.3 Spectral moments

In Chapter 4, the first two spectral moments were shown to separate effectively the voiceless fricatives [f], [s], and [ʃ].

The *first spectral moment* or *spectral centre of gravity* is a weighted average in a statistical sense and is given by

$$mom_1 = \frac{\sum \boldsymbol{f}\,\boldsymbol{I}}{\sum \boldsymbol{I}} \quad \text{Hz}, \tag{6.10}$$

where $\boldsymbol{I}$ is the amplitude (in decibels) and $\boldsymbol{f}$ the frequency (in Hertz) of the spectral components (so for a DFT applied to a 256 point time signal, $\boldsymbol{I}$ and $\boldsymbol{f}$ would be vectors of length 129). For example, in order to calculate the spectral centre of gravity of five spectral components, equally spaced between 0 and 10 kHz and with intensity values of 2, 8, 12, 0, 4 dB respectively, the calculation would be as follows:

$\boldsymbol{f}$	$\boldsymbol{I}$	$\boldsymbol{f}\,\boldsymbol{I}$
0	2	0
2500	8	20000
5000	12	60000
7500	0	0
10000	4	40000

$$\sum \boldsymbol{I} = 26 \qquad \sum \boldsymbol{f}\,\boldsymbol{I} = 120000$$

$$\begin{aligned} mom_1 &= \sum \boldsymbol{f}\boldsymbol{I} / \sum \boldsymbol{I} \\ &= 120000/26 \\ &= 4315 \quad \text{Hz.} \end{aligned}$$

The *second spectral moment* is analogous to the statistical variance, and it can give an indication of how spread or diffuse the spectrum is. For the current example, we might expect [f] to have a higher second spectral moment value than either [s] or [ʃ] since [f] has energy evenly spread throughout the spectrum, whereas for [s] and [ʃ] it is concentrated in a more compact frequency range. An appropriate formula for the second spectral moment is

$$mom_2 = \sqrt{\frac{\sum \boldsymbol{f}^2 \boldsymbol{I}}{\sum \boldsymbol{I}} - mom_1^2}, \tag{6.11}$$

where mom_1 is the first spectral moment calculated from 6.10.

6.5.4 Cepstral processing

At the end of the last chapter, we considered time-domain convolution as a possible model for the combination of the source and filter in the production of the speech. Cepstral processing is the first technique to be discussed that provides an estimated separation of the source from the filter. The motivation for this is that the harmonics that occur at pitch frequency can obscure formant analysis — therefore, a clearer estimation of the formants would be obtained if the harmonics due to the vibrating vocal folds (or the noise source due to the turbulent airstream) could be removed in some way.

In order to understand how cepstral processing works, it is first necessary to be clear about how the source and filter are represented in spectrum. As we shall see in Section 6.6, when two signals are convolved in the time domain, their DFTs are multiplied together. This means that if two signals $\boldsymbol{x}$ (the source) and $\boldsymbol{h}$ (the filter) are convolved to produce a new signal $\boldsymbol{y}$ (the output), the DFT of $\boldsymbol{y}$ is equal to the DFT of $\boldsymbol{x}$ multiplied by the DFT of $\boldsymbol{h}$. Second, when displaying dB-spectra of speech, the amplitude scale is converted into a logarithmic decibel scale. The important point, as far as cepstral analysis is concerned, is not so much the unit of the scale (decibels), but that the scale is *logarithmic*. Since the logarithm of a product is equal to the sum of logarithms ($\log(a \times b) = \log(a) + \log(b)$), and since the source and filter are multiplied when a DFT of a speech signal is calculated, it follows that a dB-spectrum of speech represents the *sum* of the source and filter components. Having recognised that a dB-spectrum (or any log-spectrum) displays the sum of the source and filter, we can now consider how they can be separated. Before doing so, one further observation must be made about the relative contributions of the source and filter to spectrum: as noted in Chapter 2 (see Figure 2.14 and Figure 2.15), the source is manifested as a rapidly oscillating component, whereas the filter is detectable as a slowly changing trend line through the oscillations attributable either to vocal fold vibration or a turbulent airstream.

At this point, we need to take a step back from the intricacies of cepstral processing and consider the following problem. Suppose we had a time-signal that had been created by summing two different sinusoids, one of which changed slowly in time and the other rapidly. Such a time signal could have been obtained by adding together two sinusoids of quite different frequencies in the manner described in Chapter 2. Suppose we now wish to separate these components. The simplest way to do this would be to apply a Fourier transform to this time signal, which would cause the two sinusoids to appear in the low and high parts of the spectrum.

This is exactly the same logic that underlies the separation of the rapidly oscillating and slowly varying dB-spectral components in cepstral processing. If we pretend that the dB-spectrum is a time-signal and apply a Fourier transform to it, then the source signal that varies rapidly should appear in the high part of the resulting spectrum, while the slowly varying spectral trend due to the filter will be manifested in the low part of the spectrum. In fact, rather than applying a Fourier transform, we apply an *inverse* Fourier transform to the dB-spectrum, which gets us from the frequency-domain back into the time-domain. Curiously, although the names imply that one is the polar opposite of the other, the DFT and IDFT are so similar that the desired effect is the same: whether we apply a DFT or an IDFT, the transformation causes a fast-changing signal to appear in the high part of the resulting spectrum (DFT) or resulting time signal (IDFT).[2]

Once an IDFT has been applied to the dB-spectrum, causing the slowly and rapidly changing spectral components to appear in two different parts of the time signal, one of them can be removed by filtering it out. Since the time-waveform only contains components due to the vocal tract filter after filtering, if we make a spectrum of this (by calculating the DFT of the waveform), the rapid oscillations due to the source should also have been removed — leaving only the trend-line attributable to the vocal tract filter.

There are obviously many steps in filtering out the source from the filter in the spectrum of a speech sound using cepstral analysis, and so it will be useful to summarise them with reference to an actual example. Figure 6.13 shows the Hamming-windowed waveform of the [æ] vowel from "stamp" in panel 1 (a Hamming or Hann window should always be applied prior to cepstral analysis). A DFT is applied to the waveform and the amplitude components of the DFT are converted into logarithmic values — simply by taking their logarithms. The logarithms are actually natural ones (to the base e) in Figure 6.13, but it would also be possible to work from a dB-spectrum (log to the base 10). Panel 2 therefore now encodes the addition of the source and filter. An IDFT is applied to the full reflected spectrum in panel 2 to derive a (reflected) time-waveform known as a *cepstrum*. The values of the cepstrum are called *cepstral coefficients* (numbered $0, 1, 2 \dots N - 1$) and the axes of the cepstrum are amplitude and time.[3] The first few cepstral coefficients, excluding the initial one at time $t = 0$, which is a measure of the energy in the signal, contain the salient information of the vocal tract filter. If the original waveform is periodic (and it obviously is, as panel 1 shows), then the source is manifested as a "spike" at pitch period duration. In panel 3, this spike is clearly visible at $n = 221$ points, which, at a sampling frequency of 20000 Hz ($T = 0.05$), corresponds to $221 \times 0.05 = 11.05$ ms — this is approximately equal to the duration of the central pitch period in panel 1. This shows therefore, that as well as providing a technique for separating the source from the filter, cepstral analysis can be used to estimate the fundamental frequency of voiced sounds (and also to estimate whether the signal is voiced, since no spike is present for aperiodic signals).

In panel 4, all the waveform including the source is attenuated to zero leaving the contribution from the vocal tract filter. Panel 5, which is known as a *cepstrally smoothed spectrum*, is derived by applying a DFT to the waveform in panel 4. A comparison of panels 2 and 5 shows how the rapidly oscillating part of spectrum has been removed, which allows the peaks in the cepstrally smoothed spectrum corresponding to the formant frequencies, to be more easily seen. In panel 6, the axes are changed, first to show the more usual decibel scale, and second to show only the lower part of the spectrum in the frequency range (0–4000 Hz). The cepstrally smoothed spectrum is shown as the solid trend line superimposed on the dB-spectrum of the waveform in panel 2. The vertical dotted lines mark the estimated first three formant frequencies.

There are various further points of detail that need to be addressed in the discussion of the cepstrum. First, the prominence of the spike in the cepstrum (panel 3) depends on how many pitch periods have been included in the original waveform (panel 1). At least two pitch periods are needed for any kind of spike

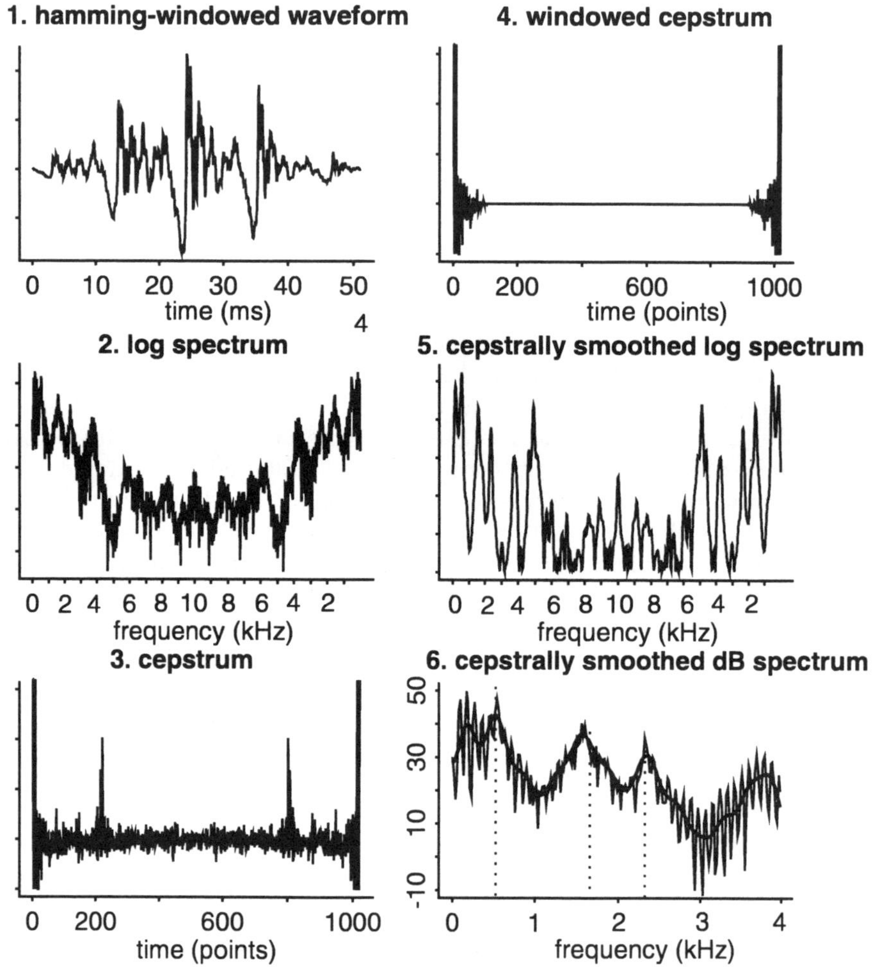

Figure 6.13: The stages in computing the cepstrum and cepstrally smoothed spectrum.

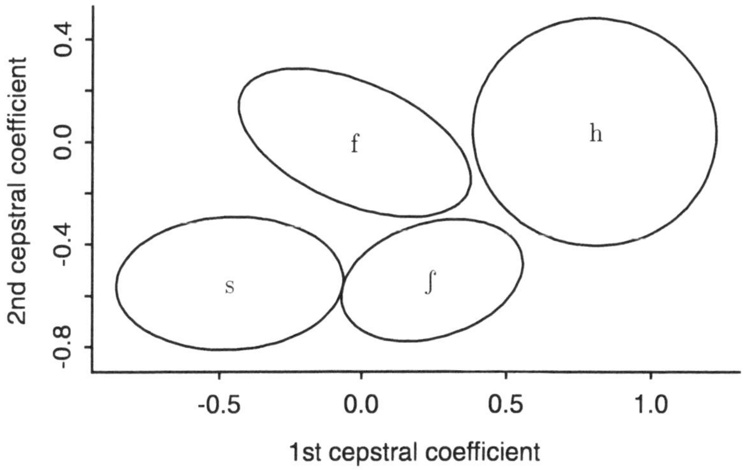

Figure 6.14: The distribution of the same 719 fricative tokens in Figure 4.27 in the plane of the first two cepstral coefficients. The ellipses include at least 95% of the tokens.

to be present at all — in the current analysis, the original waveform that was Hamming windowed in panel 1 included just over four pitch periods.

Second, since the values in the lower part of the cepstrum encode information about the vocal tract filter, they can be used to distinguish between sounds of different phonetic quality. As an example of this, cepstral coefficients were calculated for the central sections (central 512 points) on the same fricatives that were analysed for the first two spectral moments (Figure 4.27 in Chapter 4). If we examine the distribution of these fricatives in the plane of the first and second cepstral coefficients, then it is clear (Figure 6.14) that they are entirely separated — thereby confirming, at least for this small set of data, that cepstral coefficients in the bottom part of the cepstrum encode salient information about the vocal tract filter. For this reason, cepstral coefficients are often used as parameters for distinguishing between sounds in automatic speech recognition systems.

The third factor to consider is that the smoothness of the cepstrally smoothed spectrum is directly dependent on the number of points that are set at zero (panel 4) in the windowed cepstrum. The smooth spectrum in panel 5 was produced by eliminating all except the first (and last) 70 coefficients. Clearly, the part of the cepstrum including the pitch period spike must be removed, but it is important not to remove too many coefficients because otherwise two formants might be merged into a single peak.

Fourth, the peaks in the cepstrally smoothed spectrum do not necessarily correspond directly to the formant frequencies: in particular, a very low frequency peak, which is due to the fundamental frequency, may remain in the cepstrally smoothed spectrum, as the spectra in panels 5 and 6 clearly show.

Finally, in many applications of automatic speech recognition, the frequency scale of the log spectrum (panel 2 in Figure 6.13) is transformed into an auditory scale such as the mel scale, from which *mel-scaled cepstral coefficients* can be derived. A brief description and formulae for obtaining mel-scaled cepstral coefficients are presented in Davis and Mermelstein (1980) and Rabiner and Juang (1993).

6.6 Frequency-domain filtering

In Section 5.6, we considered a model of speech production in which the source signal is convolved with the filter signal. An appropriate digital model to describe this is

$$\boldsymbol{a} * \boldsymbol{y} = \boldsymbol{b} * \boldsymbol{x}, \tag{6.12}$$

where $\boldsymbol{x}$ is the source, $\boldsymbol{y}$ is the output and $\boldsymbol{a}$ and $\boldsymbol{b}$ are the weights (coefficients) of the filter. Furthermore, the weights on the output, $\boldsymbol{a}$, can be shown to represent resonances (i.e. formant frequencies in the case of speech), while the weights on the input correspond to antiformants. In order to relate filtering in the time and frequency domains we will need to develop the *z-transform*, which is closely related both to the difference equations that were discussed at the end of Chapter 5, and to the discrete Fourier transform, which was introduced at the beginning of this chapter.

6.6.1 Frequency-domain transforms

The aim in this section is to show that the frequency transformation of a digital signal can be very simply and compactly represented as the *sum of terms of a polynomial*. We will need such a representation in order to reanalyse convolution and filtering in the frequency domain which is, in turn, a prerequisite to understanding how the source and filter are combined in digital formant synthesis and separated in the linear predictive analysis of speech.

In order to explain how the frequency content of a digital signal can be expressed in a polynomial form, we will develop the following three related points. First, we will consider some general time-frequency relationships and illustrate that the spectrum of an infinitely narrow time signal, known as an *impulse*, has an infinitely broad spectrum. Second, we will show that if a signal is delayed in time, or more generally time-shifted, its phase spectrum is changed in a predictable way, but not its amplitude spectrum; from these considerations we will derive an expression for the frequency-domain representation for a *time-delayed* impulse. Finally, we will arrive at the polynomial expression of any digital signal both by observing that a digital signal can be expressed as the sum of such impulses and by invoking the property of linearity, which was mentioned in the brief discussion of linear time invariant systems in the preceding chapter.

The spectrum of an impulse

In the discussion of spectrograms in chapter 2, we noted that there was an inverse relationship between time and frequency resolution: while a wideband spectrum cannot separate two spectral components that are close together in frequency, it can be used to resolve two events that are close together in time. As discussed in Kent and Read (1992), this comes about because of the inverse relationship between a sinusoid's frequency and the duration of its pitch period. In order to register that there is a sinusoid of 10 Hz present in the spectrum, the algorithm that computes the Fourier transform has to "see" at least one pitch period. So the time-window that is used for its computation must be at least 1/10 second long. But this means that two events that are separated by less than this duration are indistinguishable. Conversely, if we want to guarantee that two temporal events appear separately in the spectrum, they must not fall within the same time window that is used to compute the spectrum. This, of course, means that the window must be very short (at least as short as the temporal separation between the events); but windows that are short in duration cannot include an entire pitch period of a low-frequency sinusoid. As discussed in Hamming (1989) and Steiglitz (1996), this inverse relationship between time and frequency resolution is the corollary in quantum mechanics of Heisenberg's uncertainty principle in which it is demonstrated that if we measure the position of a particle with great precision, we cannot define the particle's momentum with any great accuracy and *vice-versa*.

As a further example of the inverse resolution in time and frequency domains, consider that a line spectrum with a single frequency at exactly 20 Hz — that is, a spectrum with a 0 Hz bandwidth — corresponds to a 20 Hz sinusoid that repeats itself *indefinitely* i.e. one which is infinitely long in duration. If we now give the line spectrum some bandwidth, the spectrum is no longer infinitely narrow (no longer a line): in the time-domain, the signal is still a sinusoid, but it *decays* and is therefore no longer infinitely long in duration. A sinusoid that decays very rapidly indeed has a broad bandwidth with many frequency components represented in the spectrum.

The narrowest possible time signal must therefore have spectral components at *all* frequencies. In the digital domain, a time-signal that is maximally narrow in time (of zero duration) is known as a *unit impulse* (this is also defined to have an amplitude of one unit and to be positioned at time $t = 0$). The unit impulse can be defined by the equation

$$\begin{aligned} x[n] &= 1 \quad n = 0 \\ &= 0 \quad \text{elsewhere.} \end{aligned}$$

As discussed above, the unit impulse has spectral components at all frequencies. The amplitude spectrum can therefore be written in terms of k, the number of cycles as follows (we will adopt the convention of capitalising vectors when they represent frequency information):

$$X_A[k] = 1 \quad \text{for all } k,$$

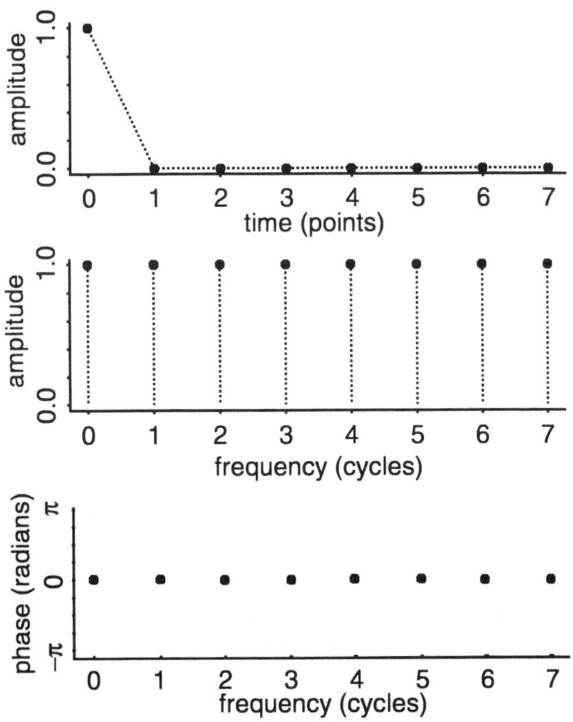

Figure 6.15: An 8-point unit impulse (*top*), its amplitude spectrum (*middle*), and its phase spectrum (*bottom*).

which is the same as saying

$$\boldsymbol{X_A} = [1, 1, 1, 1, \ldots],$$

in which $\boldsymbol{X_A}$ is a vector of the amplitude of the sinusoids at frequency 0, 1, 2, 3, 4, $\ldots k$ cycles. This is an exact summary of the information displayed in the amplitude spectrum in Figure 6.15. This figure also shows that the unit impulse has a phase of zero radians at all frequencies. The phase spectrum shown in this figure can be summarised as

$$X_\phi[k] = 0 \quad \text{for all } k.$$

But a mathematical way of combining both amplitude and phase in a *single* expression is to use *complex numbers*:

$$X[k] = 1 + 0i \quad \text{for all } k,$$

in which the real and imaginary parts of the complex number precede and follow the plus sign respectively and i is the square root of minus 1. In fact, this is exactly the way in which the DFT encodes the frequency information of a digital signal. If we ran a computer program that applied a DFT to a unit impulse, the output would be (some variation of)

$$[1+0i, 1+0i, 1+0i, 1+0i, \ldots].$$

The relationship between the complex number representation and the more familiar amplitude and phase representations is as follows. The amplitude is given by

$$X_A[k] = \sqrt{Re(X[k])^2 + Im(X[k])^2},$$

where $Re(X)$ and $Im(X)$ are the real and imaginary parts of X, respectively. In the above example, the real part of the complex number is 1 and the imaginary part is zero, so

$$\begin{aligned} X_A[k] &= \sqrt{1^2+0^2} \\ &= 1 \quad \text{for all } k. \end{aligned}$$

The relationship between the complex number and the phase requires use of the inverse tangent tan^{-1}:

$$X_\phi[k] = \tan^{-1}(Im(X[k])/Re(X[k])).$$

In this case

$$\begin{aligned} X_\phi[k] &= \tan^{-1}(0/1) \\ &= \tan^{-1}(0) \\ &= 0 \quad \text{radians (for all } k). \end{aligned}$$

The reason for the relationship between the complex number representation and the amplitude and phase of sinusoids follows from the famous identity established by the mathematician Leonhard Euler:

$$e^{ix} = \cos(x) + i\sin(x),$$

together the possibility of representing sinusoids as rotating vectors in the complex plane. This most interesting branch of mathematics is covered extensively in many other books and need not concern us here. What is important is to recognise that a complex number representation encodes both the amplitude and phase information of sinusoids and that they can be easily derived from the real and imaginary parts of the complex number as shown above.

As discussed earlier, the DFT gives us all the information we need to reconstruct the original time signal by summing the sinusoids. The information

$$X[k] = 1 + 0i \quad \text{for all } k$$

literally means that if we sum all the sinusoids at the frequencies corresponding to $k = 0, 1, 2, \ldots$ cycles, we will reconstruct the unit impulse. So, replacing the constant $2\pi/N$ by W, the unit impulse must be given by

$$x[n] = \cos(0n) + \cos(Wn) + \cos(2Wn) + cos(3Wn) + \ldots \quad (6.13)$$

In other words, a unit impulse is the sum of sinusoids at all frequencies with amplitude 1 and phase 0 radians[4]. We have now effectively carried out an inverse Fourier transform (an inverse discrete Fourier transform or IDFT in the digital domain) by converting the frequency representation X back into the time signal $\boldsymbol{x}$.

The spectrum of a time-shifted impulse

We will now examine the spectrum of a unit impulse that is *delayed* by a certain number of points such as the 4-point delayed impulse:

$$\boldsymbol{xshift} = [0, 0, 0, 0, 1, 0, 0, 0].$$

We can begin by making a good guess at what the spectrum might be based on the earlier analysis of the impulse. The spectrum is likely to consist of exactly the same sinusoids as before, but with each one *time-shifted by the same number of points.* So working from Equation 6.13, the sinusoids that make up an impulse that is shifted by four points must be:

$$\boldsymbol{xshift} = \cos(0(n-4)) + \cos(W(n-4)) + \cos(2W(n-4)) + \cos(3W(n-4)) + \ldots$$

Here there is no change in the amplitude spectrum compared with that of the impulse considered earlier (the amplitude of all the cosine terms is still 1), but the *phase* of each sinusoid is changed by an amount that is proportional to the delay. For example, the sinusoid at $k = 1$ cycle is $\cos(W(n-4)) = \cos(Wn-4W)$ i.e. the phase is changed by $-4W$ radians. The sinusoid next up in frequency, $\cos(2W(n-4)) = \cos(2Wn-8W)$, has its phase shifted by $-8W$ radians. So the phase spectrum of the unit impulse delayed by four points must be:

$$\begin{aligned} XSHIFT_\phi[k] &= [0, -4W, -8W, -12W, \ldots] \\ &= -4Wk \quad (0 \le k \le N-1), \end{aligned}$$

while the amplitude spectrum is exactly as before:

$$\begin{aligned} XSHIFT_A[k] &= [1, 1, 1, 1,] \\ &= 1 \quad \text{for all } k. \end{aligned}$$

More generally, when a unit impulse is delayed by p points, its amplitude spectrum is unchanged but its phase spectrum is changed by $-Wkp$.

In order to carry out the various filtering operations that underlie digital speech synthesis and analysis, we have to come up with an equivalent complex number representation that encodes simultaneously the amplitude and phase information of the spectrum of the delayed unit impulse. Recall from the earlier analysis that $X[k] = 1+0i$ encodes both $X_A[k] = 1$ and $X_\phi[k] = 0$. We now want a similar expression for the 4-point delayed unit impulse that simultaneously embodies $XSHIFT_A[k] = 1$ and $XSHIFT_\phi[k] = -4Wk$. It turns out that the expression is

$$XSHIFT[k] = \cos(4Wk) - i\sin(4Wk),$$

which, using the famous Euler relationship given earlier, can also be stated as

$$XSHIFT[k] = e^{-4iWk}.$$

In order to see why this gives the appropriate amplitude and phase spectra, we need to make use of two trigonometrical identities: first, $\cos^2(x) + \sin^2(x) = 1$, and second $\sin(x)/\cos(x) = \tan(x)$. We also need to bear in mind that the effect of taking the inverse tangent of a tangent is a cancellation so that $\tan^{-1}(\tan(x)) = x$. Using the relationship between complex numbers and amplitude and phase given earlier, the amplitude spectrum must be

$$\begin{aligned} XSHIFT_A[k] &= \sqrt{\cos^2(4Wk) + \sin^2(4Wk)} \\ &= \sqrt{1^2} \\ &= 1, \end{aligned}$$

and the phase spectrum is

$$\begin{aligned} XSHIFT_\phi[k] &= \tan^{-1}(-\sin(4Wk)/\cos(4Wk)) \\ &= -\tan^{-1}(\tan(4Wk)) \\ &= -4Wk, \end{aligned}$$

which therefore shows that $X[k] = e^{-4iWk}$ is an appropriate expression for the amplitude and phase of a unit impulse that is time-delayed by four points. We therefore arrive at a very compact expression for the frequency transformation of a impulse of height A units and delayed by p points:

$$X[k] = Ae^{-iWkp} \quad k \le 0 \le N-1.$$

In the above expression for $X[k]$, k is of course a vector of cycles, $k = 0, 1, 2, \ldots N-1$. It is therefore a shorthand notation for

$$X[k] = [A, Ae^{-iWp}, Ae^{-2iWp}, Ae^{-3iWp}, \ldots].$$

However, for the purposes of filtering and deriving the speech synthesis and analysis models, we can replace e^{iWk} by the single variable z and, most importantly, also turn the vector representation into a polynomial. So another even more compact representation for the frequency representation of any impulse of amplitude A time-shifted by p units is

$$X(z) = Az^{-p},$$

where $X(z)$ is now a polynomial in z: this is known as the z-transform of an impulse delayed by p points. It turns out that we can treat z like any other variable in solving equations (for example, if $z - 1 = 0$, then $z = 1$, etc.). and this is particularly useful in recasting in the frequency domain the recursive and nonrecursive filters that underlie the digital speech synthesis and analysis models. At the same time, the notation in z is also confusing because it is not immediately clear how a frequency-domain representation of a *vector* of digital samples can be reduced to what appears to be a single variable like z^{-2} (the z-transform of the unit impulse shifted by 2 units). [5] But as long as we remember that the DFT can be derived by replacing z by e^{iWk} where k is itself a *vector of the same length as the digital signal* (and W is a constant, $2\pi/N$), then the relationship should become a little clearer.

The z-transform of a digital signal

We have established in the preceding section that an appropriate frequency-domain representation for any impulse of amplitude A delayed by p points is

$$X(z) = Az^{-p}.$$

Now consider that any digital signal can be represented as the sum of delayed and weighted impulses. For example, a time signal such as

$$\boldsymbol{x} = [4, 2, 1, 3, 0, 0, 0, 0]$$

is nothing more than the sum of the impulses

$$\begin{array}{cccccccccc} & 4 & 0 & 0 & 0 & 0 & 0 & 0 & 0 \\ + & 0 & 2 & 0 & 0 & 0 & 0 & 0 & 0 \\ + & 0 & 0 & 1 & 0 & 0 & 0 & 0 & 0 \\ + & 0 & 0 & 0 & 3 & 0 & 0 & 0 & 0 \\ \hline\hline & 4 & 2 & 1 & 3 & 0 & 0 & 0 & 0 \end{array}$$

where the columns represent digital time points starting at $n = 0$. So the signal $\boldsymbol{x}$ could also be represented as

$$x[n] = 4\delta[n] + 2\delta[n-1] + \delta[n-2] + 3\delta[n-3],$$

where $\delta[n]$ is just the unit impulse

$$\delta[n] = [1, 0, 0, 0, \ldots].$$

Second, in the brief discussion of linear time invariant systems at the end of the last chapter, the property of linearity was discussed and this applies directly to the z-transform and to the Fourier transform of digital signals. Thus, where $\boldsymbol{a}$ and $\boldsymbol{b}$ are two digital signals and $Z()$ means "take the z-transform of a digital signal", then the z-transform of the sum of the signals is the same as the sum of the z-transforms of the two signals. Or, using a shorthand notation, we can write

$$Z(\boldsymbol{a} + \boldsymbol{b}) = Z(\boldsymbol{a}) + Z(\boldsymbol{b}).$$

Since we have just established that a digital signal is the sum of scaled and delayed impulses, then the z-transform of any digital signal must be the sum of the z-transforms of these impulses. But we already know that the z-transform of $A\delta[n-p]$ is Az^{-p}. It therefore follows that the z-transform of the digital signal

$$x[n] = [4, 2, 1, 3, 0, 0, 0, 0]$$

must be the polynomial

$$X(z) = 4 + 2z^{-1} + z^{-3} + 3z^{-3}.$$

In other words, we simply append increasing powers of z to successive sample values in time and sum the result. More generally, the z-transform of a digital signal $\boldsymbol{x}$ of length N is given by

$$\begin{aligned} X(z) &= x[0] + x[1]z^{-1} + x[2]z^{-3} + \ldots x[n]z^{N-1} \\ &= \sum x[n]z^{n-1}. \end{aligned}$$

By making the substitution $z = e^{iWk} = e^{2i\pi k/N}$, we arrive at a similar summation expression for the discrete Fourier transform of $\boldsymbol{x}$ — which is, of course, just the sum of the Fourier transform of scaled and delayed impulses.

6.6.2 Convolution as frequency-domain multiplication

We are finally in a position to reexpress the time-domain convolution of the input signal with the filters discussed at the end of Chapter 5 in terms of the product of their z-transforms. The basis for this is that the multiplication by z^{-p} delays a signal by p points in time. We can observe this effect by multiplying the z-transform of $\boldsymbol{x}$ above by z^{-1}, which has the effect of delaying it by one point. The resulting z-transform is

$$X(z) = 4z^{-1} + 2z^{-2} + z^{-3} + 3z^{-4}.$$

The digital value at time N of the time signal always corresponds to the coefficient of z^{N-1}. The time-signal of the above z-transform must therefore be

$$\boldsymbol{x} = [0, 4, 2, 1, 3, 0, 0, 0]$$

i.e. the original signal delayed by one time-point. This process of using z-transforms to shift digital signals in time is just a consequence of the fact that the spectra of a signal and a delayed version thereof are the same except for a change in phase. What we are effectively doing in the above z-transform operation is obtaining a spectral representation of the signal $\boldsymbol{x}$, shifting the phase of each sinusoid in that spectrum by an amount corresponding to one digital time point, and then converting the resulting spectral representation back into a time-signal: because the sinusoids have been delayed in time (= a change in phase), the resulting time signal must also necessarily be delayed in time by a corresponding amount. All this is succinctly expressed in terms of multiplication by z^{-1} (or z^{-p} if we want to delay the signal by p points).[6]

Now consider the convolution of the two signals $\boldsymbol{x}$ and $\boldsymbol{b}$ of Chapter 5, which was shown to be the same as the sum of scaled and delayed signals:

$$\begin{aligned} y[n] &= x[n]b[0] + x[n-1]b[1] + x[n-2]b[2] + \ldots + x[n-p]b[p] \\ &= x[n] * b[n]. \end{aligned}$$

Since we now know that the z-transform of $x[n-p]$ is $z^{-p}X(z)$ and since the z-transform of a sum of terms is equal to the sum of the z-transforms of those same terms, the corresponding z-transform of the above equation must be

$$Y(z) = X(z)b[0] + z^{-1}X(z)b[1] + z^{-2}X(z)b[2] + \ldots + z^{-p}X(z)b[p]$$

The terms on the right-hand side have a common factor $X(z)$, so

$$Y(z) = X(z)\left\{b[0] + z^{-1}b[1] + z^{-2}b[2] + \ldots + z^{-p}b[p]\right\}.$$

A closer inspection of the terms inside the braces shows that these are the summation of $b[p]z^{p-1}$ for increasing integer values of p: but this is by definition the z-transform of $\boldsymbol{b}$! So the z-transform of the convolution equation must be

$$Y(z) = X(z)B(z).$$

In other words, when we recast $\boldsymbol{y} = \boldsymbol{x} * \boldsymbol{b}$ in the frequency (z-transform) domain, the z-transform of the output signal is equal to the z-transform of the input signal multiplied by the z-transform of the filter.

The equivalence between time-domain convolution and z-transform multiplication can be demonstrated for the nonrecursive filter ($\boldsymbol{y} = \boldsymbol{b} * \boldsymbol{x}$) discussed in Section 5.6 in which $\boldsymbol{b} = [0.5, 0.2]$ and $\boldsymbol{x} = [4, 2, 1, 3]$. Using the methods of time-domain convolution, the output was shown to be (page 153)

$$\boldsymbol{y} = [2, 1.8, 0.9, 1.7, 0.6] \qquad (6.14)$$

The same output can be obtained by multiplying the z-transforms of the input and filter signals, i.e. since $\boldsymbol{y} = \boldsymbol{b} * \boldsymbol{n}$

$$\begin{aligned} Y(z) &= B(z)X(z) \qquad (6.15) \\ &= (0.5 + 0.2z^{-1})(4 + 2z^{-1} + z^{-2} + 3z^{-2}). \end{aligned}$$

Multiplication term by term produces

$$\begin{array}{rcccccccc} & 2 & + & z^{-1} & + & 0.5z^{-2} & + & 1.5z^{-3} & & \\ + & & & 0.8z^{-1} & + & 0.4z^{-2} & + & 0.2z^{-3} & + & 0.6z^{-4} \\ \hline Y = & 2 & + & 1.8z^{-1} & + & 0.9z^{-2} & + & 1.7z^{-3} & + & 0.6z^{-4} \end{array}$$

which is the z-transform of (6.14); therefore, multiplication of the z-transforms has produced the same results of the convolution of the source and filter in the preceding chapter.

In the case of *recursive* filters of the form $\boldsymbol{y} * \boldsymbol{a} = \boldsymbol{x}$, the z-transform is

$$Y(z)A(z) = X(z), \tag{6.16}$$

or, dividing by $A(z)$,

$$Y(z) = \frac{1}{A(z)} X(z), \tag{6.17}$$

which means that the the z-transform of the output has to be derived by (long) division. For example, for the recursive filter discussed in the preceding chapter with $\boldsymbol{a} = [1, 0.5, 0.2]$ and $\boldsymbol{x} = [4, 2, 1, 3]$, the z-transform of the output is given by

$$Y(z) = \frac{4 + 2z^{-1} + z^{-2} + z^{-3}}{1 + 0.5z^{-1} + 0.2z^{-2}} \tag{6.18}$$

The first few terms can be calculated as follows:

$$\begin{array}{r|l}
 & 4 + 0z^{-1} + 0.2z^{-2} \; etc. \\
\hline
1 + 0.5z^{-1} + 0.2z^{-2} & 4 + 2z^{-1} + \quad z^{-2} \quad + \quad 3z^{-3} \\
 & \underline{4 + 2z^{-1} + 0.8z^{-2}} \\
 & \qquad\qquad 0.2z^{-2} \quad + \quad 3z^{-3} \\
 & \qquad\qquad \underline{0.2z^{-2} \quad + 0.1z^{-3} + 0.04z^{-4}} \\
 & \qquad\qquad\qquad\qquad 2.9z^{-3} + 0.04z^{-4}
\end{array}$$

Notice that this long division could be continued indefinitely — which is compatible with the notion discussed in the preceding chapter that recursive filters, also known as infinite impulse response filters, produce an infinite number of output values.

The above long division produces a z-transform of the output:

$$Y(z) = 4 + 0.2z^{-2} + \ldots \tag{6.19}$$

of which the time signal is

$$\boldsymbol{y} = [4, 0, 0.2, \ldots], \tag{6.20}$$

which agrees with the results obtained in Section 5.6.1.

6.6.3 The impulse response of a filter

We have discussed so far that an appropriate model for speech is of the form $\boldsymbol{y} * \boldsymbol{a} = \boldsymbol{b} * \boldsymbol{x}$ in which there are weights on both the output (that encode information about the resonances) and the input (that encode information about the anti-resonances). In this section, we discuss two techniques for rearranging this convolution equation as

$$\boldsymbol{y} = \boldsymbol{h} * \boldsymbol{x}. \tag{6.21}$$

The time-signal $\boldsymbol{h}$ is called the *impulse response* of the filter (or the impulse response of the vocal tract filter in the case of speech analysis). In digital speech synthesis and analysis, $\boldsymbol{h}$ represents the contribution to the speech signal from the vocal tract without (as far as possible) any contribution from the source: its amplitude spectrum therefore encodes all the information about resonances and antiresonances that depend, in turn, on the supralaryngeal vocal tract shape.

One way to obtain $\boldsymbol{h}$ is to reexpress the convolution equation with weights on both the output and input signal in terms of z-transforms. In this case, we have

$$Y(z)A(z) = B(z)X(z).$$

Dividing by $A(z)$

$$\begin{aligned} Y(z) &= \{B(z)/A(z)\}\, X(z) \\ &= H(z)X(z), \end{aligned}$$

where $H(z) = B(z)/A(z)$ is the z-transform of the impulse response $\boldsymbol{h}$. For example, if for a filter $\boldsymbol{y} * \boldsymbol{a} = \boldsymbol{b} * \boldsymbol{x}$, we have

$$\begin{aligned} \boldsymbol{a} &= [1, 0.5, 0.2] \\ \boldsymbol{b} &= [1, -0.2] \\ \boldsymbol{x} &= [4, 2, 1, 3] \end{aligned}$$

The z-transform of the impulse response is

$$\boldsymbol{H} = \frac{1 - 0.2z^{-1}}{1 + 0.5z^{-1} + 0.2z^{-2}} \tag{6.22}$$

We could now obtain $\boldsymbol{h}$ by the method of long-division discussed earlier. Once $\boldsymbol{h}$ has been calculated, the output $\boldsymbol{y}$ could be obtained by convolving it with the source, i.e. $\boldsymbol{y} = \boldsymbol{h} * \boldsymbol{x}$.

Another way of obtaining $\boldsymbol{h}$ is to pass a unit impulse through the equivalent difference equation. Following the principles discussed in Chapter 5, the convolution $\boldsymbol{y} * \boldsymbol{a} = \boldsymbol{b} * \boldsymbol{x}$ with the coefficients as defined above can be reexpressed as the difference equation:

$$\begin{aligned} y[n] &= -a[1]y[n-1] - a[2]y[n-2] + b[0]x[n] + b[1]x[n-1] \\ &= -0.5y[n-1] - 0.2y[n-2] + x[n] - 0.2x[n-1]. \end{aligned}$$

In order to pass an impulse through the system defined by this equation, $\boldsymbol{x}$ becomes $[1, 0, 0, 0, \ldots]$. For the present example, the first few terms are calculated as follows:

$$\begin{aligned} h[0] &= -0.5h[-1] - 0.2h[-2] + x[0] - 0.2x[-1] \\ &= 0 - 0 + 1 - 0 \\ &= 1 \end{aligned}$$

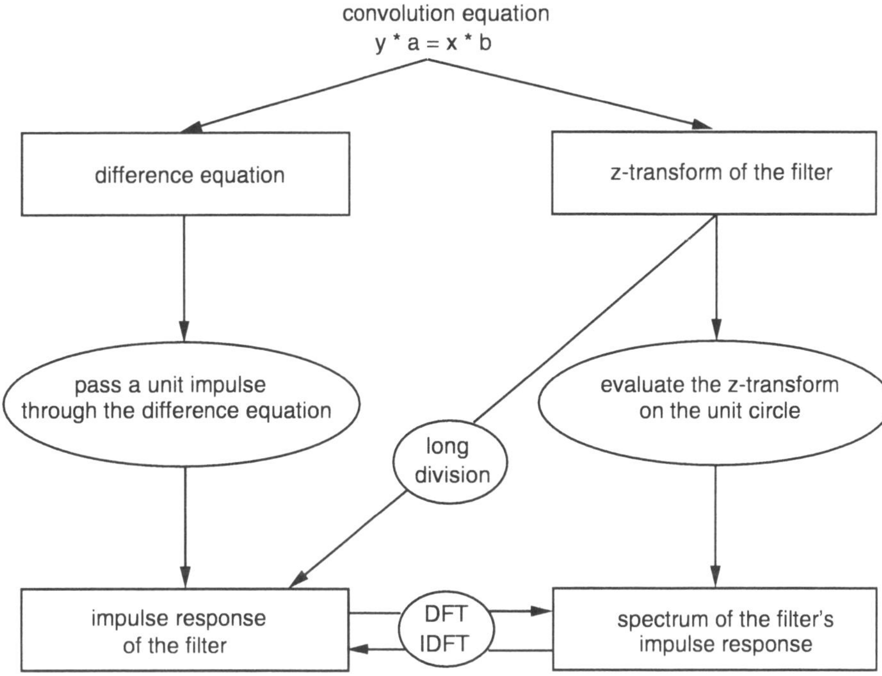

Figure 6.16: The impulse response of the filter can be obtained either by time-domain methods (*left*) or by frequency-domain methods (*right*).

$$\begin{aligned} h[1] &= -0.5h[0] - 0.2h[-1] + x[1] - 0.2x[0] \\ &= -0.5 - 0 + 0 - 0.2 \\ &= -0.7 \end{aligned}$$

$$\begin{aligned} h[2] &= -0.5h[1] - 0.2h[0] + x[2] - 0.2x[1] \\ &= 0.35 - 0.2 + 0 + 0 \\ &= 0.15. \end{aligned}$$

Once the impulse response $\boldsymbol{h}$ has been obtained, the spectrum of the vocal tract filter (in the case of speech) is produced by applying a DFT to this time signal. The result is a smooth spectrum showing the peaks and troughs that are due to the formants and antiformants, respectively, with a negligible contribution from the source.

A summary of the ways in which the impulse response of the filter can be obtained is given in Figure 6.16. The convolution equation with weights on both the input and output signals is essentially a difference equation as discussed at the end of Chapter 5. The impulse response of the filter can be obtained by passing a unit impulse through this difference equation in the manner described above. Alternatively, the filter that underlies the convolution equation can be expressed as a z-transform. By making the substitution $z = e^{iWk}$ ($W = 2\pi/N$), a procedure which in engineering terms is known as evaluating the z-transform on the unit-circle, we obtain the spectrum of the impulse response, which, if plotted, would show smooth peaks and dips of the resonances and antiresonances with no contribution from the input (source signal). As Figure 6.16 shows, the time-signal $\boldsymbol{h}$ and the spectrum of the impulse response are related to each other by the discrete Fourier transform and the inverse discrete Fourier transform, respectively.

The same figure can be used to show that there are at least two different ways of combining a input (source) signal with a filter to obtain an output signal (this is the operation that underlies digital formant synthesis to be discussed in the next chapter): either the source signal $\boldsymbol{x}$ can be passed through same difference equation that is used to obtain the filter's impulse response, or else, if the impulse response $\boldsymbol{h}$ has already been obtained, then the output (speech output if we are dealing with formant-based speech synthesis) is obtained from the convolution equation $\boldsymbol{y} = \boldsymbol{x} * \boldsymbol{h}$. This synthesis equation has a direct corollary in the frequency domain, as the preceding discussion on z-transform and cepstral processing suggests (Figure 6.17). If a frequency-domain transformation of the impulse response has been obtained as a complex number representation that encodes both amplitude and phase information, then the frequency-domain transform (also in terms of complex numbers) $Y[k]$ is given by the product of $H[k]$ and $X[k]$. If amplitude and phase values have been obtained from the complex number representation, then the amplitude spectrum $Y_A[k]$ is similarly given by the product of $H_A[k]$ and $X_A[k]$, while the phase spectrum $Y_\phi[k]$ is given by the sum of $H_\phi[k]$ and $X_\phi[k]$. Finally, following the discussion on cepstral processing, if the raw amplitude-spectrum values are rescaled logarithmically to obtain a dB-spectrum, then the dB-spectrum of the output signal is given by the sum (not the product) of the dB-spectrum of the impulse-response and the dB-spectrum of the input (source) signal, as Figure 6.17 shows.

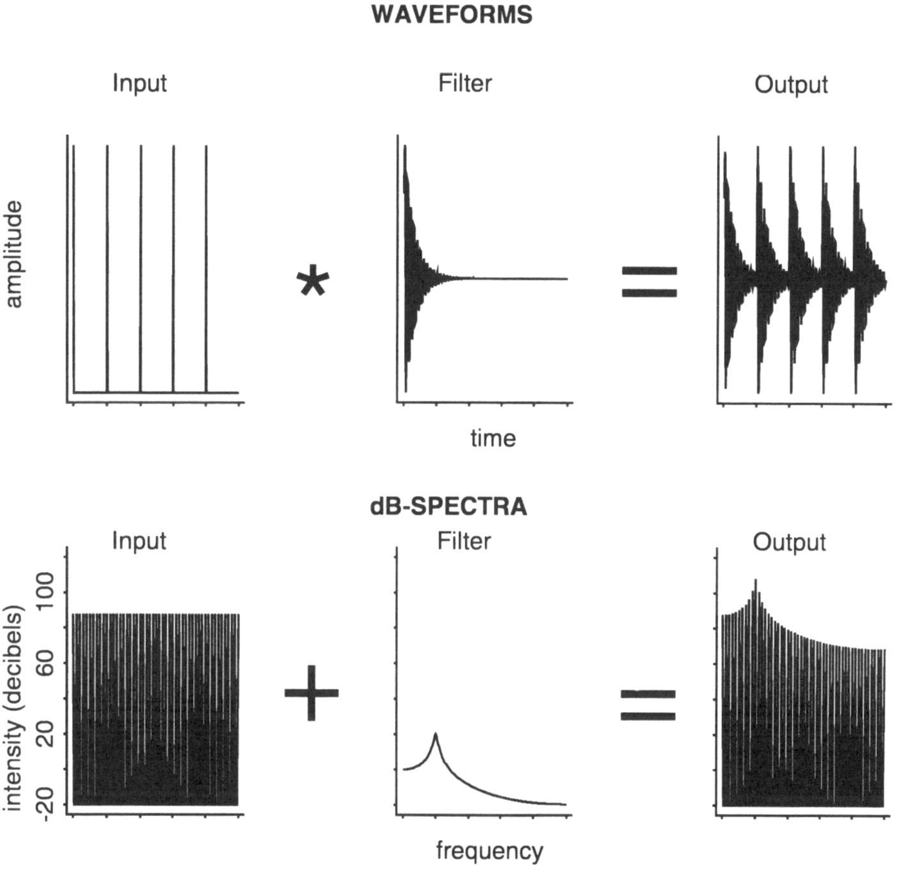

Figure 6.17: *Top row*: an input signal that is convolved with the impulse response of a filter to obtain an output signal (the interval on the time axis is 100 points). *Bottom row*: their corresponding dB spectra. In order to obtain the dB-spectrum of the output, the dB-spectrum of the input is added to the dB-spectrum of the filter (the horizontal axis extends from 0 Hz to the Nyquist frequency).

Notes

1. When a signal is delayed in the sense of Chapter 5, we lose one data point: thus if $x[n] = [10, -2, 0, 3, 1, 0, -1, 8]$, $x[n-1] = [0, 10, -2, 0, 3, 1, 0, -1]$ and the last value is dropped off. Consequently, the spectrum of the difference, $x[n]$ - $x[n-1]$, does not have a smooth rise of 6 dB/octave. There are three solutions to this problem. (1) For speech signals, the length of the signal is of course much longer than eight data points and so the fact that one data point is lost has a virtually negligible effect on spectral preemphasis. (2) If zeros are appended to the digital time signal, all the original values are preserved resulting in the expected smooth preemphasis boost. (3) For very short signals that are not padded with zeros, a smooth preemphasis boost is produced if the signal is delayed "circularly" i.e. in the above case, we subtract (a scaled version of) $x[n-1] = [8, 10, -2, 0, 3, 1, 0, -1]$ in which the last sample value, $x[N-1]$ becomes the first.

2. Indeed the formulae for the DFT and IDFT are nearly identical: the differences amount to a scale factor and a change in the sign of exponent (negative for the DFT, positive for the IDFT).

3. We avoid the term *quefrency* (in place of *time*) as well as many of the other anagrammatic terms that are sometimes associated with cepstral analysis.

4. In fact, the right-hand-side of this equation has to be scaled by N, the number of data points in the signal. So an N point unit impulse is exactly reconstructed from

$$x[n] = \frac{\cos(0n) + \cos(Wn) + \cos(2Wn) + \cos(3Wn) + \ldots \cos((N-1)Wn)}{N}$$

Note that the first term, $cos(0n)$, reduces to the vector $[1, 1, 1, 1, 1, 1, 1, 1]$ assuming $N = 8$.

5. We will therefore give the following simple example. (This can be done on a calculator with cosine and sine functions; and since a DFT is the sum of the DFT of impulses of this kind so — in theory at least — could the DFT of any digital signal). Consider the case of an 8-point impulse in which the height of the impulse is $A = 5$ and the delay is $p = 2$. The time-signal is

$$\boldsymbol{x} = [0, 0, 5, 0, 0, 0, 0, 0],$$

and its z-transform must therefore be

$$X(z) = 5z^{-2}.$$

This, amongst other things, can be thought of as a shorthand notation for the DFT, which can be obtained by making the substitution $z = e^{iWk}$ where W is a constant, $2\pi/N$, and $k = 0, 1, 2 \ldots 7$:

$$\begin{aligned} X[k] &= [5, 5e^{-iWp}, 5e^{-2iWp}, 5e^{-3iWp} \ldots 5e^{-7iWp}] \\ &= [5, 5e^{-2iW}, 5e^{-4iW}, 5e^{-6iW} \ldots 5e^{-14iW}]. \end{aligned}$$

Each of these terms reduces to a complex number. For example, the second term

$$\begin{aligned} 5e^{-2iW} &= 5(\cos(4\pi/8) - i\sin(4\pi/8)) \\ &= 5(0 - i) \\ &= 0 - 5i. \end{aligned}$$

This complex number encodes both the amplitude and phase of the sinusoid at $k = 1$ cycle. The amplitude is

$$\begin{aligned} X_A[1] &= \sqrt{0^2 + (-5)^2} \\ &= 5, \end{aligned}$$

and the phase

$$\begin{aligned} X_\phi[1] &= \tan^{-1}(-5/0) \\ &= \tan^{-1}(-\inf) \\ &= -\pi/2. \end{aligned}$$

So the sinusoid corresponding to $X[1] = 5e^{-2iW}$ is

$$\begin{aligned} s[n] &= 5\cos(2\pi kn/N - \pi/2) \\ &= 5\cos(2\pi n/8 - \pi/2) \\ &= 5\cos(0.25\pi n - \pi/2). \end{aligned}$$

Therefore eight sinusoids are derived from $X[k]$, which, when summed, would reconstruct exactly the original scaled, delayed impulse.

6. And in fact, following on from the discussion in the remainder of this section, multiplication by z^{-p} is exactly the same as the convolution with a unit impulse shifted by p points. So we could also get from

$$\boldsymbol{x} = [4, 1, 2, 3, 0, 0, 0, 0]$$

to

$$[0, 4, 1, 2, 3, 0, 0, 0]$$

by convolving $\boldsymbol{x}$ with $b[n] = [0,\ 1]$. This is because $\boldsymbol{x} * \boldsymbol{b}$ is given by

$$x[n]b[0] + x[n-1]b[1] = x[n-1] = [0, 4, 1, 2, 3, 0, 0, 0].$$

7

Digital Formant Synthesis

The attempt to build a talking machine has a long history and can even be traced back to a time before the beginning of the Christian era (Linggard, 1985). The first complete talking machine is due to von Kempelen (1791) and is described in a book of over 400 pages that also reports on the twenty or so years of experimentation that were needed to build the device (interesting historical accounts of the development of speech synthesis are given in Dudley & Tarnóczy, 1950; Flanagan, 1972; Linggard, 1985; see also Klatt, 1987; and Flanagan & Rabiner, 1973). It was not until the 20th century that speech synthesis became a widespread research endeavour. Part of the reason for this is that with the invention of the telephone, there was an increasing need to find a way of reducing the data in speech transmission without degrading significantly its quality; and this was one of the principal motivations that led to the invention of the first electronic speech synthesis system capable of synthesising whole utterances which was demonstrated publicly at the New York World's Fair in 1939 and in San Francisco in 1940 (Dudley, 1939; Dudley et al., 1939). Another reason was that mechanical devices that model the vocal tract accurately enough to produce intelligible speech are very difficult to construct; and the advent of electronic instrumentation at the beginning of this century provided a way of synthesising speech without having to copy the action of the vocal organs in detail. Some landmarks in the development of speech synthesis systems in the 1950s include the pattern playback system of the Haskins Laboratories (Cooper, Liberman, & Borst, 1951), the Parametric Artificial Talker (PAT) by Lawrence (1953) and the Orator Verbis Electris (OVE) system developed by Fant (1953). In more recent times, major advances in the development of text-to-speech systems have been made both in the development of the MITtalk text-to-speech system developed over a number of years by Dennis Klatt at MIT (Allen, Hunnicutt, & Klatt, 1987; Klatt, 1980, 1982, 1987; Klatt & Klatt, 1990), which can synthesise intelligible and natural English speech in different voices and from an unrestricted vocabulary, and the KTH synthesis-by-rule system developed at Stockholm (Carlson & Granström, 1975, 1976; Carlson, Granström, & Hunnicutt, 1982).[1]

An initial distinction can be made between synthesisers that are copy, or *analysis/resynthesis*, systems and those that are based on some form of *synthesis-by-rule*. The purpose of analysis/resynthesis systems is usually to parameterise a given utterance in such a way to achieve a data-reduced version while compromising minimally on intelligibility. The different types of speech coding (reviewed extensively in O'Shaughnessy, 1987, and Owens, 1993), which may be in either the time-domain (e.g., a linear predictive coding system: Atal & Hanauer, 1971; Markel, 1972a; Markel & Gray, 1976), or the frequency domain (e.g., a channel vocoder: Holmes, 1980) are examples of analysis/resynthesis. Forerunners of modern analysis/resynthesis systems include Dudley's (1939) vocoder and the Pattern Playback system used in the classic speech perception experiments at the Haskins Laboratories, which converts painted spectrographic patterns into an electrical signal using an optical scanning device (Cooper et al., 1951; Liberman et al., 1954).

Synthesis-by-rule systems on the other hand do not in any sense copy a single utterance and are designed to generate novel utterances and sounds. Within this category, an approximate distinction can be made between *waveform-concatenation* and *parametric* systems. The former category would include synthesisers that store some form of acoustic template. For example, in a word-based concatenation system, the templates would be words (or parameterised forms of words) which are then concatenated to produce a form of continuous speech. A significant drawback with this kind of synthesis strategy is that it does not model the effects of utterance-level prosody resulting in a very unnatural speech quality. Another is that the vocabulary is limited by the templates that are stored.

Diphone synthesis (Charpentier & Stella, 1986; Dixon & Maxey, 1968; Isard & Millar, 1986; O'Shaughnessy, Barbeau, Bernardi, & Archambault, 1988) is another example of a concatenation system in which the units are not words but diphones that usually includes all pairings of phonemes extending between their temporal midpoints. The main purpose of using diphones is to include the transitions between abutting sounds are thereby encode much of the variability that is due to coarticulation in speech production. A variation on the diphone which is sometimes used in speech synthesis systems is the demisyllable (Fujimura & Lovins, 1978).

Parametric synthesis-by-rule systems are of two main kinds: *articulatory synthesisers* in which speech articulation is modelled directly (Dunn, 1950) and resonance or *formant synthesisers* in which the resonances of the vocal tract are modelled rather its shape (Flanagan, 1957; Lawrence, 1953). Focusing first on articulatory synthesisers, in Chapter 3 we saw how the vocal tract can be modelled as a set of interconnecting cylinders from the mouth to the lips. The resonances of the vocal tract can be calculated from the vocal tract transfer function which is defined as the ratio of the volume-velocity of airflow at the lips to that at the glottis: this transfer function is primarily dependent on the cross-sectional area of the cylinders that are used to model the vocal tract. The basis for articulatory synthesis is that, because there is a direct relationship between acoustic and electrical quantities (for example, volume-velocity and pressure can be directly translated into current and voltage), the transfer-function of the interconnecting cylinders can be exactly and equivalently represented in terms of an electrical transmission line in which each cylinder is represented as an equivalent electrical circuit. The transfer function and resonances can then be determined from the ratio of the input current (at the "glottal" end) to the output current (at the "lip" end). One of the first articulatory synthesisers based on an equivalent representation of 25 electrical sections is described in Dunn (1950): in this system, static vowels could be synthesised by modelling the principal constriction (the greatest point of articulatory narrowing) in terms of varying the electrical quantity *inductance*. Notable articulatory synthesis systems that followed include those of Stevens, Kasowski and Fant (1953), in which the vocal tract was modelled as 35 interconnecting electrical circuits, and Rosen (1956) which allowed dynamic variation in the synthesis of sounds (for example, in order to synthesise diphthongs). Some more recent examples of articulatory synthesis systems include e.g., Mermelstein (1973) and Rubin, Baer, and Mermelstein (1981).[2]

The remainder of this chapter will focus on the digital implementation of a formant-based system: one of the main motivations for doing so is that, in order to understand the fundamentals of the linear predictive coding of speech, it is first necessary to consider how synthetic speech output can be obtained by passing a source through a set of digital resonators. In the next chapter, we will see that linear predictive coding reverses this process by estimating the parameters of equivalent formant resonators that could have given rise to the digital speech signal.

7.1 Core structure of a formant synthesiser

Since the existence of the first dynamically controllable formant synthesiser (Fant, 1953; Lawrence, 1953), there has been a basic distinction between those of a serial, or *cascade*, structure and those of a *parallel* structure. In a cascade synthesiser, it is necessary to specify only the formant centre frequencies and bandwidths, but not the formant levels (or peak amplitudes of the formants). This type of synthesiser can be used for synthesising both oral and nasal sonorants (Klatt, 1980). A parallel synthesiser has the additional complexity that the formant levels, as well as the centre frequencies and bandwidths must be specified, but this additional complexity has been shown to be necessary to synthesise adequately many consonants. Essentially, consonants — and in particular fricatives — have antiresonances in the low part of the frequency spectrum that are difficult to model using a cascade system. The solution to this problem, which the parallel structure presents, is to simulate the effects of the antiresonances by having direct control over the formant levels.

The cascade system is most closely based on the acoustic theory of speech production (Fant, 1960), which shows that the vocal tract transfer function can be represented as the product of (an infinite number of) "filters" that give rise to the resonance frequencies. The cascade synthesiser is a model of this system to the extent that there is one filter per resonance whose coefficients are convolved with each other in the time-domain or multiplied together in the frequency domain.

There is a direct link between the cascade and parallel systems in that the

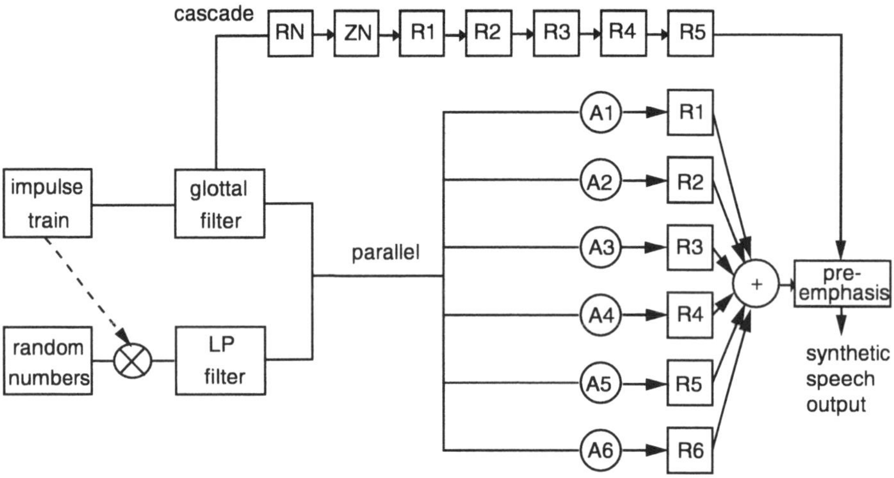

Figure 7.1: A combined cascade and parallel model of speech synthesis (adapted from Klatt, 1980). The top branch shows the cascade system consisting of a nasal resonator (*RN*), a zero resonator (*ZN*), and five resonators connected in series. The lower branch shows six resonators arranged in parallel whose levels can be set by *A1* ... *A6*. The output from the cascade and parallel branches is passed through a first difference (preemphasis) filter to synthesise the 6 dB/octave rise. The system is excited by an impulse train passed through a lowpass filter for voiced speech, by random numbers for voiceless consonants, or by an impulse train modulated by random numbers for voiced fricatives.

product of the filter coefficients (cascade) in the frequency-domain can be re-expressed as a summation (parallel). The summation essentially implies that the filters are independent of each other, and it is because of this independence that the formant levels can be, to a large extent, independently controlled in the parallel system.

Figure 7.1 shows a (somewhat simplified) flow-diagram of the combined cascade and parallel system in Klatt (1980). The input to the system is a *source* consisting of either an impulse train, which controls the fundamental frequency in voiced sounds or random numbers for voiceless sounds. These two components are used to generate three principal voicing states for synthesising sounds. First, for fully voiced sounds (i.e. all sonorants), the impulse train is sent, via a glottal filter, through the cascade system. Second, for voiceless sounds the output of the random number generator is sent through the parallel system. The third major possibility arises in the synthesis of fully voiced fricatives in which the output from the random number generator is modulated (multiplied) by a rectangular wave repeated at pitch frequency and sent through the parallel configuration. Three other possibilities for generating voiceless and voiced aspiration and for producing the low-frequency "voice bar" that is typical in voiced oral stops are also discussed in Klatt (1980).

The cascade system consists of seven formant filters connected in series: five of these are used for F1-F5 to synthesise vowels; the additional two include a formant and an antiformant filter for synthesising nasal consonants and nasalised vowels. As stated earlier, there are two parameters to set for each filter arranged in cascade: the formant *centre frequency* and formant *bandwidth*. In the parallel system, the *amplitude level* of the formant must also be specified. Finally, the synthetic speech that is generated from either the cascade or parallel systems is sent through a first difference filter to simulate the 6 dB/octave rise due to sound pressure radiation from the lips.

7.2 Digital considerations

A synthesiser can be implemented either in hardware or digitally, although a digital system has the advantage that the modules can be easily changed (see Linggard, 1985; Rabiner, 1968, and Witten, 1982, for a discussion of the relationship between digital and analogue speech synthesis models). In a digital implementation of a speech synthesiser, consideration must be given to the *sampling frequency*, which in turn has implications for the number of resonance filters that are included in the system. Since the cues to speech sounds are usually distributed in the range 0-5000 Hz, a sampling frequency of 10000 Hz is generally adequate for high quality speech synthesis. From theoretical considerations of vocal tract modelling, it can be shown that the average spacing between formants is $c/2L$, where c is the speed of sound and L the total length of the vocal tract. For an adult male vocal tract of presumed length 17 cm (and assuming that c is approximately 34000 cm/s), formants occur, therefore, on average at intervals of $34000/34 = 1000$ Hz. Accordingly, for a digital synthesiser with a frequency resolution of 5000 Hz, at least 5 resonance (formant) circuits are needed to synthesise vowels. In analogue synthesisers, a fairly complicated correction factor is needed to take account of the influence of resonances at a higher frequency range (Fant, 1960), but because of the inherent properties of digital frequency transformations, this is not needed in digital systems (Gold & Rabiner, 1968; Rabiner, 1968; Holmes, 1983).

7.3 Periodic excitation

The synthesis of glottal waveforms is complicated both by the lack of experimental data from natural speech and because of the large degree of phonatory voice quality variation both within and across talkers (Abercrombie, 1967; Laver, 1980). We will first discuss a relatively simple model of the glottal waveform and then consider some of its shortcomings for the synthesis of voiced speech.

One of the simplest ways of synthesising voiced speech is to use an *impulse train* consisting of scaled impulses at pitch frequency. The spacing between the pulses is used to control the pitch period and therefore the fundamental frequency: Figure 7.2 shows an impulse train appropriate for the synthesis of five pitch periods at a sampling frequency of 10000 Hz and corresponding to a

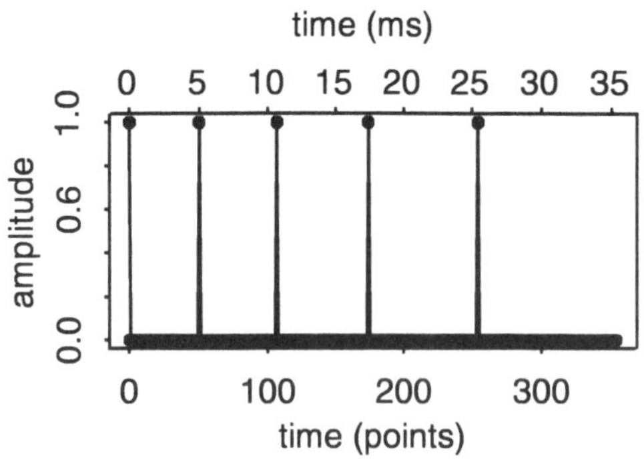

Figure 7.2: An impulse train for synthesising a linearly decreasing fundamental frequency.

linearly falling fundamental frequency from 200 Hz to 100 Hz.

Although even this very basic waveform can by itself generate speech that is perceptually voiced (at normal fundamental frequencies), it clearly bears little resemblance to a real glottal waveform and consequently the synthetic speech suffers from a lack of naturalness. A simple approximation to the glottal waveform can be obtained by passing the impulse train (set $\boldsymbol{x}$ equal to the impulse train in Equation 7.1 below) through a second-order recursive filter, of the form

$$y[n] - 2ry[n-1] + r^2 y[n-2] = x[n] \tag{7.1}$$

in which $r = e^{-1/tc}$ and tc is the time-constant that defines the time at which the glottal pulse obtains a maximum value (and also the rate at which the pulse decays). An example of the waveform for a single pitch period of 100 Hz and $tc = 10$ points is shown in Figure 7.3 together with the corresponding spectrum, which has a downward slope of approximately 12 dB/octave. The time waveform is in fact also defined by

$$\boldsymbol{y}[n] = nr^{n-1} \quad n \geq 0. \tag{7.2}$$

Equation 7.1 can be recast in z-transform terms, $Y(z) = H_g(z)X(z)$, where $H_g(z)$ is the z-transform of the glottal filter. From Equation 7.1, the z-transform of the glottal filter can be seen to be

$$\begin{aligned} H_g(z) &= \frac{1}{1 - 2rz^{-1} + r^2 z^{-2}} \\ &= \frac{1}{(1 - rz^{-1})^2} \end{aligned} \tag{7.3}$$

Two important observations can be made about this synthetic glottal waveform. First, compared with natural glottal waveforms, the synthetic waveform in Figure 7.3 is time-reversed (compare Figure 7.3 with Figure 3.4 of Chapter 3);

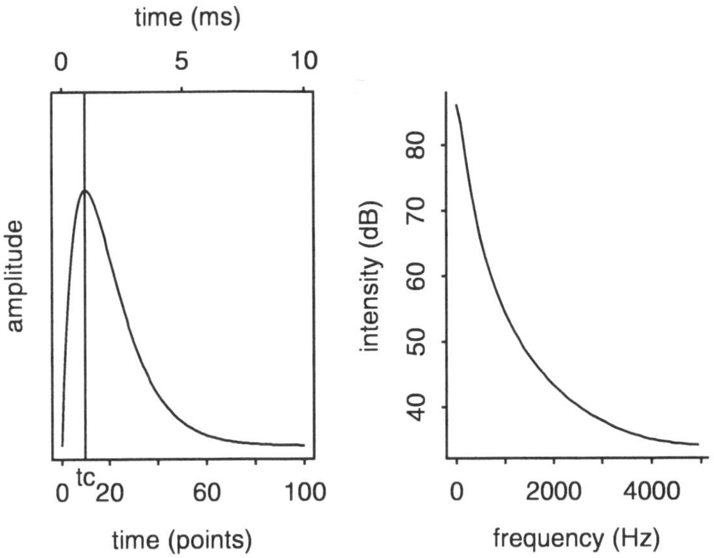

Figure 7.3: *Left*: A synthesised glottal pitch period of 100 Hz produced by convolving an impulse with the coefficients of a second-order (low-pass) filter. The time constant, tc, is 10 points. *Right*: The spectrum of this waveform which slopes downwards at roughly 12 dB/octave.

however, this has been shown to make little perceptual difference (Holmes, 1973, 1983). Second, if we wish to ensure that the maximum value of the pulse (at $x[tc]$) is always attained at a constant fraction of the pitch period duration, then the time-constant, and consequently r in Equation 7.1, needs to varied in relation to the fundamental frequency. For example, to ensure that the maximum is always 1/10th of the pitch period duration (as in Figure 7.3), the time-constant would have to be reset for each pitch period to $f_s/(10f_0)$, where f_s is the sampling frequency (in Hz) and f_0 the fundamental frequency (in Hz).

Shaping a pulse train to generate glottal waveforms using the filter described by Equation 7.1 is appealing both because it is a simple operation and because it generates perceptually quite adequate voicing waveforms. This technique was used in the synthesiser described in Klatt (1980); however, it has many shortcomings from the points of view of synthesising more natural glottal waveforms. A more recent implementation of the voicing source (Klatt & Klatt, 1990), based on the model discussed in Rosenberg (1971), allows specification of the opening and closing phases (see Section 3.2.1). Parameters are also included to tilt the spectrum of the glottal waveform, include aspiration noise, introduce frequency jitter to the glottal waveform, and allow for double (diplophonic) pulsing. The effects of the coupling between the glottal source and the vocal tract are also modelled by taking account of the effects of additional tracheal resonances and anti-resonances in the glottal waveform. (See also Chapter 3 for further references on modelling the glottal waveform).

7.4 Formant filter

A second-order recursive filter is also appropriate for synthesising formants. The general form of the equation is very similar to that of Equation 7.2:

$$\boldsymbol{y}[n] - 2r\cos(\omega)\boldsymbol{y}[n-1] + r^2\boldsymbol{y}[n-2] = A\boldsymbol{x}[n]. \tag{7.4}$$

In this case, $r = e^{-(bw/2)}$, bw is the formant bandwidth in radians and ω is the formant centre frequency (also in radians). The coefficient of the input, A, is a constant which, when equal to the coefficients of the output ($A = 1 - 2r\cos(\omega) + r^2$), normalises the resulting spectrum to 0 dB at a frequency of 0 Hz. The conversion from a frequency in Hertz to an equivalent radian frequency can be made using the formulae discussed in Chapter 6. Once again Equation 7.4 can be recast in z-transform terms in the form $Y(z) = H_{fn}(z)X(z)$, where $H_{fn}(z)$ is the z-transform of the nth formant filter:

$$H_{fn}(z) = \frac{A}{1 - 2r\cos(\omega)z^{-1} + r^2z^{-2}} \tag{7.5}$$

The impulse response of a single formant with centre frequency and bandwidth at 2000 Hz and 200 Hz, respectively, is shown in Figure 7.4; the corresponding spectrum is shown in Figure 7.5. The impulse response of a formant is in fact defined by

$$h_{fn}[n] = \frac{A}{\sin(\omega)}r^n\sin(\omega(n+1)) \quad n \geq 0, \tag{7.6}$$

where A, r and ω are as defined above. A closer examination of the impulse response, $\boldsymbol{h_{fn}}$, shows that it is the product of a gain factor ($A/\sin(\omega)$), a decaying exponential (r^n) that depends on the bandwidth, and a sinusoid ($\sin(\omega(n+1))$), whose frequency is equal to the formant centre frequency (see Figure 7.4).

When formant filters are arranged in cascade, their z-transforms are *multiplied* together. Since the z-transform for any single formant is defined by a second-order difference equation, it follows that n formants in cascade are defined by a difference equation of order $2n$ (or of a z-transform of order $2n$). For example, when two formants are arranged in cascade, the z-transform will be of the form

$$H_{f1}(z)H_{f2}(z) = \frac{A}{1 + a[1]z^{-1} + a[2]z^{-2} + a[3]z^{-3} + a[4]z^{-4}}$$

where $\boldsymbol{a}$ are the coefficients that depend on the combined effects of each of the formants' centre frequencies and bandwidths as determined by Equation 7.5.

Three main effects of changing the formant centre frequencies and bandwidths, which are described in detail in Klatt (1980) and Fant (1960), are shown in Figure 7.6. First, when the bandwidth of any formant is doubled, its peak dB level falls by 6 dB: this is shown in the left panel, in which the bandwidth of

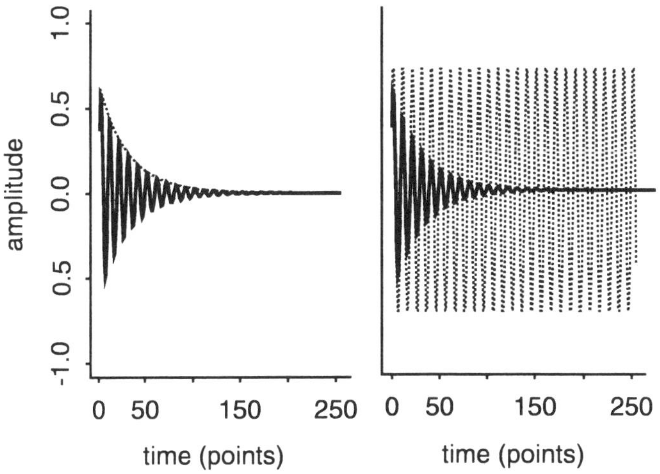

Figure 7.4: *Left*: The impulse response for a formant frequency filter. The rate of decay (dotted line) is directly proportional to the formant bandwidth. *Right*: The rate of oscillation of the impulse response depends on the formant centre frequency — a sinusoid at the same frequency as the formant centre frequency is superimposed as a dotted line.

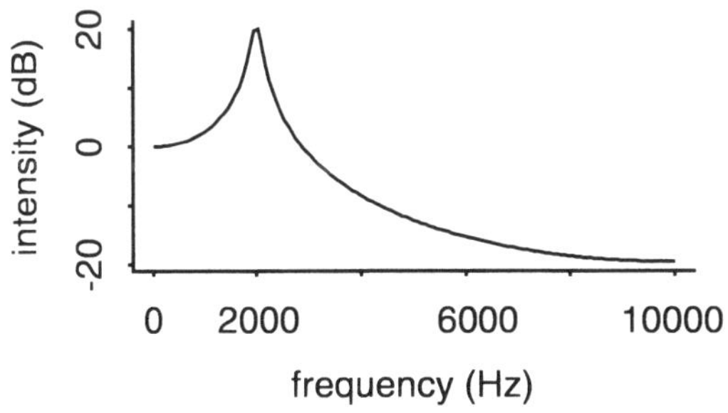

Figure 7.5: The spectrum of the impulse response of the formant shown in Figure 7.4.

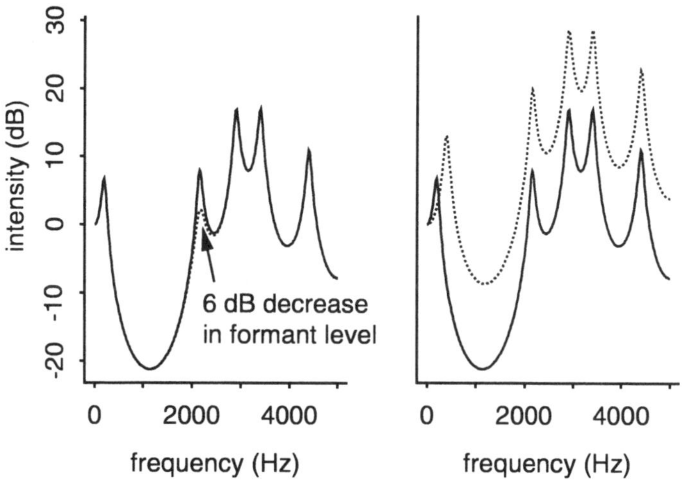

Figure 7.6: *Left*: The bandwidth of the second formant is doubled (dotted line) causing a 6 dB drop in the level of that formant. *Right*: The dotted line shows the effect of doubling the centre frequency of the first formant.

the second formant (at 2150 Hz) is doubled from 100 Hz to 200 Hz producing a drop in 6 dB of that formant's level. Second, when a formant centre frequency is raised/decreased, the dB level of that formant and all higher formants are increased/decreased. For a doubling of frequency, the formant level is increased by 6 dB and that of higher formants by 12 dB: this effect is shown in the right panel of the same figure in which the first formant frequency at 200 Hz is doubled to 400 Hz producing the corresponding rises both to its own level (6 dB) and to those of higher formants (12 dB). Third, when two formants approach each other, their levels are raised. This is shown in Figure 7.7 in which the interval between the centre frequencies of two formants is decreased by 400 Hz causing the levels of both formants to rise.

The cascade model of formant resonators described above is entirely appropriate for the synthesis of oral sonorants (vowels and approximants). However, as discussed in Chapter 3, the spectrum of nasal consonants includes one major anti-formant due to the effects of the side-branching resonator of the oral tract. An anti-resonance is the mirror image of a resonance and is defined by a *non-recursive* filter (weights on the input):

$$y[n] = \frac{x[n] - 2r\cos(\omega)x[n-1] + r^2x[n-2]}{A} \tag{7.7}$$

where r, ω, and A are as defined earlier. The spectrum of an anti-resonance for a centre frequency of 1000 Hz and bandwidth of 400 Hz is shown in Figure 7.8.

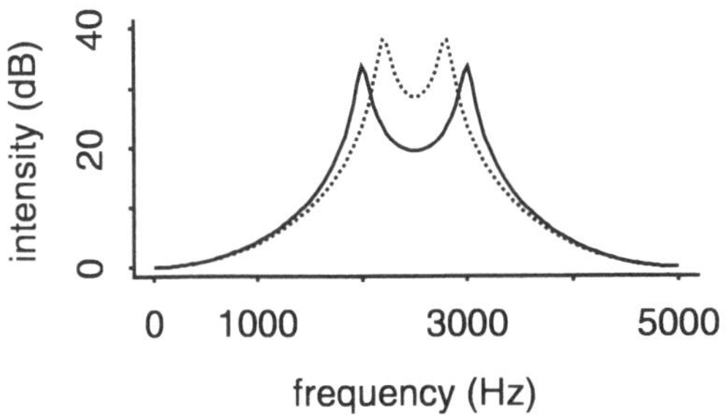

Figure 7.7: The effect on the levels of the formant frequencies when their centre frequencies move closer together.

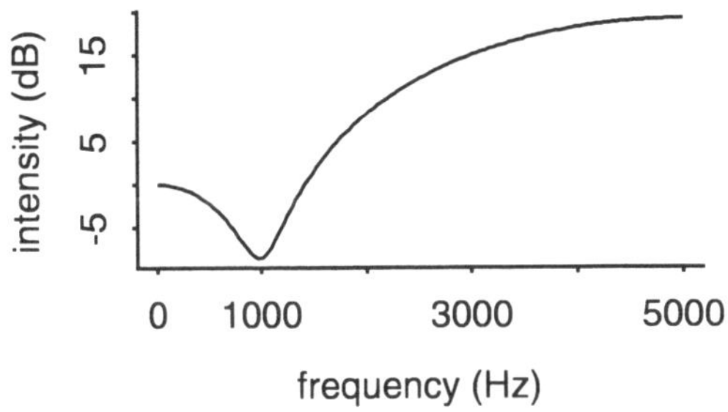

Figure 7.8: The spectrum of (the impulse response of) an antiformant with centre frequency 1000 Hz and bandwidth 400 Hz.

The z-transform of this antiformant filter, H_{fzero}, is given by

$$H_{fzero} = \frac{1 - 2r\cos(\omega)z^{-1} + r^2 z^{-2}}{A} \tag{7.8}$$

i.e. the coefficients of z are all in the numerator. The introduction of an anti-resonance has two main effects on the spectrum. First, it introduces a dip into the spectrum at the anti-resonance frequency; and second, it changes the resonance frequencies (which is sometimes described as changing the *balance* of the spectrum — Atal, 1985). This is shown in Figure 7.9, in which the introduction of a single anti-resonance of centre frequency 350 Hz and bandwidth 150 Hz lowers the level of the first formant frequency and raises the levels of the higher formant frequencies. From a perceptual point of view, this change in spectral balance is at least as important (as a cue for nasalisation) as the actual frequency location of the antiresonance itself (Carlson, Granström, & Klatt, 1979; Malme, 1959).

In nasal consonants, the resonances depend on the combined nasal-pharyngeal tract, while a major anti-resonance is introduced because of the side-branching oral cavity. Since the nasal-pharyngeal tract is longer than the oral tract (ap-

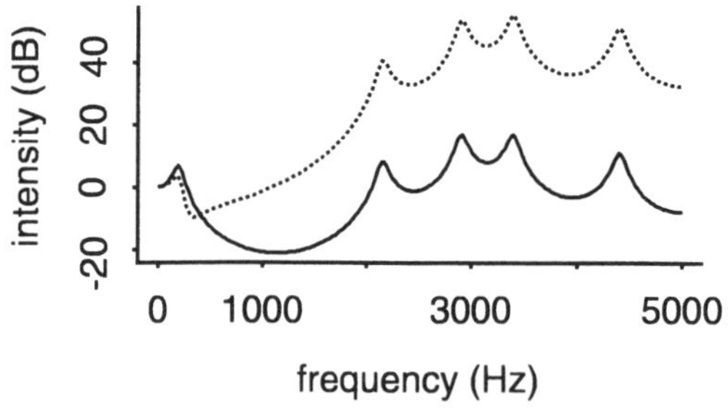

Figure 7.9: The dotted line shows the resulting spectrum when an antiformant of centre frequency 350 Hz is introduced into the solid line spectrum.

proximately 20 cm as opposed to 17 cm for an adult male), the average spacing between resonances is less: as discussed earlier, this average spacing is given by $c/2L$ (c is the speed of sound and L the tract length), so for a nasal-pharyngeal tract, we can expect a spacing between (nasal) formants of 34000/40 or 850 Hz. Since nasal consonants are characterised by $5000/850 \approx 6$ nasal resonances in the 0-5000 Hz range, their synthesis requires one extra resonance compared with that of vowels. In the cascade configuration of the Klatt (1980) system described earlier, there are six formant filters and one antiformant filter and one of the formant filters is used for synthesising the additional (low frequency) resonance needed for nasal consonants. The same configuration can be used to synthesise *oral* sonorants by setting the centre frequency and bandwidth of one of the formant filters equal to that of the antiformant filter so that they cancel each other out (leaving therefore five resonances). For synthesising nasal consonants, the additional resonance in the Klatt (1980) system is set to a low centre frequency (270 Hz) to model the prominent low frequency nasal formant that characterises nasal consonants. The anti-resonance frequency, which in nasal consonant production depends on the length of the side-branching oral cavity, could be varied for synthesising different places of articulation; another way of synthesising nasal consonant place of articulation differences is to vary the nasal resonance *bandwidths* (Fujimura, 1962). An example of the spectrum of the filter for [n] (following closely the specifications given in Klatt, 1980, is shown in Figure 7.10.

7.5 Combining the source with the filter

As discussed earlier, the essential relationship between source, filter, and output in a cascade model of speech synthesis is

$$Y(z) = H(z)X(z)$$

For oral sonorants $H(z)$ is defined as

$$H(z) \quad = \quad H_g(z)\, H_{f_1}(z)\, H_{f_2}(z)\, H_{f_3}(z)\, H_{f_4}(z)\, H_{f_5}(z)$$

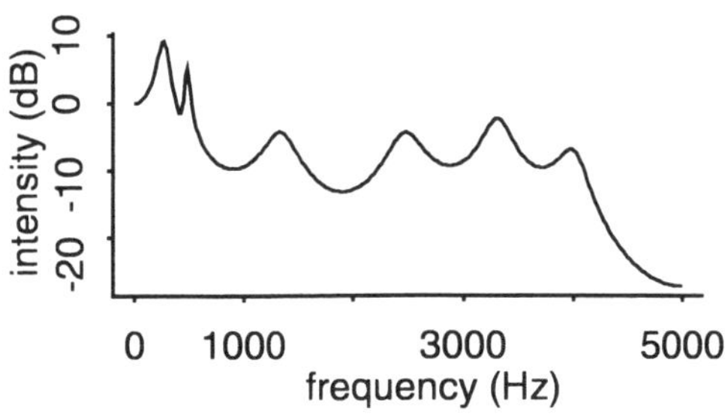

Figure 7.10: The spectrum for a synthetic [n].

$$= H_g(z) \prod_{n=1}^{5} H_{f_n}(z), \qquad (7.9)$$

where $H_{f_n}(z)$ is the z-transform of the nth formant frequency filter and H_g(z) is the z-transform of the glottal filter. Since each formant frequency and the glottal filter are second-order recursive, it follows that the relationship between output and input (for a five-formant cascade system) is defined by a 12th-order recursive filter of the form:

$$y[n] - a[1]y[n-1] \ldots - a[12]y[n-12] = b[0]x[n]. \qquad (7.10)$$

A synthetic speech waveform could be produced by passing a source (either an impulse train or random numbers as in Figure 7.1) through the above filter in the manner described in Chapter 5. For nasal consonants, there are 14 coefficients on the output because of the inclusion of an additional filter to model the nasal formant and two coefficients on the input ($b[0]x[n] + b[1]x[n-1] + b[2]x[n-2]$) if a single anti-formant filter is included.

The resulting synthetic waveform slopes downwards at approximately 12 dB/octave in voiced speech (due to the effects of the glottal filter). In order to provide a 6 dB/octave boost (producing an overall trend of -6 dB/octave) to model the effect of sound radiation from the lips, the output waveform could be first-differenced as discussed in Section 6.4.4. However, another exactly equivalent way of producing an approximate 6 dB/octave lift is to include a term $(1 - az^{-1})$ in the z-transform of the vocal tract filter, i.e.

$$H(z) = H_g(z)H_p(z) \prod_{n=1}^{5} H_{f_n}(z)$$

for oral sonorants where $H_p(z) = 1 - az^{-1}$ (a is a constant between 0.9 and 1 as discussed in Section 6.4.4). A spectrum of a synthetic [i] vowel in which an impulse train corresponding to a fundamental frequency of 100 Hz was passed through the above filter (including the glottal filter and preemphasis effects) is shown in Figure 7.11.

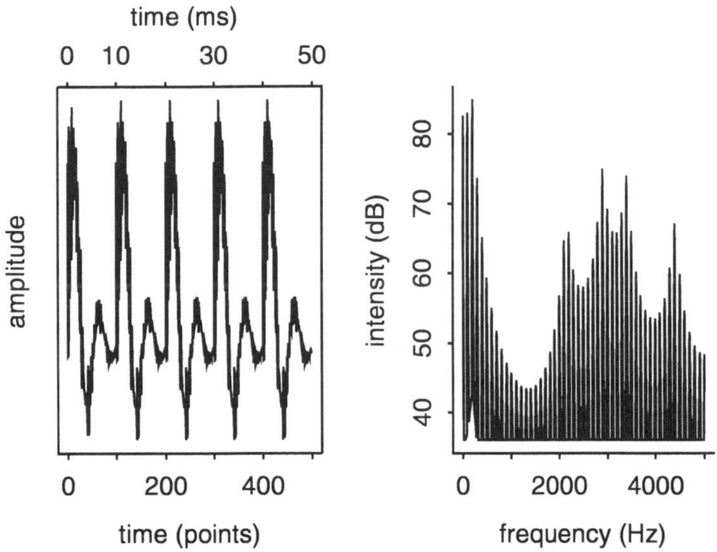

Figure 7.11: Waveform of a synthetic [i] (*left*) and corresponding spectrum (*right*) for a fundamental frequency of 100 Hz passed through the filter shown as the solid line in the left panel of Figure 7.6.

7.6 Parallel structure

In the cascade model described so far, there is no direct control over the peak amplitude levels of the formants: the user specifies a set of formant centre frequencies and bandwidths, and the formant levels emerge appropriately from the filter function. Providing the centre frequencies (and to a lesser extent) the bandwidths are appropriately specified for a particular sound, the formant levels will also be appropriate.

This is only true, however, for oral and nasal sonorants. In the case of fricatives and the fricated part of the release of oral stops and affricates, the amplitude level in the lower part of the spectrum cannot be correctly set just by specifying the centre frequencies and bandwidths of formants. This is because in fricated sounds, the source is at the place of articulation of the consonant and the cavity behind the source causes anti-resonances in the spectrum that are difficult to model using a cascade configuration.

The alternative is to specify the filter function in such a way that there is direct control, not only over the centre frequencies and bandwidths, but also over the levels. The basic structure for achieving this is a parallel configuration (Holmes, 1973, 1983) in which the filters for the separate formants are *added* rather than multiplied i.e. a parallel configuration has the form:

$$H_{fnp}(z) = H_{f1p}(z) + H_{f2p}(z) + H_{f3p}(z) + H_{f4p}(z) + H_{f5p}(z), \qquad (7.11)$$

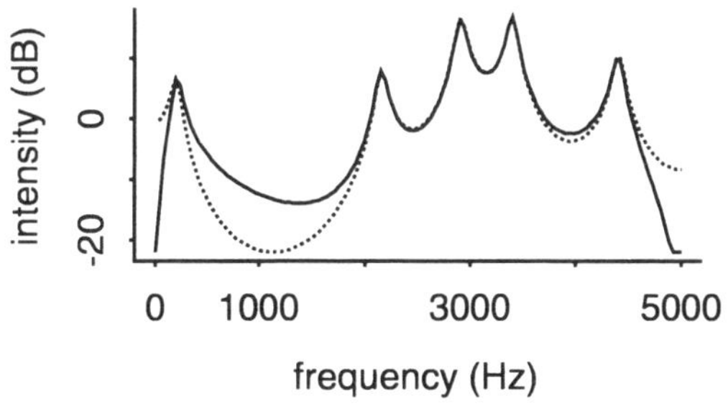

Figure 7.12: The approximation of the output from parallel bandpass filters (*solid line*) to the spectrum from a cascade model (*dotted line*) for an [i] vowel.

where $H_{fnp}(z)$ is the z-transform of the nth formant filter used in a parallel configuration. The object now is to manipulate (the numerator of) each of the terms in Equation 7.11 such that the parallel model is as close as possible to the theoretically correct cascade model (theoretically correct in the sense that the cascade configuration is a more appropriate model of the vocal tract filter in speech production). Unfortunately, the manipulation to convert multiplication into a summation such that the cascade and parallel models produce nearly identical output is very complicated, not so much in a mathematical sense (it can be done using partial fraction expansion) but more by the number and different kinds of operations that are necessary to make this possible (see, e.g., Holmes, 1983 in which the cascade model is completely abandoned in favour of a parallel configuration).

An approximation to the cascade output can be obtained by defining each of the terms in Equation 7.11 as a *bandpass filter* (Linggard, 1985). This implies that the z-transform for each formant filter has the same denominator but that there is an extra z^{-2} term in the numerator and a different coefficient.[3] Specifically, for a formant in a parallel configuration, the z-transform (assuming a bandpass filter) is

$$H_{fp}(z) = \frac{A_p(1 - z^{-2})}{1 - 2r\cos(\omega)z^{-1} + r^2 z^{-2}}$$

where r and ω are as defined earlier and $A_p = 0.5(1 - r^2)$. The effect of the coefficient in the numerator is to set the formant centre frequency to 0 dB. The dB level of the formant centre frequency can be varied by multiplying the numerator by $10^{I/20}$, where I is the desired level in decibels. The output from each filter should be combined in opposite signs (negate the output from the filter for odd formant numbers as described in Weibel, 1955; Fant, 1960; Flanagan, 1972). An example of the extent to which the parallel configuration approximates the cascade configuration for the [i] vowel shown earlier is given in Figure 7.12. Further details on the refinements that are necessary for accurate parallel synthesis are presented in Holmes (1983, 1985).

Further reading

This chapter has presented only a sufficient overview of formant synthesis and speech synthesis in general to provide the necessary background for an understanding of the linear predictive coding of speech discussed in the next chapter. Above all, we have not discussed any details of the top-end of a text-to-speech system nor the details of phoneme synthesis by rule (Holmes, Mattingley, & Shearme, 1964). There are numerous publications that deal with these and many other aspects of speech synthesis in detail. See, in particular, Klatt (1987) and more recently (Bailly et al., 1992) and van Heuven and Pols (1993). A presentation of an alternative to the standard synthesis-by-rule paradigm, based on Firthian phonology, is given in Local (1994). Many other references and examples of speech synthesis systems can be found on the World Wide Web in the addresses given in the footnotes of this chapter. A summary of articulatory speech modelling that forms the basis for articulatory speech synthesis is given in Gabioud (1994).

Notes

1. There are many other different kinds of speech synthesis systems available. The most important of these are discussed in the review article in Klatt (1987) which also includes recordings of them. There is also much material available on the WWW currently on the *comp.speech* web page (`http://www.speech.cs.cmu.edu:80/comp.speech`).

2. These publications provide some of the background to the Haskins Laboratory articulatory synthesis system — an excellent demonstration of this (and also of the Pattern Playback system) can be found on their WWW site: `http://www.haskins.yale.edu`.

3. This causes a zero to be introduced at frequencies of zero Hertz and the Nyquist frequency.

8

LINEAR PREDICTION OF SPEECH

The technique of linear prediction has been available for speech analysis since the late 1960s (Itakura & Saito, 1973a, 1970; Atal & Hanauer, 1971), although the basic principles were established long before this by Wiener (1947). Linear predictive coding, which is also known as autoregressive analysis, is a time-series algorithm that has applications in many fields other than speech analysis (see, e.g., Chatfield, 1989).

The underlying motivation for the linear predictive analysis of speech is that it provides a decomposition of a signal into its source and filter components. From this point of view, LPC can be interpreted as the reverse of digital formant synthesis discussed in the preceding chapter: in synthesis, the object is to convolve a source with a set of filter coefficients to obtain a (synthetic) output signal; in LPC, a natural speech signal is decomposed into a source signal and a set of filter coefficients which, when combined in a synthesis model, reconstruct exactly the original signal.

Since the source and filter coefficients are separated in LPC, it follows that LPC can be used to provide a smoothed spectrum of a speech signal and this is one of the main applications of LPC speech analysis. As well as providing a smoothed spectrum, the formant frequencies and bandwidths can be extracted from the LPC coefficients, thereby providing the basis for formant tracking of speech signals. In conjunction with autocorrelation analysis, LPC is also fundamental to some procedures for tracking the pitch of voiced signals (Markel, 1972b).

Another application of LPC is in analysing the shape of the vocal tract that could have produced a speech signal. Such an analysis is possible, at least for some sounds, because LPC coefficients can be directly related to *cross-sectional areas* of a lossless tube model. Early analyses of this kind are discussed in Wakita (1972). In fact, the coefficients that provide the link between LPC and area functions (known as reflection coefficients) also provide the basis for *LPC analysis-resynthesis*, allowing speech to be coded in a data-reduced form (Atal & Hanauer, 1971; Markel, 1972a; Makhoul, 1973; Wakita, 1979).

In the following sections, we begin by presenting an overview of the LPC algorithm itself and then consider how it can be applied to spectral analysis, pitch tracking, vocal tract modelling, and speech synthesis.

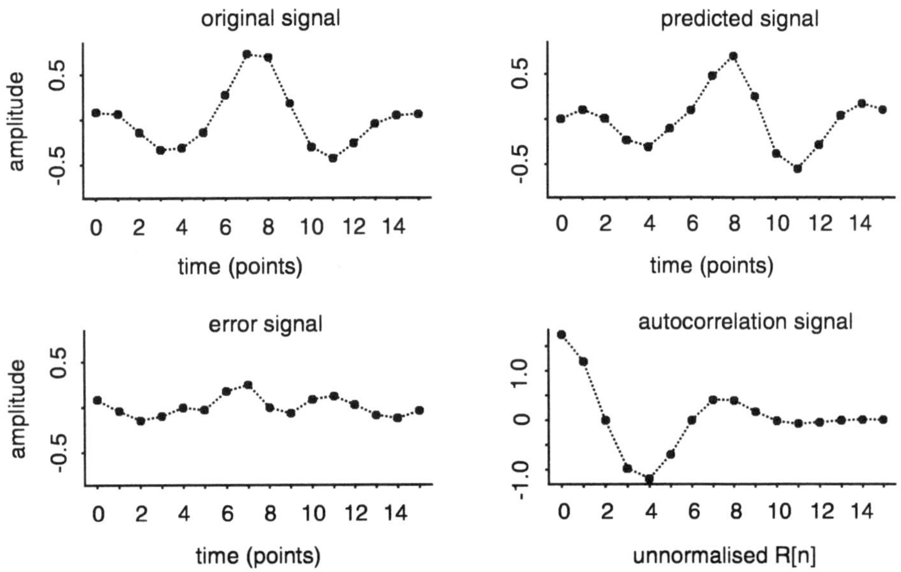

Figure 8.1: An example of a second order LPC analysis. *Top left*: The signal to be analysed. *Top right*: The predicted signal. *Bottom left*: The error signal, which is equal to the difference between the original and predicted signals. *Bottom right*: The short-time (unnormalised) autocorrelation of the original signal.

8.1 LPC and its relationship to digital speech

LPC is a time domain technique that models a signal as a linear combination of weighted delayed values. For a speech signal $\boldsymbol{y}$, linear predictive coding is used to calculate a set of coefficients $\alpha[1] \ldots \alpha[p]$ such that

$$y[n] = \alpha[1]y[n-1] + \alpha[2]y[n-2] + \ldots + \alpha[p]y[n-p] + \epsilon[n]. \tag{8.1}$$

We can begin by making two observations about the model defined by Equation 8.1. First, a decision must be made about the number of coefficients that are needed in order to model accurately $\boldsymbol{y}$ by its weighted past values. (A model with p coefficients, $\alpha[1] \ldots \alpha[p]$ is often known as a pth order LPC model; the coefficient $\alpha[0]$, which is the weighting on $y[n]$, is always assumed to be 1). This is a very important consideration in speech analysis that will be returned to later. Second, it is (almost) never possible to model *exactly* $y[n]$ by weighted delayed values: there is a discrepancy, or *error signal* (denoted by $\epsilon[n]$ in Equation 8.1). In LPC, the object is to find a set of coefficients such that $\boldsymbol{\epsilon}$ is as close to zero as possible.

Figure 8.1 shows an example of a 2nd-order LPC analysis applied to a signal $\boldsymbol{y}$ of length 16. In this example the object is to calculate $\alpha[1]$ and $\alpha[2]$ such that the mean square of $\boldsymbol{\epsilon}$ is minimised in

$$y[n] = \alpha[1]y[n-1] + \alpha[2]y[n-2] + \epsilon[n]. \tag{8.2}$$

The way in which the coefficients are actually calculated will be left until a later section. For the present, we will note that once they have been determined, the predicted signal, $\hat{\boldsymbol{y}}$, is given by

$$\hat{y}[n] = \alpha[1]y[n-1] + \alpha[2]y[n-2] \tag{8.3}$$

and the error signal is the difference between the original and the predicted signal (Figure 8.1)

$$\epsilon[n] = y[n] - \hat{y}[n], \tag{8.4}$$

or equivalently, the original signal is the sum of the predicted signal and the error signal:

$$y[n] = \hat{y}[n] + \epsilon[n]. \tag{8.5}$$

8.1.1 Relationship to speech production

The production of speech can be modelled by passing a periodic or noise source through the vocal tract filter. As we have seen earlier, this process can be described by the convolution equation

$$\boldsymbol{y} * \boldsymbol{a} = \boldsymbol{b} * \boldsymbol{x}, \tag{8.6}$$

where $\boldsymbol{x}$ is the source, $\boldsymbol{a}$ and $\boldsymbol{b}$ are coefficients on the input and output signals that encode the frequencies of the vocal tract resonances and anti-resonances, and $\boldsymbol{y}$ is the output signal. For speech sounds that have no anti-resonances (for example, voiced oral vowels), $\boldsymbol{b} = 1$ and the equation reduces to

$$\boldsymbol{y} * \boldsymbol{a} = \boldsymbol{x}. \tag{8.7}$$

Furthermore, since convolution is equivalent to a linear combination of scaled delayed values (see Chapter 5), Equation 8.7 can also be expressed as

$$y[n] + a[1]y[n-1] + a[2]y[n-2] + \ldots + a[k]y[n-k] = x[n] \tag{8.8}$$

or rearranging the terms

$$y[n] = -a[1]y[n-1] - a[2]y[n-2] - \ldots - a[k]y[n-k] + x[n]. \tag{8.9}$$

Expressed in words, in an all-pole model of speech, i.e. for a speech signal whose spectrum is uniquely determined by resonances, the current value $y[n]$ is entirely determined by the sum of scaled past values and the input value. From this point of view, if we start with any speech signal $\boldsymbol{y}$ and wish to estimate the coefficients $\boldsymbol{a}$ that encode information about the vocal tract filter, then linear predictive coding is a suitable technique to provide this estimation, precisely because it models a signal as the weighted sum of past values plus an error term (compare Equations 8.1 and 8.9).

It is immediately apparent that one of the limitations of linear predictive coding is that, while it can provide an estimate of the coefficients of the *output* signal that determine the pole (resonance) frequencies, it does not model those on the *input* signal $\boldsymbol{b}$ that determine the frequency locations of the zeros (anti-resonances). In this sense, the LPC coefficients are those of an all-pole model that provides most accurate results when it is applied to speech signals such as voiced oral vowels that have no zeros. This does not necessarily mean, however, that the LPC model is inapplicable for analysing consonants that have zeros. As discussed in Atal (1985), a zero introduces two kinds of effects into a spectrum, a local dip and a change in the spectrum's spectral balance; while LPC cannot model the dips that are introduced by zeros, it can provide a good approximation to the change in spectral balance.

In comparing the LPC Equation 8.1 with the one that defines the all-pole model of speech production in Equation 8.9, it is clear that, apart from a change in sign, a direct correspondence can be established between the coefficients of LPC model, $\alpha[n]$, and the coefficients of the vocal tract filter, $a[n]$. But this assumption is valid only to the extent that the error signal $\boldsymbol{\epsilon}$ represents a realistic and appropriate model for the source signal in speech production, $\boldsymbol{x}$. But what are the conditions that allow this equivalence to be made?

In order to answer this question, we must first recall that LPC fits a set of coefficients under the condition that the mean of $\boldsymbol{\epsilon}^2$ is minimised. It can be shown that, under these conditions, the spectrum of $\boldsymbol{\epsilon}$ is approximately flat i.e. unchanging for increasing frequency (Markel & Gray, 1976). Now there are only two types of time signal that have a flat spectrum: an impulse train and white noise (i.e. a signal generated from random numbers). From the point of view of modelling the production of speech, this is indeed fortunate because as we saw in speech synthesis, an impulse train and random numbers represent an entirely appropriate source model for the synthesis of voiced and voiceless speech, respectively.

If LPC calculates a set of predictor coefficients under the assumption that the source corresponds either to an impulse train or random numbers, it must mean that the predictor coefficients $\alpha[1] \ldots \alpha[p]$ represent not only the spectral contribution from the vocal tract but also from the glottal flow and radiation at the lips. Consider that in speech synthesis, the z-transform of the output signal Y(z) is related to the filter and source by

$$Y(z) = H(z)X(z)$$

and

$$H(z) = H_p(z)H_g(z)\prod_{n=1}^{N} H_{fn}(z).$$

In speech analysis, LPC decomposes the output signal $Y(z)$ into a filter $H(z)$ and a source $X(z)$, but there is no further differentiation of $H(z)$ into the coefficients that are attributable to the formant frequencies ($\prod H_{f_n}(z)$), the lip-radiation/preemphasis effect ($H_p(z)$), and the glottal filter ($H_g(z)$). Therefore another limitation of LPC is that the coefficients are only an approximate model of the vocal tract shape because they are merged with the spectral contributions from the glottal wave shape and the lip-radiation effect. But despite this limitation, LPC can provide a very accurate spectral representation for most speech sounds, as discussed in Section 8.3.

8.2 Techniques for calculating the LPC coefficients

There are at least three techniques for calculating the predictor coefficients: from the *autocorrelation method*, from the *covariance method*, and from reflection coefficients using a *lattice filter* (Makhoul, 1977). Since all of these techniques have been dealt with rigorously in many previous publications (Atal, 1985; Markel & Gray, 1976; Makhoul, 1975; Rabiner & Schafer, 1978; O'Shaughnessy, 1987; Owens, 1993; Parsons, 1987; Wakita, 1973; Witten, 1982), we will not discuss them in detail here nor give any derivations or proofs. The reader is referred to any of the above publications for these.

We will, however, review briefly the autocorrelation method using a recursive procedure (Makhoul, 1975; Markel & Gray, 1976), both because it is a frequently used technique, and because it gives some insight into the relationship between autocorrelation coefficients, LPC coefficients, and another set of coefficients known as reflection coefficients.

The technique for obtaining LPC coefficients using the autocorrelation method involves calculating four quantities from each other recursively. These are

$\boldsymbol{R}$ A vector of short-time autocorrelation coefficients (of length equal to the length of the signal). These are the same coefficients that were used in Chapter 5 both to estimate whether the signal is voiced and to estimate the fundamental frequency of voiced signals.

$\boldsymbol{\alpha}^{(p)}$ A vector of LPC coefficients of length $p+1$. The superscript denotes the *order* of the LPC-model that is calculated. For example, for a 4th-order LPC model, there are 5 coefficients: $\alpha^{(4)}[0] = 1$, $\alpha^{(4)}[1]$, $\alpha^{(4)}[2]$, $\alpha^{(4)}[3]$, $\alpha^{(4)}[4]$ ($\alpha^{(p)}[0]$ is always 1).

$\boldsymbol{k}$ A vector of partial-correlation (PARCOR) coefficients of length p where p is the order of the LPC model. These coefficients can be further used to calculate the relative areas of the equivalent lossless tube model on which LPC is based. They are discussed in further detail in the section on area functions.

$E^{(n)}$ The sum of the error signal squared associated with a nth-order LPC model. For example, if $\boldsymbol{\epsilon}$, the error signal, for a second-order LPC-model calculated on a signal of length 4 is $\boldsymbol{\epsilon} = [3, 0, -1, 4]$, then $E^{(2)} = 9 + 0 + 1 + 16 = 26$.

There are two main stages in calculating these variables: an initialisation stage and then a recursive stage. The equations are as follows:

Initialisation stage:

1 $\alpha^{(1)}[1] = k[1] = R[1]/R[0]$

2 $E^{(1)} = (1 - k[1]^2)R[0]$

Steps 3–6 represent the recursive stage (for $2 \leq j \leq p$, where p is the order of the LPC-model to be calculated):

Calculation of PARCOR coefficients of order j $(2 \leq j \leq p)$

3 $k[j] = \frac{R[j] - \sum_{i=1}^{j-1} \alpha^{(j-1)}[i]R[j-1]}{E^{(j-1)}}$

Calculation of LPC coefficients

4 $\alpha^{(j)}[j] = k[j]$

5 $\alpha^{(j)}[i] = \alpha^{(j-1)}[i] - k[j]\alpha^{(j-1)}[j-i] \quad (1 \leq i \leq j-1)$

Calculation of error term

6 $E^{(j)} = (1 - k[j]^2)E^{(j-1)}$

As an example, the recursive procedure above is used to calculate a second-order LPC model for the signal shown in Figure 8.1. Once the short-time autocorrelation coefficients have been obtained (Figure 8.1, bottom right panel), the LPC coefficients are calculated as follows:

First-order LPC-coefficients

$$\begin{aligned} R[0] &= 1.732 \quad \text{(Figure 8.1)} \\ R[1] &= 1.186 \quad \text{(Figure 8.1)} \end{aligned}$$

$$\begin{aligned} \alpha^{(1)}[1] &= k[1] = R[1]/R[0] \\ &= 0.685 \end{aligned}$$

$$\begin{aligned} E^{(1)} &= (1 - k[1]^2)R[0] \\ &= (1 - 0.685^2) \times 1.732 \\ &= 0.919 \end{aligned}$$

Second-order coefficients

$$k[2] = \frac{R[2] - \alpha^{(1)}[1]R[1]}{E^{(1)}}$$

$$
\begin{aligned}
&= \frac{-0.007 - (0.685 \times 1.186)}{0.919} \\
&= -0.892
\end{aligned}
$$

$$
\begin{aligned}
\alpha^{(2)}[2] &= -0.892 \\
\\
\alpha^{(2)}[1] &= \alpha^{(1)}[1] - k[2]\alpha^{(1)}[1] \\
&= 0.685 + 0.892 \times 0.685 \\
&= 1.296 \\
\\
E^{(2)} &= (1 - k[2]^2)E_1 \\
&= (1 - (-0.900)^2) \times 0.982 \\
&= (1 - 0.796) \times 0.919 \\
&= 0.187
\end{aligned}
$$

For a second-order LPC model, the predicted signal (top right panel of Figure 8.1) is therefore given by

$$\hat{y}[n] = 1.296y[n-1] - 0.892y[n-2],$$

while the sum of $\epsilon^2 \approx E^{(2)} = 0.187$. Higher-order LPC coefficients can be similarly calculated using the above equations.

8.3 Analysis of the error signal

The error signal can provide information about the pitch of the signal as well as evidence of the signal's voicing status. Figure 8.2 shows different representations of the error signal for [ɛ] and [ʃ], both produced by an Australian English female talker in the context of isolated words (sampling frequency = 20000 Hz). The [ɛ] vowel was Hamming windowed (row 1, left panel) and analysed using a 16th-order LPC model; the [ʃ] was also Hamming windowed and analysed using a 10th-order LPC model. The error signals for both sounds (second row) were obtained in the manner described in Section 8.1.

For voiced speech, the error signal often shows peaks at the beginning of each period (second row, left panel), whereas no such peaks are evident for voiceless speech (second row, right panel). The error signal by itself could therefore be used to estimate the fundamental frequency of voiced signals, although the peaks are often a good deal less clear for nasal and lateral consonants (Makhoul & Wolf, 1972). Another way of estimating the fundamental frequency is to carry out an autocorrelation analysis of the error signal: for voiced speech, a peak in the autocorrelation function is expected corresponding to the duration of the pitch period (note the peak at 4 ms in the left panel of row 3), whereas no significant peak is expected in the error signal's autocorrelation function for voiceless speech. In a well-known procedure (the SIFT algorithm) developed by

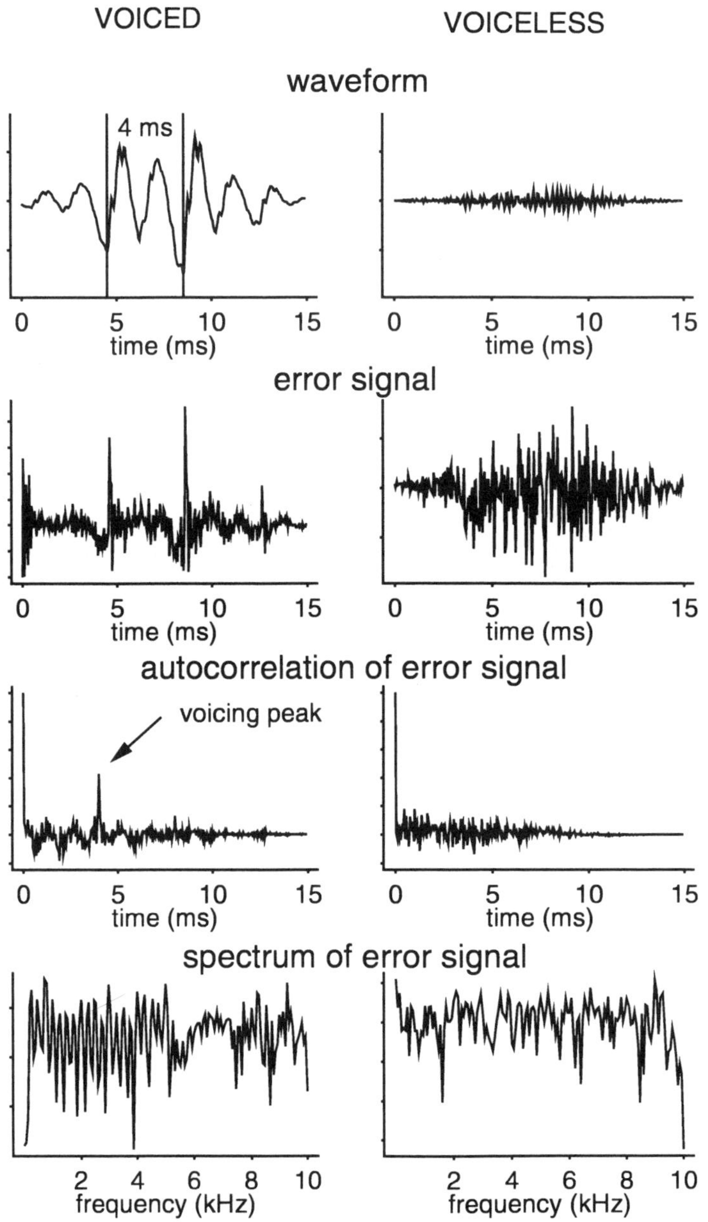

Figure 8.2: *Left column*: Voiced speech. *Right column*: Voiceless speech. *Top row*: Hamming windowed signals. *Second row*: The error signal produced by LPC analysis. *Third row*: The short-time autocorrelation of the error signal — note the peak in the voiced signal at roughly 4 ms. *Bottom row*: The spectrum of the error signal in row 2.

Markel (1972b), various other types of processing are applied (including low-pass filtering and down-sampling) to obtain a more distinct peak in the error signal's autocorrelation function.

The panels in the fourth row show the spectrum of the error signal for [ɛ] and [ʃ]. Compatibly with the earlier discussion, both are relatively "flat" i.e. there are no significant changes in the amplitude level for increasing frequency. The spectrum of voiced error signals is expected to show the influence of the fundamental frequency as harmonics that occur at pitch frequency (for the present example, these occur at approximately 250 Hz, as shown in the left panel, row 4 of Figure 8.2).

As discussed in the preceding section, an important parameterisation of the error signal is $E^{(p)}$, the sum of error signal squared. This quantity is used not only in the recursive procedure for deriving the LPC and reflection coefficients from each other (see Section 8.3), but also to normalise the amplitude level of an LPC-smoothed spectrum relative to that of the original signal (see Section 8.5). An analysis of $E^{(p)}$ can also show how the size of the error signal is progressively reduced for an increasing order of LPC model (Atal & Hanauer, 1971). This is shown in Figure 8.3 for the same [ɛ] and [ʃ] sounds analysed with an increasing number of LPC coefficients. As expected, the size of the error signal is reduced as progressively more coefficients are fitted to the signal; however, for this voiced sound ([ɛ] produced by a female talker), a plateau is reached at somewhere between 10 and 15 coefficients, while for the voiceless [ʃ], there is only a very slight decrease in $E^{(p)}$ for $n > 5$. The figure therefore supports earlier studies that have shown that voiceless sounds can be modelled with a smaller number of coefficients than voiced sounds. It is also clear from the same figure that the error signal is larger for the voiceless than the voiced sounds: this is because it is more difficult to predict a sample from previous samples in aperiodic signals.

8.4 LPC-smoothed spectra and formants

We have seen that when a source, $\boldsymbol{x}$, is passed through an all-pole filter — i.e. one that is characterised only by resonances — the relationship between input and output is of the form

$$y[n] + a[1]y[n-1] + a[2]y[n-2] + \ldots + a[p]y[n-p] = x[n].$$

Essentially, linear predictive coding provides an estimate of the coefficients $(a[1] \ldots a[p])$ under the assumption that $\boldsymbol{x}$ is either an impulse train in periodic sounds or white noise in aperiodic sounds. As discussed in Chapter 6, if the coefficients of the filter are known, the spectrum of the filter can be easily calculated. Therefore, since LPC is essentially an estimate of the filter coefficients in an all-pole model, these can in turn be used to provide a smoothed spectrum of the speech signal — that is, a spectrum that is of the filter only in the absence of any contribution from the source. The smoothed spectrum is derived from

$$\frac{\sqrt{E^{(p)}}}{1 - \alpha[1]z^{-1} - \alpha[2]z^{-2} \ldots - \alpha[p]z^{-k}}$$

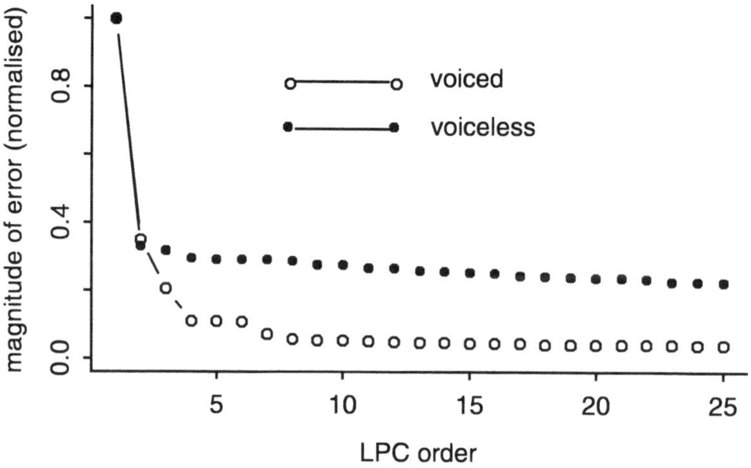

Figure 8.3: The magnitude of the sum of the squared error, $E^{(p)}$, for increasingly higher-ordered LPC analyses of a voiced and voiceless segments.

in which $\alpha[1] \ldots \alpha[p]$ are the LPC-coefficients and $E^{(p)}$ is the sum of the error signal squared. It can be shown that this numerator coefficient provides the optimal alignment in amplitude between the smoothed spectrum and a DFT of the original speech signal (Rabiner & Schafer, 1978).

Before deriving a smoothed spectrum from the LPC-coefficients, we should remind ourselves of the nature of the spectral information that they encode. The coefficients model the spectral information not only from the vocal tract filter but also from the glottal flow and the lip-radiation effect. What we would like, however, is for our smoothed spectrum to represent (as far as possible) the spectral contribution due to the vocal tract filter by itself, since this is, of course, more directly related to the vocal tract shape and therefore to the formant frequencies and bandwidths. We therefore need to consider if there is a way of eliminating the spectral contribution from the other two factors. A possible way of achieving this is to boost voiced speech signals by +6dB/octave (preemphasis) before the LPC coefficients are calculated. The motivation for this is as follows. The glottal filter contributes roughly a −12dB/octave effect, while the contribution from lip-radiation is +6dB/octave resulting in a net −6dB/octave trend. Therefore by preemphasising the speech signal, this spectral trend, which is caused by the combined glottal and lip-radiation effects, is largely eliminated. From another perspective, we saw in the previous chapter that the z-transform of the glottal filter is $1/(1 - rz^{-1})^2$ while the z-transform of the lip-radiation effect is $1 - az^{-1}$ (a is usually close to 1). The resulting filter is therefore

$$\frac{1 - az^{-1}}{(1 - rz^{-1})^2}$$

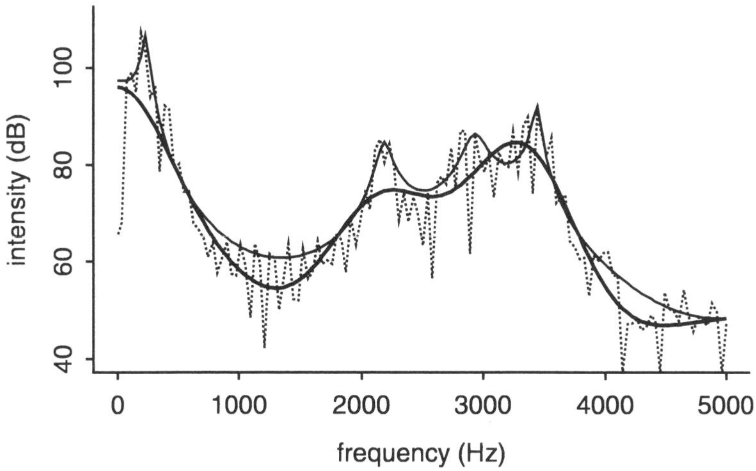

Figure 8.4: An LPC-smoothed spectrum (*solid line*) line and a cepstrally smoothed spectrum (*solid bold line*) superimposed on a DFT spectrum (*dotted line*) of an [i] vowel.

On the assumption that r and a are approximately equal, the combined effect is

$$\frac{1}{1 - az^{-1}}$$

Therefore, the z-transform of the signal should be multiplied by a factor $(1 - az^{-1})$ to remove the remaining combined effect. The +6dB/octave (preemphasis) lift should be applied only to voiced signals, however, because in voiceless signals, the source, which has a downward sloping trend of −6dB/octave, is (approximately) cancelled by the lip-radiation effect (see Chapter 3).

An LPC-smoothed spectrum for an [i] vowel is shown as a bold line in Figure 8.4 superimposed on the DFT-spectrum. The smoothed spectrum was obtained by preemphasising (first differencing) the speech signal and applying a Hamming window to the first differenced data (the order in which these two operations is applied makes little difference — Markel & Gray, 1976). A 20th-order LPC model was then calculated for the first differenced and Hamming windowed signal. Also shown for comparison is a cepstrally smoothed spectrum of the same sound, calculated using 40 cepstral coefficients.

The smoothness of the LPC-spectrum is dependent on the number of coefficients that are used to calculate the spectrum (fewer coefficients imply a smoother spectrum). As a general indication, for voiced speech, the number of coefficients in LPC analysis should be at least equal to the sampling frequency in kHz for a typical adult male vocal tract (for example, 20 coefficients for a sampling frequency of 20000 Hz); the motivation for this is discussed in the section on reflection coefficients and vocal tract modelling below. For voiceless

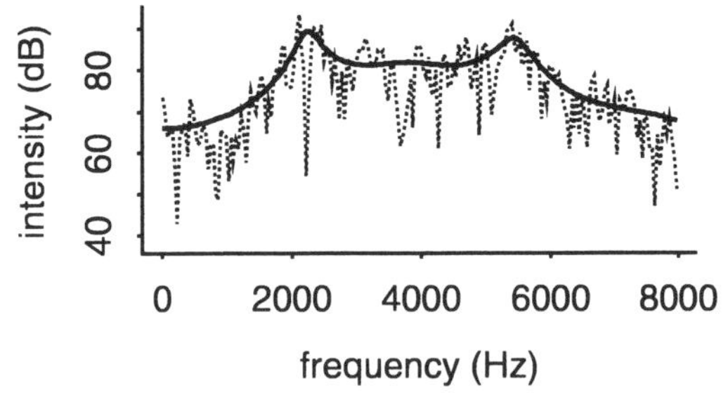

Figure 8.5: An LPC-smoothed spectrum superimposed on a DFT spectrum of [ʃ].

sounds, and in particular for voiceless fricatives, it is often possible to model adequately the resonances with far fewer coefficients (Markel & Gray, 1976; see also the discussion on the error signal in the preceding section).

Figure 8.4 highlights one of the important differences between an LPC-smoothed and a cepstrally-smoothed spectrum: an LPC-smoothed spectrum fits the peaks of original spectrum (i.e. the resonances) better than the valleys (Markel & Gray, 1976). This is because the LPC-spectrum is based on an all-pole model (i.e. one in which there are no zeros, or anti-resonances, and therefore one in which the valleys are not explicitly modelled). Second, the number of possible peaks in an LPC-smoothed spectrum is directly dependent on the number of coefficients that are used to calculate it: there can only be $l/2$ spectral peaks where l is the number of LPC-coefficients. Therefore, since $l = 20$ coefficients were used in the present example, the LPC-spectrum in Figure 8.4 can have at most 10 peaks between 0 Hz and the Nyquist frequency (10000 Hz). For cepstrally-smoothed spectra, however, there is no such relationship between the number of coefficients and peaks.

Figure 8.5 shows an LPC-smoothed spectrum superimposed on a DFT of a voiceless sound [ʃ]: the spectra were obtained from the central 512 samples of this sound and the LPC-order was 8, resulting in at most 4 peaks in the spectral range up to the Nyquist frequency (10000 Hz). As the figure shows, two of these peaks show up clearly at around 2200 Hz and 5500 Hz; the other two are not distinct, either because they have very large bandwidths or because they occur at 0 Hz. We will return to this point again below.

As well as providing a smoothed spectrum, LPC coefficients can be used to estimate automatically the formant centre frequencies and bandwidths. The technique is essentially the reverse of one discussed in the preceding chapter on digital formant synthesis, in which we saw that it is possible to convert a formant centre frequency and bandwidth (ω and bw in radians) into the coefficients of a second-order recursive filter, i.e.,

$$y[n] - 2r\cos(w)y[n-1] + r^2y[n-2] = x[n]$$

where $r = e^{-bw/2}$. Recall that the number of coefficients is twice the number of formant frequencies that are synthesised: so for 5 formants, there are 10 coefficients on delayed output samples. This is directly related to the observation made earlier that a maximum of $l/2$ peaks can be derived from a LPC-model of order l (so 5 peaks for a 10th-order LPC model).

The process of converting formant centre frequencies and bandwidths into coefficients is reversible, so that given a set of coefficients such as

$$y[n] + a[1]y[n-1] + a[2]y[n-2] + a[3]y[n-4] \ldots = x[n] \tag{8.10}$$

it is possible to extract the formant centre frequencies and bandwidths that gave rise to them. The process is a little involved algebraically and requires finding the roots of the z-transform of the filter. Specifically, if we represent Equation 8.10 in z-transform terms, we have

$$Y(z) = X(z)/A(z)$$

where $A(z)$ is given by

$$A(z) = 1 + a[1]z^{-1} + a[2]z^{-2} + a[3]z^{-3} + a[4]z^{-4}$$

The resonance frequencies can be found by calculating where $A(z) = 0$ (because then $Y(z) = X(z)/A(z)$ becomes infinitely large as a result of dividing by zero); and in order to find where $A(z) = 0$ we have to factor the polynomial in order to find the roots of $A(z)$. For filters that encode resonance frequencies and bandwidths, the roots occur in *complex conjugate pairs*, i.e. they are of the form $a + ib$ and $a - ib$ where i is the square root of minus 1: this follows from Euler's identity, which can be used to show that the filter for a single formant, $H(z) = 1/(1 - 2r\cos(\omega)z^{-1} + r^2z^{-2})$, is equivalent, in factored form, to $H(z) = 1/((1 - re^{-i\omega}z^{-1})(1 + re^{i\omega}z^{-1}))$. Finally, for a root $a + ib$, the bandwidth bw is given by:

$$bw = 2\log(a)$$

and the formant centre frequency ω by

$$\omega = \log(b),$$

where *log* is the natural (to base e) logarithm and ω and bw are radian frequencies. These can be converted into Hertz using the formula given earlier (Equation 6.3, page 160).

The important point to note from the algebra is that, since it is possible to compute $l/2$ centre frequencies and bandwidths from l output coefficients of a recursive filter, they can also be directly estimated from the LPC coefficients, which are in the form

$$y[n] = \alpha[1]y[n-1] + \alpha[2]y[n-2] + \ldots + \alpha[k]y[n-k],$$

by finding the roots of the z-transform of the filter and then converting them into centre frequencies and bandwidths using the formulae given above.

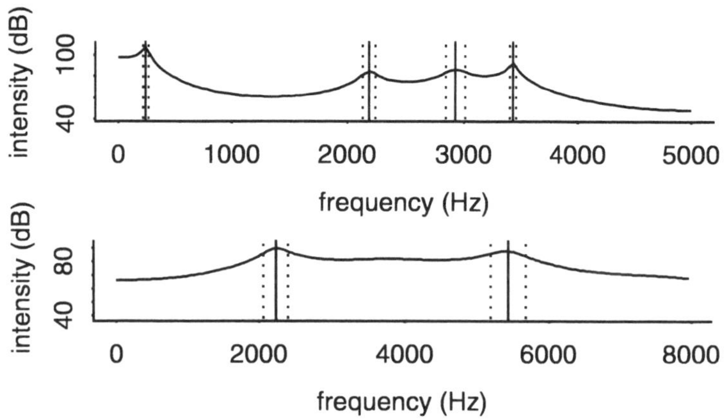

Figure 8.6: LPC smoothed spectra and superimposed formant centre frequencies (solid lines) and bandwidths (dotted lines) calculated from the roots of the LPC polynomial for [i] (*top*) and for [ʃ] (*bottom*).

Some of the centre frequencies calculated from LPC coefficients do not always correspond to real formant frequencies, either because they occur at 0 Hz, or simply because there is no visible peak in the spectrum at which the peak is calculated (this can happen when the bandwidth is absurdly large). Therefore, it is almost always necessary to discard some of the peaks after they have been extracted from the roots of the polynomial in z.

Examples of the calculated centre frequencies and bandwidths for the two sounds given earlier are shown in Figure 8.6. In both cases, spurious peaks, which could not possibly correspond to real formant frequencies, have been discarded. The resulting analysis shows four calculated formant centre frequencies in the 0–5000 Hz range for the [i] vowel and two resonances for the fricative [ʃ] (two were discarded because of very large bandwidths).

There has been a considerable discussion in the literature on the accuracy of the centre frequencies and bandwidths obtained from an LPC model. There is some agreement that the bandwidths can be quite unreliable (Rabiner & Schafer, 1978, p. 450), because LPC does not model the vocal tract losses and these have their greatest influence on the bandwidths.

While preemphasising the (voiced) signal before LPC analysis can help to reduce the combined spectral contribution of the glottal flow and lip-radiation, the factor $1/(1-rz^{-1})^2$ is in particular a highly simplified representation of the glottal filter: consequently, the LPC coefficients are still likely to represent a blending of the vocal tract and source filtering effects. Additionally, as discussed earlier, LPC is an all-pole model, and so it cannot model directly the anti-resonance frequencies of many consonants.

Nevertheless, in spite of these limitations, LPC has been shown to be a very useful technique for formant estimation and usually forms the core of routines

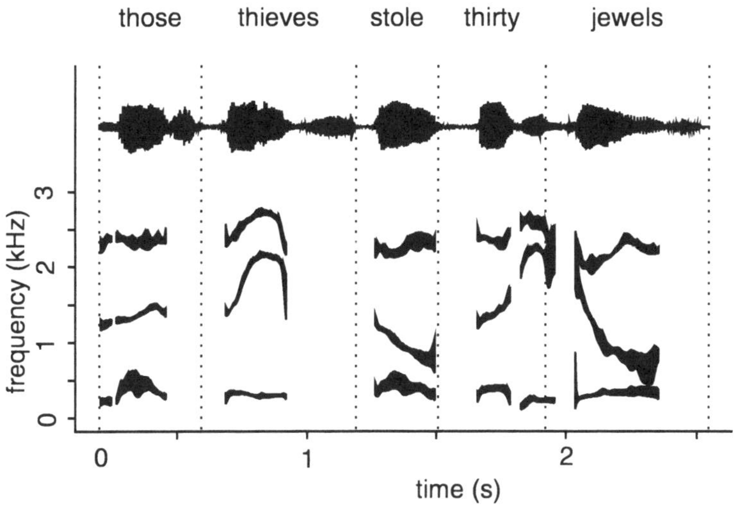

Figure 8.7: Automatically tracked F1-F3 showing the bandwidths on either side of the centre frequency for a sentence produced by an adult male talker.

for tracking automatically formants in the changing acoustic speech signal.

An example of three automatically tracked formants (using the formant tracking system in the speech signal processing package *Waves*[1]) for the sentence "those thieves stole thirty jewels" produced by an adult male Australian talker is shown in Figure 8.7. This type of plot is typically derived by obtaining the formants from the roots of LPC coefficients, in the manner described earlier. Some further processing is also included to remove calculated pole frequencies that would be unlikely to correspond to formants. The LPC coefficients are calculated by windowing the data, often using a window length of roughly 10-15 ms (see also Chandra & Lin, 1974, 1977), since this represents durations for which the dynamic change in the vocal tract shape is presumed to be minimal.

Figure 8.7 shows both the formant centre frequency as well as the bandwidth frequency for the predominantly voiced parts of the speech signal. The centre frequencies are more or less appropriate for this utterance (other than an obvious error in F1 at the onset of "jewels"), but the bandwidths are less convincing: note in particular the way in which F2 and F3 overlap at the end of "thirty" and at the onset of "jewels" because of the very large bandwidth values at these time points.

8.4.1 LPC-derived cepstral coefficients

Once a smoothed LPC-spectrum has been obtained, it is possible to obtain *LPC-derived cepstral coefficients* following the procedures outlined in Chapter 6: that is, by taking the logarithm of the smoothed LPC spectrum and then applying an inverse discrete Fourier transform to obtain a time signal, the LPC-

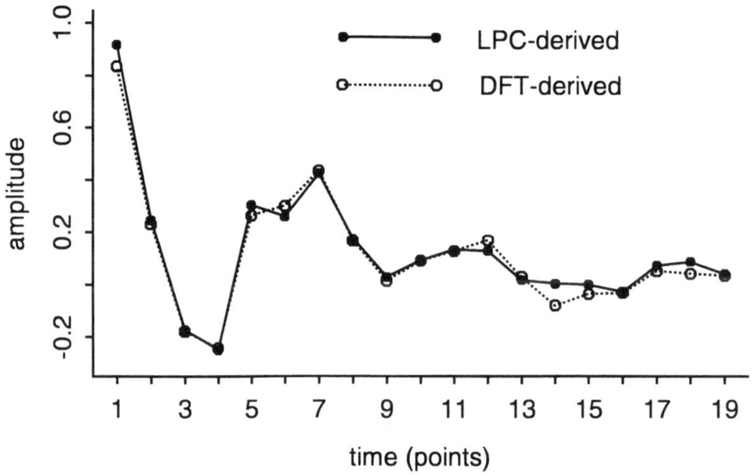

Figure 8.8: LPC-derived and DFT-derived cepstral coefficients calculated for the same [i] vowel whose spectrum is shown in Figure 8.4.

derived cepstral coefficients. These can also be obtained directly from the LPC coefficients using a recursive relationship, rather than from the (log of) the LPC-smoothed spectrum: formulae are given in Rabiner and Juang (1993) (see also Loizou, Dorman, & Spanias, 1995, for a recent comparison with mel-scaled cepstral coefficients in the recognition of nasal consonants). One of the motivations for obtaining LPC-derived cepstral coefficients is that they have been shown to result in slightly improved machine recognition scores compared with those obtained from LPC coefficients (Rabiner & Juang, 1993).

Finally, as Figure 8.8 shows, it should be emphasised that, although similar, LPC-derived cepstral coefficients are *not* equivalent to the DFT derived cepstral coefficients discussed in Chapter 6.

8.5 Area functions and reflection coefficients

Linear predictive analysis can also give some insight into the shape of the vocal tract that produced the acoustic speech signal (Wakita, 1972, 1973; Wakita & Gray, 1974, 1975). This follows from the fact there is a direct correspondence between the all-pole LPC model and the transfer function (a characterisation the input-output relationships) of a wave as it travels through a series of connected cylinders of equal length and different cross-sectional areas. The cylinders are also assumed to be lossless, which implies that none of the input energy of the sound wave is dissipated in the cylinders themselves. As far as speech production is concerned, a model of the vocal tract as a set of lossless tubes is a gross simplification since, as discussed in Chapter 3, it is known that energy is lost due to various factors such as heat conduction, friction, and vocal-tract

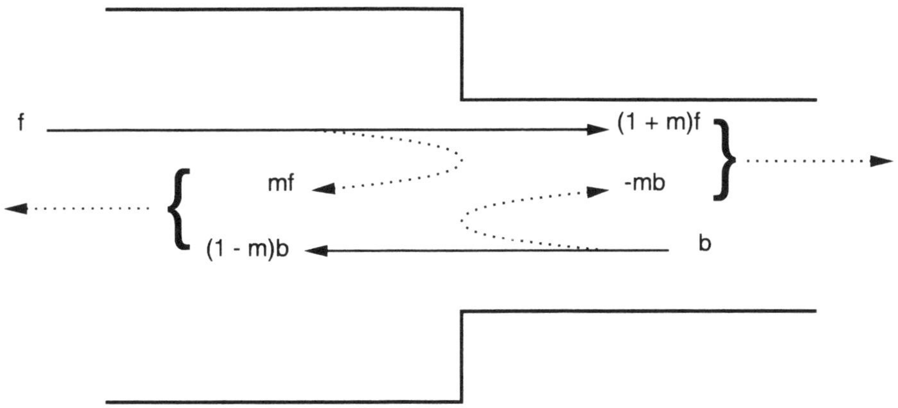

Figure 8.9: The junction between two lossless tubes showing how a forward travelling wave ***f*** and a backward travelling wave ***b*** are reflected and propagated depending on the reflection coefficient, m, between the cylinders (adapted from Linggard, 1985).

wall vibration (Fant, 1960). Some further limitations of the lossless model are discussed below. In spite of these limitations, a lossless tube model derived from LPC-coefficients can sometimes give an approximation to the vocal tract shape, at least for oral vowels.

The link between linear predictive analysis and the lossless tube model is via *reflection coefficients* that are equal to the partial-correlation (PARCOR) coefficients (Itakura & Saito, 1973b) discussed earlier. In the lossless tube model, the reflection coefficients determine both the portion of the travelling wave that is reflected at the junction between cylinders and, more importantly for the present discussion, the ratio of the areas between the connecting cylinders starting at the lips and working towards the glottis. We shall deal briefly with the reflection of the travelling wave and then focus on how the tubes' cross-sectional areas can be calculated.

In the lossless tube model of the vocal tract, a wave, parameterised either as air pressure or volume-velocity variations, travels from one end of the connecting cylinders (the glottal end) to the other (the lip end). At each junction between the cylinders, part of the wave is propagated forward, and part is reflected back. The proportion of the wave that is propagated and reflected depends on the reflection coefficient between each pair of abutting cylinders; the reflection coefficient is, in turn, entirely a function of the ratio of the abutting cylinders' cross-sectional areas.

Since a portion of the forward travelling wave is reflected at each junction between the cylinders, a backward travelling wave is created from the lip to the glottis which is also subject to propagation and reflection at the cylinder boundaries. Figure 8.9 shows in further detail how the forward and backward travelling waves are propagated and reflected at the junction between two cylinders. The travelling waves can be thought of as air pressure variations, as in Linggard (1985). At the junction between the cylinders, the forward travelling wave $\boldsymbol{f}$ divides into a propagated part and a reflected part. If the reflection coefficient has a value of m, it can be shown that the propagated part is given by $(1+m)\boldsymbol{f}$ and the reflected part by $m\boldsymbol{f}$. The backward travelling wave, which passes through the cylinders from the lips to the glottis, also divides into a propagated and reflected part, but here the sign of the reflection coefficient is reversed giving $(1-m)\boldsymbol{b}$ for the propagated part, and $-m\boldsymbol{b}$ for the reflected part. There are now two outputs from the abutting cylinders, a new forward travelling wave, $(1+m)\boldsymbol{f}-m\boldsymbol{b}$ (the sum of the propagated forward wave and the reflected backward wave) and a new backward travelling wave, $(1-m)\boldsymbol{b}+m\boldsymbol{f}$ (the sum of the propagated backward wave and the reflected forward wave). The new forward and backward travelling waves enter the next cylinders to the right and left, respectively, and are once again propagated and reflected depending on the value of the reflection coefficients at the junction with the next cylinders.

If the reflection coefficients are known, then the relative cross-sectional areas of the cylinders can be calculated. The relationship is (Markel & Gray, 1976, p. 70)

$$m[i] = \frac{A[i-1] - A[i]}{A[i-1] + A[i]} \tag{8.11}$$

where $m[i]$ is the reflection coefficient between the abutting cylinders and where i extends from 1 to p, and p is the order the of the model (the tube $A[0]$ represents the tube of infinite area just in front of the lips; $A[p]$ represents the area of the tube which behind the glottis). Rearranging the above equation, the cross-sectional area $A[i-1]$ nearer the lips can be calculated from the adjacent tube's area $A[i]$ (nearer the glottis), and from the reflection coefficient $m[i]$ between them:

$$A[i-1] = \frac{A[i](1+m[i])}{1-m[i].} \tag{8.12}$$

As stated earlier, the reflection coefficients are equal to the PARCOR coefficients (Markel & Gray, 1976, p. 76) and so $m[i]$ can be replaced with $k[i]$ in Equation 8.12, where $k[i]$ is the ith partial-correlation coefficient. Since the PARCOR coefficients are a by-product of the derivation of LPC coefficients using the recursive method described earlier, it follows that linear predictive analysis can be used to calculate the cross-sectional areas of the lossless cylindrical tubes.

It is clear that Equations 8.11 and 8.12 only allow a calculation of *relative* cross-sectional areas: the relationship between the areas is recursive, and a value of $A[p]$ must be assumed in order to start the recursion. A common assumption is to set this cross-sectional area of the cylinder at the glottal end equal to unity, on the assumption that there is little change to the glottal area in speech production compared with other sections of the vocal tract. Under this assumption, the cross-sectional area of tube $P-1$, which is the first tube

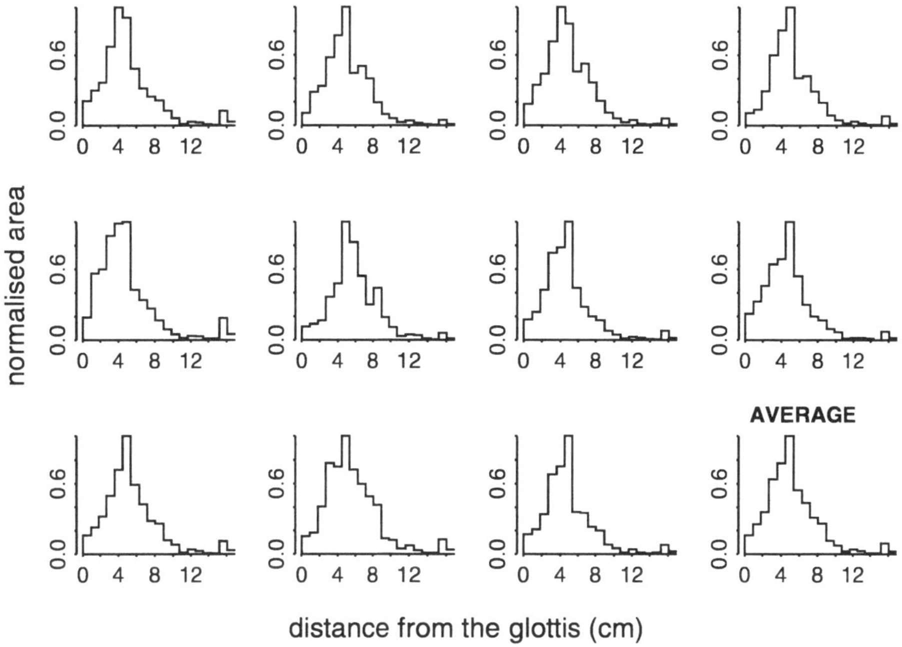

Figure 8.10: Area functions calculated for eleven [i] vowels. The bottom right panel shows the average cross-sectional area for these eleven tokens. Note the estimated constriction is at roughly 11-12 cm which is not inappropriate for a palatal constriction.

beyond the glottis is given by

$$A[p-1] = \frac{1 \times (1 + m[p])}{1 - m[p]}$$

where $m[p]$ is the pth reflection coefficient (assuming an LPC analysis of order p which derives p LPC and p PARCOR/reflection coefficients). Successive areas are calculated recursively from Equation 8.12; the final area to be calculated, $A[0]$, represents the (theoretically infinite) cylindrical area in front of the lips.

Figures 8.10 and 8.11 show cross-sectional areas that were calculated for eleven [i] and eleven [ɔ] vowels taken from isolated productions of Australian monosyllables of the form /CVd/ produced by a male talker. The speech data were sampled at 20000 Hz and the middle 512 points (25.6 ms) were extracted from each vowel. A Hamming-window was applied to each of the central sections, the data were preemphasised, and a 20th-order LPC analysis was applied to the data to extract 20 LPC and 20 PARCOR coefficients for each vowel. The 20 PARCOR coefficients were then used to estimate the cross-sectional areas of the corresponding lossless tube models in the manner described above. All calculated cross-sectional areas were normalised relative to the tube with the

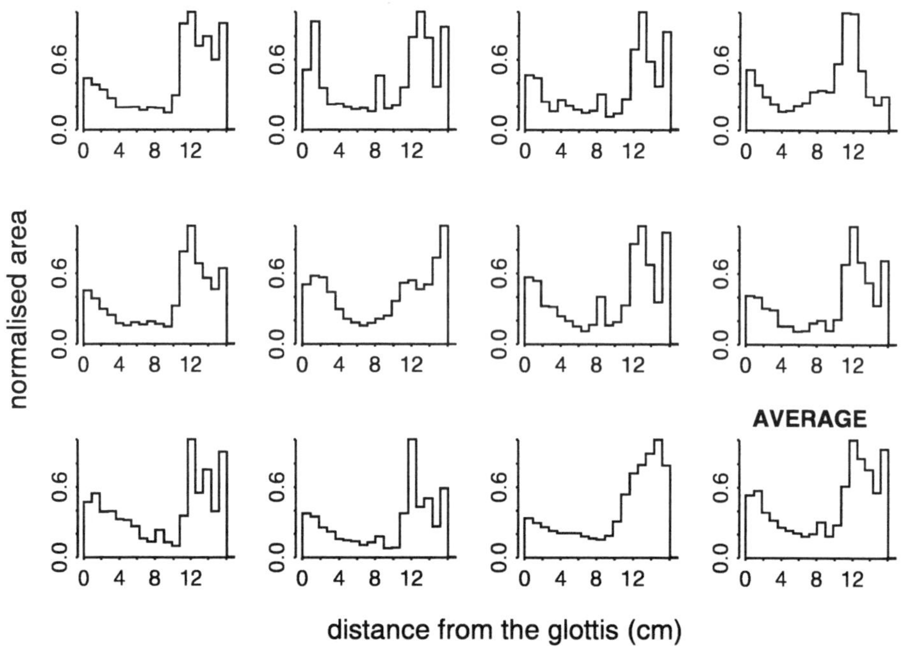

Figure 8.11: Area functions calculated for eleven [ɔ] (as in "hoard") vowels. The bottom right panel shows the average cross-sectional area for these eleven tokens. In this case, the principal constriction is further back (nearer the glottis) than for [i] in Figure 8.10.

largest area which was set equal to unity.

A comparison between Figures 8.10 and 8.11 shows differences that are in accordance with the production characteristics of these two vowels: [i] has a constriction that is further forward in the mouth (nearer the lips) than [ɔ] and appropriately the minimum and maximum constrictions of these two vowels are in different locations. (This is more clearly shown by a comparison of the bottom right panels of Figures 8.10 and 8.11 which show average cross-sectional areas for these two vowels).

8.5.1 Number of cylinders

The number of cross-sectional areas M that is calculated from reflection coefficients is constrained by the sampling frequency of the speech waveform (f_s in Hz), the presumed total length of the vocal tract (L in cm), and the estimated speed of sound in air (c in cm/s) (Wakita, 1973; Markel & Gray, 1976). The relationship is

$$M = \frac{2Lf_s}{c} \tag{8.13}$$

If we assume, as in much of the literature, that the length of an adult male vocal tract is 17 cm and that the speed of sound in air is 34000 cm/s, then M, the number of cylindrical sections, is equal to the sampling frequency in kHz. For example, where $f_s = 20000$ Hz,

$$\begin{aligned} M &= \frac{34 \times 20000}{34000} \\ &= 20. \end{aligned}$$

In all of the displays so far (see Figures 8.10 and 8.11), 20 cross-sectional areas are calculated because the sampling frequency of the speech signal was 20 kHz (and L and c were presumed to have the above values).

8.5.2 Limitations of the lossless tube model

As discussed in Sondhi (1979), it must be emphasised that the areas that are recovered from the equivalent lossless tube model are at best a gross approximation to the actual cross-sectional area of the vocal tract and at worst provide misleading and inaccurate information. This is for several reasons, including the following. First, as stated earlier, there are many different kinds of losses in the vocal tract due, for example, to heat conduction through the vocal tract walls, which influence both the formant centre frequencies and in particular the bandwidths, but which cannot be taken into account in deriving the reflection coefficients from a lossless tube model. Second, the lossless model assumes plane wave propagation, which is generally valid only for frequencies below roughly 4000 Hz; consequently, area functions cannot be accurately recovered for many consonants that have significant acoustic information above this frequency. Third, since the model does not explicitly model anti-resonances, it is not possible to produce an accurate representation of the shape of the vocal tract for nasal or most lateral consonants. Fourth, the coefficients that are estimated using linear predictive coding model not only the vocal tract filter, but also the glottal flow and lip radiation effects. For this reason, factors that have a negligible perceptual effect, such as preemphasising a signal to boost the spectrum by +6dB/octave, can have a major influence on the calculated area-functions (see, e.g., Wakita, 1972, 1973; Markel & Gray, 1976; Sondhi, 1979; Wakita, 1979).

8.5.3 Calculating reflection coefficients

In the discussion so far, it has been assumed that the PARCOR coefficients (and therefore the reflection coefficients) have to be calculated from LPC coefficients. It is, however, possible to calculate the reflection coefficients directly from the speech waveform using a computational model that encodes the propagation and reflection of air pressure variations in lossless cylindrical tubes discussed earlier (Itakura & Saito, 1970, 1973b). The corresponding computational model is often known as a *lattice model* because the forward and backward travelling

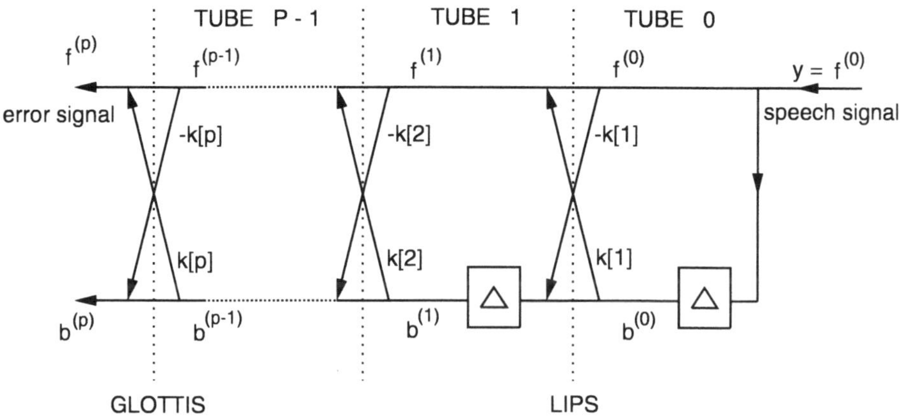

Figure 8.12: An analysis lattice model that can be used to calculate the PARCOR coefficients for a speech waveform, $\boldsymbol{y}$. The box marked Δ denotes that the waveform is delayed by one sample. $\boldsymbol{f^{(p)}}$ $\boldsymbol{b^{(p)}}$ are the forward and backward waveforms in tube P, respectively.

waves cross-over and influence each other at various stages in the calculation (Figure 8.12). We will consider first the *analysis* model, in which the object is to calculate the reflection coefficients directly from a speech waveform. The reverse of this model can be used for speech synthesis by combining a glottal waveform with an existing set of reflection coefficients: this is discussed in the next section. The input to the analysis model is the waveform $\boldsymbol{y}$ at the lip-end of the system. The object is to calculate one reflection coefficient at each cylindrical junction recursively. As the model shows (Figure 8.12), there are two waveforms, a *forward prediction waveform*, and a *backward prediction waveform*, which are propagated through the lattice from the lips to the glottis.

There are two principal stages in the analysis model: calculating the PARCOR coefficient at any junction, and then using that calculated PARCOR coefficient to determine the forward and backward prediction waveforms that leave that cylinder and enter the next cylinder.

The PARCOR coefficient for any cylindrical junction, which also represents a correlation coefficient between the forward and backward prediction waveforms (Markel & Gray, 1976), is given by

$$k[i] = \frac{\sum_{n=0}^{N-1} f^{(i-1)}[n]b^{(i-1)}[n-1]}{\sqrt{\sum_{n=0}^{N-1} f^{(i-1)}[n]^2 \sum_{n=0}^{N-1} b^{(i-1)}[n]^2}} \tag{8.14}$$

where $1 \leq i \leq p$, $i = 1$ corresponds to the lip end of the model, p is the order of the model and $\boldsymbol{f^{(i)}}$ and $\boldsymbol{b^{(i)}}$ are the forward and backward prediction waveforms in tube i. Notice that in calculating any reflection coefficient from Equation 8.14, the backward prediction waveform must first be delayed by one sample.

Once $k[i]$ has been calculated at any junction, the forward and backward waveforms that are input into the next cylinder (towards the glottis) are given by

$$f^{(i)}[n] = f^{(i-1)}[n] - k[i]b^{(i-1)}[n-1] \quad (8.15)$$
$$b^{(i)}[n] = b^{(i-1)}[n-1] - k[i]f^{(i-1)}[n] \quad (8.16)$$

and these can then be fed back into Equation 8.14 to determine the next reflection coefficient and so on. Assuming $\boldsymbol{y}$ is the signal to be analysed, the process is started as follows:

$$f^{(0)}[n] = y[n]$$
$$b^{(0)}[n] = y[n-1].$$

The forward prediction waveform at the glottal end, $\boldsymbol{f}^{(p)}$, is the error signal and is approximately equal to the error signal derived from linear-predictive analysis in Section 8.3 (the backward prediction waveform at the glottis, $\boldsymbol{b}^{(p)}$, does not have a useful interpretation). Therefore, the lattice analysis model decomposes a waveform $\boldsymbol{y}$ into a set of reflection coefficients and a source signal, where once again the source signal is modelled either as an impulse train (voiced speech) or as white noise (voiceless speech).

It is also possible to obtain the LPC-coefficients from the reflection coefficients using the same recursive relationship discussed in Section 8.3; the LPC coefficients can then be used to estimate the smoothed spectrum and the centre frequencies and bandwidths of the formant frequencies.

8.6 Speech synthesis from LPC-parameters

We have so far discussed in this chapter procedures for estimating the vocal tract shape and formant frequencies from a natural speech signal. In this final section, we will briefly consider the reverse of this process in which the object is to synthesise a speech signal starting either with a stylised area function for a vowel, or with a set of reflection coefficients that have been obtained by analysing the signal following the procedures of the preceding sections. Figure 8.13 provides an overview of relationship between the various components in both the analysis and synthesis of speech using LPC-based parameters. The core of the model includes reflection coefficients and LPC coefficients that can be derived from each other. An area function can be computed from the reflection coefficients and *vice-versa*; and formant frequencies can be estimated from LPC coefficients.

The left side of the model (*analysis*) shows the two options discussed in this chapter for analysing a natural speech signal into a set of coefficients and an error signal. With regard to speech synthesis, a source signal can be combined either with reflection coefficients in a lattice-synthesis filter or with LPC coefficients in a recursive filter. The model also shows that the coefficients could be obtained in three different ways: first by specifying an area function; second by decomposing

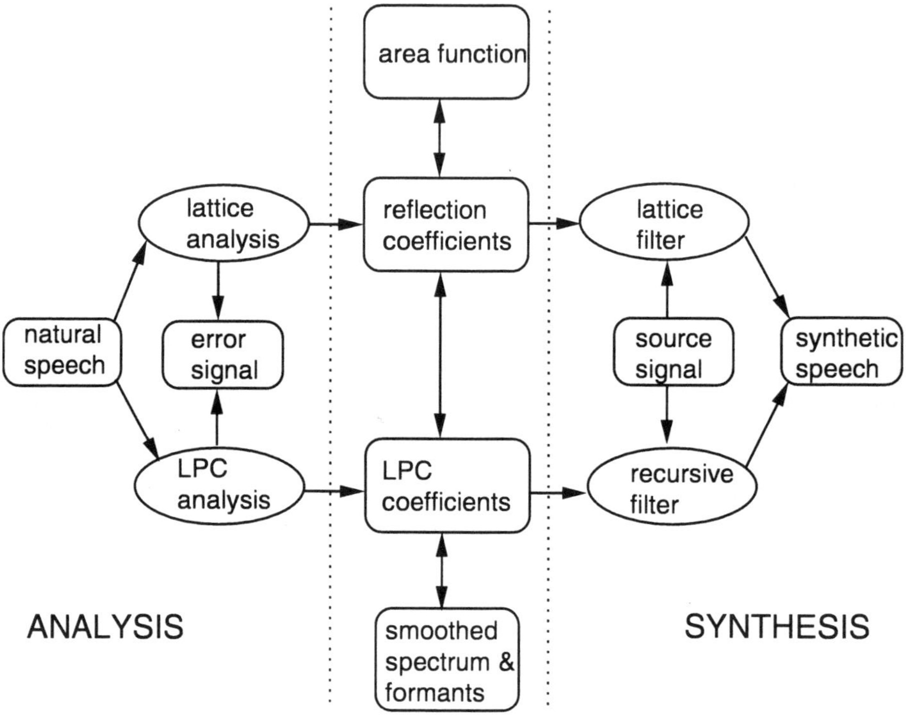

Figure 8.13: The relationship between analysis and synthesis stages in an LPC-based model. The left part of the model represents the decomposition of a natural speech signal into an error signal and LPC/reflection coefficients from LPC/lattice analyses. An area function can be derived from reflection coefficients (and *vice-versa*) and a smoothed spectrum and formant frequencies from LPC coefficients. In speech synthesis, synthetic speech can be obtained either by combining a source and reflection coefficients in a lattice model, or by combining a source and LPC coefficients in a recursive (IIR) filter.

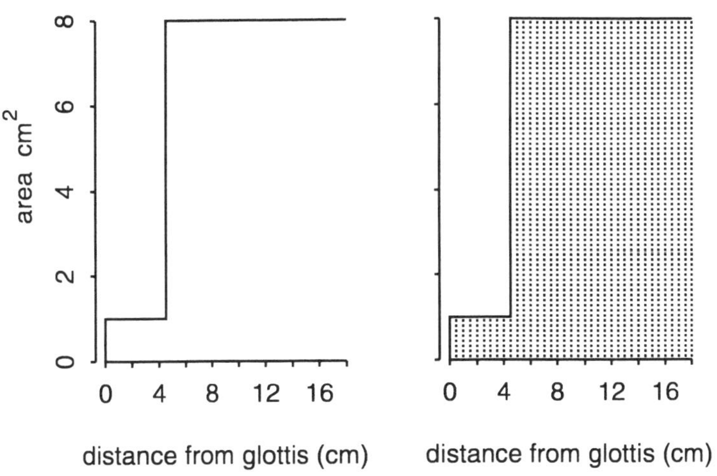

Figure 8.14: A twin-tube model for [æ] and its equivalent representation at short cylinders of equal length (*right*).

natural speech into a set of coefficients and an error signal (the left side of the model) — this amounts to performing analysis-resynthesis; and third, by converting a set of formant frequencies and bandwidths into coefficients of a recursive filter. Since this third possibility is essentially formant synthesis of the kind described in the preceding Chapter, we will discuss further only the first two possibilities below.

8.6.1 Synthesis from an area function

An all-pole lossless tube model can be converted into either an equivalent spectral representation or, if combined with a source, into synthetic speech via reflection coefficients and LPC coefficients. Consider as an example a twin-tube representation of [æ] (Figure 8.14) which, following Fant (1960), could be modelled with cross-sectional areas 1 cm^2 and 8 cm^2 and lengths 4.5 cm and 13.5 cm (thus total length 18 cm). The reflection coefficients can be derived from the area function using the relationship in Equation 8.11, but two additional factors must be considered: the number of cylinders in the model and the boundary conditions at the lips and glottis.

With regard to the number of cylinders, it is first necessary to convert the twin-tube model into an equivalent number of cylinders of equal length (Figure 8.14). The motivation for a larger number of equal length cylinders is to provide a sufficient frequency range for representing the spectra of vowels. As discussed earlier, the number of sections, total vocal tract length and sampling frequency are constrained by the relationship given in Equation 8.13. Therefore, for a Nyquist frequency of 4000 Hz (which is a suitable frequency range for modelling vowels), the sampling frequency must be at least 8000 Hz which implies a minimum of 8 cylindrical sections.

Once the twin-tube model has been represented as a number of cylinders of equal length, the reflection coefficients are calculated using Equation 8.11. Notice that, since a reflection coefficient is obtained from the area ratios between two abutting cylinders, the absolute values of the cylinders' cross-sectional areas are unimportant (the same reflection coefficient is obtained from two abutting tubes of areas (1) 1 cm^2, 8 cm^2 and (2) 10 cm^2, 80 cm^2). From Equation 8.11, it is also evident that two abutting tubes of the *same* cross-sectional area have a reflection coefficient of zero, thus correctly implying that the wave is not reflected at the junction between such cylinders. The reflection coefficients for the multiple-tube representation of the twin-tube model are therefore all zero, except at the single junction where the area changes from 8 to 1.

The first reflection coefficient $m[1]$ defines the reflection of the wave at the open end (lip-end) of the model which, in Figure 8.12, is denoted by tube 0. Under the assumption that this tube is of infinite area (Wakita, 1973), the corresponding reflection coefficient, $m[1]$, evaluates to 1, since the effect of $A[1]$ in the numerator and denominator of Equation 8.11 is negligible; it can be seen from the same equation that when tube 0 is defined to have an area less than infinity, $m[1]$ is is correspondingly less than 1.

The last reflection coefficient, $m[p]$, defines the reflection of the wave at the glottal end of the system. In order to simulate a closed tube at the glottal end with no losses, the area of tube P (Figure 8.12) is zero: this results in a reflection coefficient $m[p] = 1$. Therefore the reflection coefficients for a lossless twin-tube model open at the lip end (to a tube of infinite area) and closed at the other has reflection coefficients $[1, 0, 0, 0, \ldots, m, 0, 0, 0, \ldots, 1]$ where the number of zero values depends on the number of equal length sections used to represent the twin-tube model, and m is the reflection coefficient at the junction where the areas change.

It can be shown that the last reflection coefficient $m[p]$ controls the *losses* and therefore the *bandwidths* of the resulting spectrum of the system (Wakita, 1973). As stated earlier, $m[p] = 1$ implies a completely lossless model (when $m[p] = 1$, the bandwidths are 0 Hz): as $m[p]$ tends to zero, losses are introduced into the system and the bandwidths are broadened. This is shown for a straight-sided single tube of length 17 cm (modelled as 17 tubes of equal area and each of length 1 cm) in Figure 8.15 in which, under one condition, $m[p] = 1$ (the lossless case) and under another, $m[p] = 0.4$ (implying losses): for the first condition, the bandwidths are as close to zero as the digital spectrum will allow, while for the second condition, the bandwidths are considerably broader.

We have jumped a step in passing from reflection coefficients to the spectrum: the former must first be converted into LPC coefficients from a smoothed spectrum which can be derived using the procedures discussed earlier. The conversion from reflection coefficients to LPC coefficients is easily accomplished using the recursive relationship discussed in Section 8.3. Additionally, once the reflection coefficients have been obtained, then the sound corresponding to the area function could be synthesised by combining it with a source, as discussed in the next section.

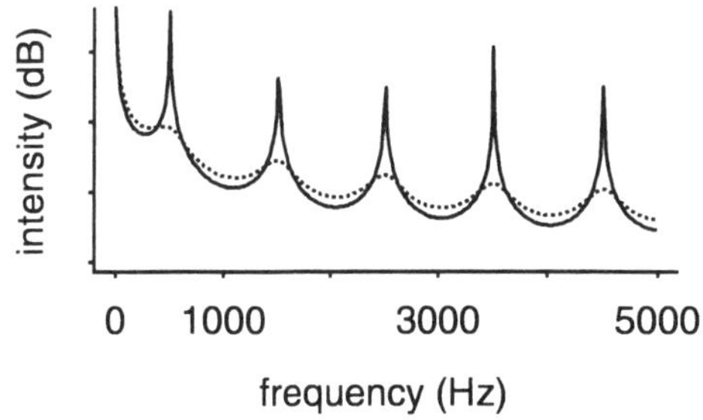

Figure 8.15: The spectrum for a straight-sided tube of length 17 cm under lossless conditions (*solid line*) in which the last reflection coefficient $m[p] = 1$ and in which losses are simulated by changing $m[p]$ to 0.4 (*dotted line*).

8.6.2 Synthesis

Waveforms can be synthesised by combining reflection coefficients in an appropriate filter with a chosen source. The coefficients themselves might have been derived by first analysing a natural waveform (this would then be LPC analysis-resynthesis); or they could be obtained from area functions in the manner described in the preceding section.

Resynthesis from LPC coefficients follows exactly the principles discussed in Chapter 7 for passing a source through a recursive filter. This is necessarily so because LPC coefficients are simply the estimated weights on the delayed output values. Therefore, to synthesise an output signal $\boldsymbol{y}$ from LPC coefficients, our chosen source, $\boldsymbol{x}$, is passed through the filter:

$$y[n] - \alpha[1]y[n-1] - \alpha[2]y[n-2] - \ldots = x[n].$$

Notice that if $\boldsymbol{x}$ is set equal to $\boldsymbol{\epsilon}$ (the error signal), then the output signal, which was decomposed into the LPC coefficients and the error signal, is exactly reconstructed. The alternative to the above is to obtain the impulse response of the filter (using the methods discussed in Chapter 6) and then pass the source through the non-recursive filter,

$$\boldsymbol{y} = \boldsymbol{h} * \boldsymbol{x},$$

where $\boldsymbol{h}$ is the impulse response obtained from the LPC coefficients.

A problem with the above method of synthesis from LPC-coefficients is that there is no guarantee that the filter will be *stable*: the output might grow, rather than decay exponentially, producing undesirable noise during synthesis. On the other hand, it can be shown that reflection coefficients are always guaranteed to lie within the range ± 1, which in turn ensures stability of the filter (Wakita, 1973; Markel & Gray, 1976). Therefore, synthesis using reflection coefficients is the preferred method of generating an output signal.

The synthesis-lattice model, which is based directly on the propagation of airflow from the glottal to the lip end discussed earlier, has one input, a source,

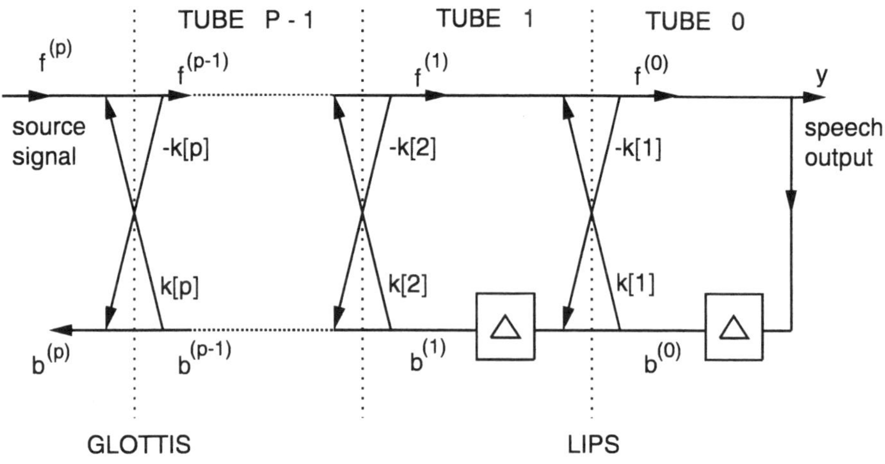

Figure 8.16: A lattice-synthesis structure. The component Δ denotes a delay by one sample.

and two outputs: a forward signal to the lip-end of the model and a backward signal from the lips to the glottis (Figure 8.16). Compatibly with the preceding discussion, the reflection coefficient in the path towards the forward signal is opposite in sign to the one in the cross-over path to the backward signal. As well as showing the direction in which the signals travel, there are also delay elements to show that the previous, and not the current, value of the waveform must be used in any subsequent calculations. Since this is an all-pole filter, in which the current sample is determined from a combination of the source and weighted preceding samples, the output must be calculated *iteratively* from previous output values.

LPC analysis-resynthesis from reflection coefficients is an important technique in speech coding in which the object is to compress the speech signal as far as possible but with a minimum loss of intelligibility. An LPC-analysis of a signal can be used to estimate the fundamental frequency, the likelihood that the signal is voiced, and also the filter coefficients of the vocal tract. Consequently, a signal of 512 points could be compressed to around 10–20 values (e.g., 10–15 for the coefficients, 1 for the fundamental frequency, 1 for the amplitude of the signal, 1 for the voicing decision) and then reconstructed synthetically using the techniques discussed in this section.

Notes

1. *Waves* is a trademark of Entropic Inc.

9 Classification of Speech Data

We have seen in previous chapters that a given acoustic parameter, such as zero-crossing-rate, the first formant frequency, or root-mean-square energy, can be used to differentiate between classes of speech sounds. In many cases, it takes a combination of parameters to separate different classes of speech sounds. For example, vowels of different phonetic quality require at the very least not just the first formant frequency but a combination of the first two formants for their separation; similarly, classes of fricatives such as [s], [ʃ], and [θ] can be quite effectively separated in a plane of two parameters based on the first two cepstral coefficients, as we saw in Chapter 6.

As a first approximation, we can get a good idea of how well a parameter set differentiates classes of speech sounds by looking at the speech data on the chosen parameters, but ultimately, this form of analysis is limited for two main reasons. First, we can usually only examine the distribution of up to two (as for the ellipses of the fricatives in Chapter 6), and at most three parameters at a time. However, since there is no reason to presuppose that the distinction between different classes of speech sounds must be based on no more than three parameters, we must be prepared to consider an alternative kind of analysis when we want to assess how effectively sounds are separated in four, or more than four, dimensions. Another reason is that a visual examination of the data is never very precise. Consider as an example, an analysis to assess whether vowels are more effectively separated in the plane of F2 – F1 vs. F1 in mels compared with F2 vs. F1 in Hertz. We can certainly make ellipse plots and compare by eye the extent to which the ellipses in each plot overlap; but if we want an exact measure in the form of "5% more [ʊ] vowels are confused with [u]" in the Hertz compared with the mel plot, then we have to consider statistical models that can be fitted to the data to provide this form of analysis.

In this chapter, we will consider the kinds of statistical models that are commonly used for classifying speech data. A *parameter* in this context is a numerical measure for an utterance that has been calculated from the sampled speech data for a token. Speech tokens can be defined by multiple parameters that include measures of, for example, the zero-crossing-rate, fundamental frequency, formant frequencies, and so on. In order to characterise a speech sound, or *token*, a number of these parameters are usually collected together into a *parameter vector*. Typical examples of a parameter vector might be the first three formant frequencies at the acoustic vowel target or a combination of the RMS and zero-crossing density at the midpoint of each fricative. In the both cases, each sound can be considered to be represented by a point, or parameter-vector, in an n-dimensional space (three dimensions in the first example, two dimensions in the second). Parameter vectors need not be derived from a single point in time; towards the end of the chapter we will discuss some methods for encoding the time-varying nature of speech for classification purposes.

Each token also belongs to a particular *label type*: for example, if we are comparing 20 [i], 20 [ɛ], 20 [æ] and 20 [a] tokens in the F1/F2/F3 space defined above, then there are four label types each with 20 tokens. Furthermore, since formants are presumed to provide an effective separation between vowels of different quality, a three-dimensional plot of this data should show four fairly distinct, nonoverlapping clouds of points, one for each label type. We will be concerned then with quantifying the extent to which the "clouds" overlap and therefore the extent to which the tokens of different label types are confused with each other in a given space.

9.1 Speech spaces and distance measures

To most people, a ***space*** corresponds to something physical; to specify your position in space you might give a longitude and latitude or say, perhaps, that you are six metres from the door and two metres from the hat stand. The mathematical idea of a space is similar but more general than the physical space we are used to. A good example of a space is a two-dimensional graph, for example of F1 vs. F2 for some vowel data (Figure 9.1); here we would talk about the position of each vowel token in the F1/F2 space, meaning its position on the graph or its F1/F2 coordinates. In fact, a two-dimensional space is often called a ***plane***, as in the F1/F2 plane. So, for example, the graph shows an [ɪ] vowel at F1=475 Hz, F2=2590 Hz, and an [ɛ] vowel at F1=510 Hz, F2=2447 Hz.

A two-dimensional space is defined by a pair of parameters, such as the first two formant frequencies or duration and pitch frequency. Similarly, a three-dimensional space is defined by three parameters: the first three formants, duration, pitch and rms energy, etc. Three-dimensional spaces can still be pictured: they correspond to our normal surroundings with dimensions of length, breadth, and height. Since spaces of four or more dimensions have no physical analogy, they cannot be visualised but can still be defined and manipulated mathematically. One useful manipulation is to calculate the distance between two points in a space.

Normally, to find the distance between two points in the real world we might use a tape measure. However, if the points are too far apart, we can still calculate the distance between them if we know their coordinates in space: for example, if we know their latitude and longitude. In this case, the distance between points $\boldsymbol{p}$ and $\boldsymbol{q}$ can be calculated using Pythagoras's theorem, which relates the lengths of the sides of a right-angle triangle (Figure 9.2) as follows:

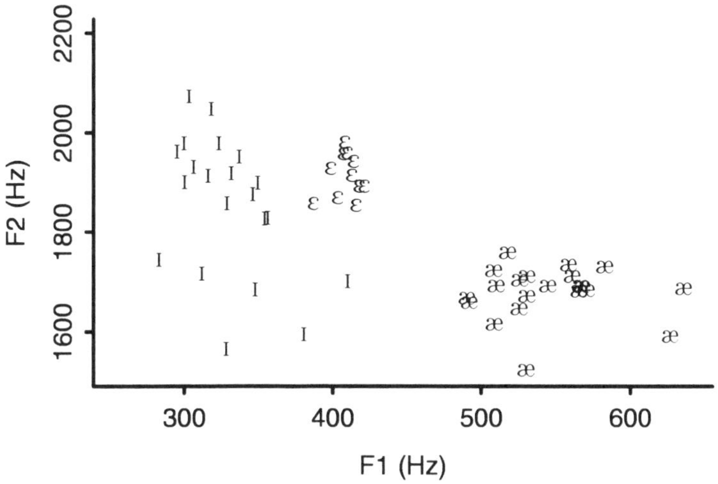

Figure 9.1: The location of a number of [æ], [ɪ], and [ɛ] vowels in the space defined by the first two formant frequencies.

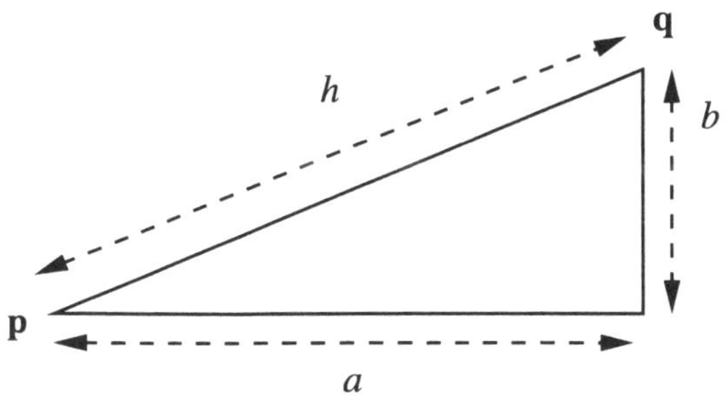

Figure 9.2: The distance between two points p and q can be calculated using Pythagoras's theorem, which relates the sides of a right-angled triangle.

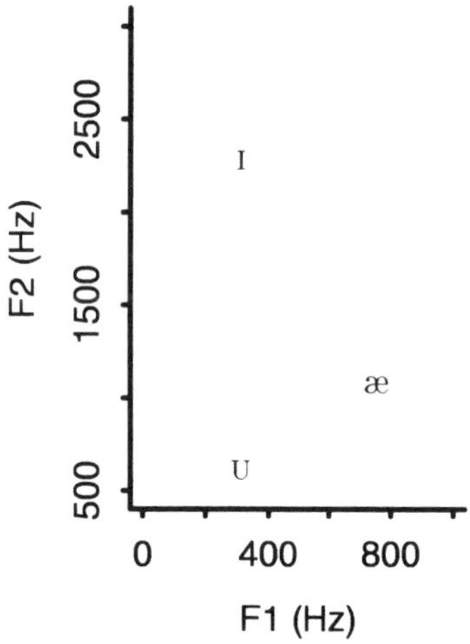

Figure 9.3: Three vowels on the F1/F2 plane.

$$h = \sqrt{a^2 + b^2}, \tag{9.1}$$

where h is the length of the longest side (or hypotenuse) and a and b are the lengths of the two shorter sides. If we know the latitude and longitude of each point, we can calculate the east-west (a) and north-south (b) distances between them by simple subtraction. If we denote the points $\boldsymbol{p}$ and $\boldsymbol{q}$ by a pair of coordinates (p_x, p_y) and (q_x, q_y), then the lengths of the sides of the triangle are

$$a = p_x - q_x \tag{9.2}$$
$$b = p_y - q_y \tag{9.3}$$

and the distance between the two points is (by Equation 9.1):

$$h = \sqrt{(p_x - q_x)^2 + (p_y - q_y)^2} \tag{9.4}$$

As an example, consider the plot of F1 vs. F2 in Figure 9.3: it shows a typical [ɪ] vowel token at coordinates (300, 2300), an [æ] vowel token at (750, 1100), and an [ʊ] vowel token at (310, 630). We can find the distance between the [ɪ] and the [æ] on this plot using Equation 9.4:

$$\begin{aligned} Distance &= \sqrt{(300 - 750)^2 + (2300 - 1100)^2} \\ &= 1281.6. \end{aligned}$$

Similarly, the distances between [ɪ] and [ʊ] and [ʊ] and [æ] can be calculated as 1670.0 and 643.8, respectively. These distances on the F1/F2 plot can be taken as a very crude indication of the *similarity* of these vowels: we might conclude that this [ʊ] token is more like the [æ] than the [ɪ], since it is closer to it on the plot and thus has a smaller distance measure.

This method of calculating distances can be generalised to any number of dimensions by replacing Equation 9.4 with the following:

$$h = \sqrt{(p[1] - q[1])^2 + (p[2] - q[2])^2 + \ldots + (p[N] - q[N])^2}$$

or more succinctly,

$$h = \sqrt{\sum_{n=1}^{N} (p[n] - q[n])^2}. \tag{9.5}$$

Here $\boldsymbol{p}$ and $\boldsymbol{q}$ are parameter vectors with N elements each (and $p[n]$ is the nth element of $\boldsymbol{p}$). The distance h is called the *Euclidean distance* between parameter vectors $\boldsymbol{p}$ and $\boldsymbol{q}$.

Using this formula we could calculate the distance between the same vowel tokens displayed in Figure 9.3 but this time in the "space" defined by the first *three* formants. Table 9.1 gives the formant values for each vowel; the distance between the [ɪ] token and the [æ] token is

$$\begin{aligned} h &= \sqrt{(300 - 750)^2 + (2300 - 1100)^2 + (3100 - 2470)^2} \\ h &= 1428.0. \end{aligned}$$

Table 9.2 shows a *distance matrix* that summarises the distances between each pair of vowels. Note that since the distance from [ɪ] to [æ] is the same as the distance from [æ] to [ɪ], only one half of the matrix is filled in. Comparing these distances to those calculated earlier from just two parameters, the distances are greater in all cases, but the difference is larger between [ɪ] and the other two vowels since its third formant is so high compared with the other two.

In summary then, a set of n *parameters*, such as the first three formant frequencies, defines an n-dimensional space or n-space. Individual speech tokens are characterised by a set of values for each parameter, called a *parameter vector*, so an [ɪ] vowel might have a corresponding parameter vector [300, 2300, 3100]. A parameter vector defines a point in n-space, and we can calculate the distance between pairs of points. These distances can be taken as a measure of the similarity of the tokens associated with the points.

	F1	F2	F3
[ɪ]	300	2300	3100
[æ]	750	1100	2470
[ʊ]	310	630	2380

Table 9.1: Formant values for three vowels.

9.2 Distributions of speech sounds

As discussed earlier, an effective set of parameters for speech tokens makes tokens of the same type, for example [ɛ] vowels, have similar parameter values while those for different types, [æ] vowels, will have more distinct parameter values. In terms of parameter spaces as discussed in the last section, this means that tokens of the same type should cluster together in space and that the clusters for distinct types should not be close together. We might hope that tokens of the same type would have the same parameter vector, for example that all [ɪ] vowels would have F1 at 300 Hz and F2 at 2300 Hz: clearly this is not the case since there is a good deal of acoustic variation in the parameter vectors of speech tokens of any type. The reasons for this are well known and relate to the inherent variability in speech, caused by, for example, coarticulation, speaker identity, and speaking rate.

Given that such variation exists, the task of assigning an unknown token to one of a number of label types is complicated. There will rarely be an exact match between the parameter vector of the unknown token and that of a known token. A method needs to be developed to measure the similarity between the unknown and all of the tokens belonging to a type: if this is available then the token can be assigned to the type to which it is most similar. The problem then becomes one of measuring similarity between a parameter vector (the unknown token) and a collection of parameter vectors (the members of a type). The key to this problem lies in the theory of *Bayesian probability*, which provides a way of characterising collections of points (tokens) and relating unknown points to these collections.

9.2.1 Bayesian probability

When we make observations of real-world events we are *sampling a population*, for example, the population of all phonetic tokens in a performance of *Twelfth Night*. Statistics and probability theory provide tools for characterising and reasoning about such populations and samples. This section introduces some of the terms and equations used in probability theory that are relevant to the problem of classifying speech tokens.

Given a population of phonetic tokens, there is associated with each phonetic type t_i a *prior probability* $P(t_i)$ that a token of that type will be chosen if a

	[ɪ]	[æ]	[ʊ]
[ɪ]	-	1428.0	1818.6
[æ]	-	-	650.0
[ʊ]	-	-	-

Table 9.2: Distance matrix for three vowels, each cell shows the distance (as calculated from Equation 9.5) between a pair of vowels.

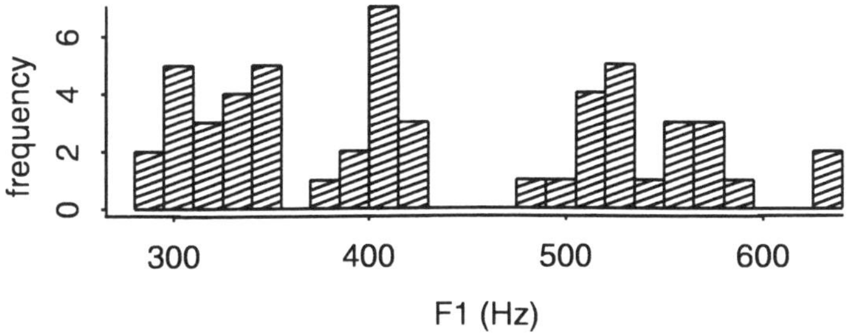

Figure 9.4: The distribution of F1 for a sample of vowel tokens.

random sample is taken from the population. This prior probability reflects the relative proportion of phonetic tokens in the population or in our example, the relative number of occurrences of each phonetic type in *Twelfth Night*. Thus [z] should have a higher prior probability than [ʤ] because it has a much higher frequency in English. Values of $P(t_i)$ could be estimated by counting the relative numbers of tokens of each type in the population and dividing by the size of the population as a whole:

$$P([\mathrm{z}]) = \frac{\text{Number of tokens of [z]}}{\text{Total number of phonetic tokens}}$$

The value of $P([\mathrm{z}])$ then measures the probability of observing *any* [z] token independent of its acoustic properties. Similarly, we can measure the probability of observing some set of acoustic properties independently of the tokens that they are observed in: this is the prior probability of a parameter p_j and is written as $P(p_j)$. This measures how often this parameter value occurs in the population; for example, how often an F1 value of 300 Hz is observed. In most cases the parameter is a continuous variable, and so the prior probability for that parameter is a continuous curve, called a *probability density*, showing the relative occurrence of any parameter value in the population. Such a continuous probability density can be estimated by counting the number of times that values of F1 fall within certain ranges: for example, 0-100 Hz, 100-200 Hz and so on. This is equivalent to plotting a histogram and Figure 9.4 shows such an example for a population of vowel tokens.

A second kind of probability measure is the probability of observing a given parameter value given that a token is known to be of type t_i (for example, the probability of observing a high F1 value, for a token of type [ɪ]). This quantity is called the *class conditional probability* and is denoted as $P(p_j|t_i)$, that is, the probability of observing a particular value of parameter p_j for a token of type t_i (the vertical bar is read as "given" such that $P(p_j|t_i)$ reads 'the probability of p_j given t_i). An example of a conditional probability is the probability of observing a high F1 value for a token of type [ɛ]. This can be estimated by:

$$P(\text{F1=High}|[\varepsilon]) = \frac{\text{Number of tokens of type [}\varepsilon\text{] with F1=High}}{\text{Total number of tokens of type [}\varepsilon\text{]}}$$

Again, the value of p_j is often a continuous quantity, in which case $P(p_j|t_i)$ is a probability density showing the relationship between the conditional probability and the value of the parameter. The prior probability for a parameter value $P(p_j)$ can be calculated by summing the conditional probabilities for the parameter over all types, multiplied by the prior probability for the type:

$$P(p_j) = \sum_k P(p_j|t_k)P(t_k). \tag{9.6}$$

The third kind of probability measure to be introduced is the *posterior probability* that a token belongs to type t_i given a value for parameter p_j, $P(t_i|p_j)$. Note the difference between this quantity and the conditional probability $P(p_j|t_i)$: the latter assumes that we know that a token belongs to type t_i, whereas the posterior probability, which assumes that we know only what the value of parameter p_j is for a token, tells us the probability that this token belongs to type t_i. It is this quantity that is most useful to us in classifying speech tokens. Unfortunately, it is difficult to estimate the posterior probability from observed data; it could be estimated by

$$P([\varepsilon]|F1 = High) = \frac{\text{Number of tokens of type [}\varepsilon\text{] with F1=High}}{\text{Total number of tokens with F1=High}}$$

In a real situation, it is difficult to estimate reliably the total number of tokens in the population with F1=High since any realistic sample will be too small. Fortunately, there is a way of calculating $P(t_i|p_j)$ from the prior probability and the conditional probability known as *Bayes's theorem*:

$$P(t_i|p_j) = \frac{P(p_j|t_i)P(t_i)}{P(p_j)} \tag{9.7}$$

All of the quantities on the right-hand side can be reliably estimated by sampling a population of speech tokens. Bayes's theorem forms the basis of the procedure for assigning unknown tokens to one of a number of classes.

In all of the examples in this section we have talked about probabilities relating to a single parameter, p_j, being the jth parameter from a parameter vector. In many cases we are interested in the probabilities relating to the whole parameter vector for example, the conditional probability of observing a particular set of F1, F2 and F3 values in a token of type [ɛ]. Probability theory tells us that if the parameters are *independent* that is, the values of F1, F2, and F3 are not systematically related to each other, then the conditional probability of the combined parameter vector $P(F1, F2, F3|[\varepsilon])$ is the product of the conditional probabilities for each parameter alone. If the parameters are not independent, then their combination is more complicated; however, it is just as easy to estimate the distribution for a parameter vector as for a single parameter and doing so avoids the question of the independence of the parameters. The preceding discussion, of how probabilities can be combined, applies to parameter vectors as well as single parameters.

9.2.2 The Gaussian distribution

In order to make an accurate estimate of the conditional probability of any token given its type ($P(token|type)$), it would be necessary to have a very large sample of tokens of that type. In reality, we often deal with small samples, from ten to a few hundred, which are unlikely to give accurate estimates of the probability distribution for the type: a histogram of the parameter values gives only a crude approximation to the true distribution which is more likely to be a smooth curve. In these cases, the conditional probability can be worked out only if some assumptions are made about the shape of the distribution. The most often made assumption is that the probability distribution follows the *normal* or *Gaussian distribution*; this assumption is valid if the data is a *random sample* from a *randomly distributed* variable. Fortunately, most variables used in speech research (such as formant frequencies or filter bank amplitudes) are randomly distributed, and if care is taken to select a random sample from the population, the assumption of a normal distribution is reasonable (O'Shaughnessy, 1987).

With only one parameter, a normal distribution is entirely characterised by two quantities: the mean and standard deviation of the sample. The *mean*, μ, is the average value of the parameter and is often called the *expected value* of the parameter for the type. The *variance*, σ^2, is a measure of the spread of the data and is defined as

$$\sigma^2 = \frac{1}{N}\sum_{j=0}^{N}(v_j - \mu)^2, \tag{9.8}$$

where v_j is the value of the parameter for token j, and N is the number of tokens. The square root of the variance, σ, is called the *standard deviation*. Given these quantities, the probability of a parameter value v for a token of type t using the Gaussian probability density is given by

$$P(v|t) = \frac{1}{\sigma\sqrt{2\pi}}e^{-\frac{(v-\mu)^2}{2\sigma^2}} \tag{9.9}$$

As an example we can calculate the probability density for a sample of [ɪ] vowels using vowel duration as the single parameter. Based on a sample of 42 vowels from a database of connected speech produced by a male talker, the mean duration of [ɪ] tokens was 117.66 ms and the standard deviation σ was 48.79 ms. Based on these quantities we can calculate the conditional probability for two values of duration, 60 ms and 117 ms, given the type [ɪ] as follows (the factor $1/\sigma\sqrt{2\pi}$ is constant for all of these calculations and comes out to $1/(48.79 \times \sqrt{2\pi}) = 1/122.3$):

$$\begin{aligned}
P(60|[\text{ɪ}]) &= \frac{1}{122.3} \times e^{-\frac{(60-117.66)^2}{2\times 48.79^2}} \\
&= \frac{e^{-0.698}}{122.3} \\
&= 0.004 \\
P(117|[\text{ɪ}]) &= \frac{1}{122.3} \times e^{-\frac{(117-117.66)^2}{2\times 48.79^2}}
\end{aligned}$$

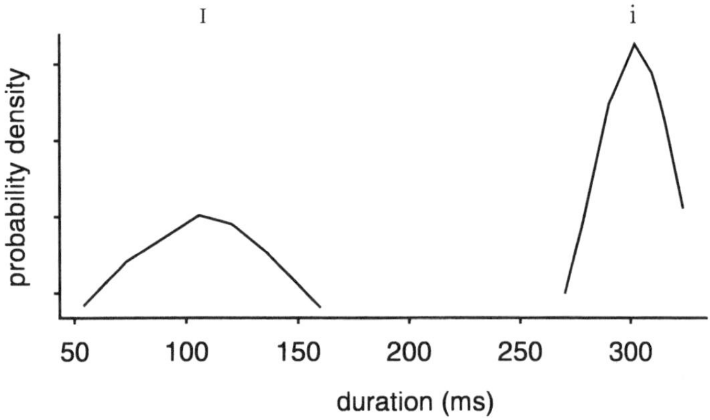

Figure 9.5: The probability density for two vowel types, [ɪ] ("hid" *left*) and [i] ("heed", *right*) on the parameter *duration*.

$$
\begin{aligned}
&= \frac{e^{-9.15\times10^{-5}}}{122.3} \\
&= 0.0082
\end{aligned}
$$

note that, as would be expected, the probability of the parameter value close to the mean (117 ms) is larger than that some distance from the mean (60 ms). This density can be plotted as a continuous curve for a range of values of the parameter, approximating the familiar bell-shaped curve of the Gaussian distribution. Figure 9.5 shows the distribution of two vowels on the parameter of vowel duration.

When there is more than one parameter, the mean becomes a vector of values known as the *centroid* or centre of gravity of the data. The variance becomes the *covariance matrix*, which measures not only the variation in each parameter but also the degree to which parameters vary together — called the *covariance* or *correlation coefficient of the parameters*. For a two-dimensional example, the covariance matrix is

$$
\boldsymbol{W} = \begin{bmatrix} \sigma_1^2 & \sigma_{12} \\ \sigma_{21} & \sigma_2^2 \end{bmatrix}
$$

The diagonal elements of the covariance matrix (σ_i^2) are the variances for each parameter as defined in Equation 9.8. The off diagonal elements are the covariances between parameters and are defined by

$$
\sigma_{ij} = \frac{N\sum_{k=0}^{N} v_k[i]v_k[j] - \sum_{k=0}^{N} v_k[i]\sum_{k=0}^{N} v_k[j]}{N(N-1)} \tag{9.10}
$$

where $v_k[i]$ is the ith element of the parameter vector for the kth token and N is the number of tokens in the data. It can be seen from this equation that the covariance matrix must be symmetric, that is, $\sigma_{ij} = \sigma_{ji}$, since swapping i and j in the equation has no effect on the result.

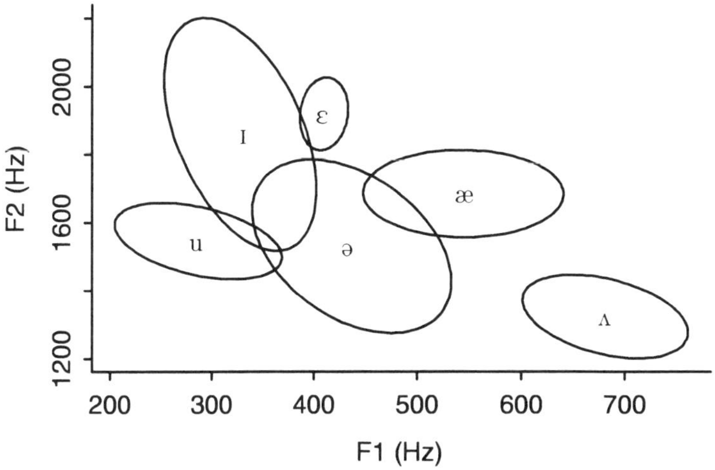

Figure 9.6: The distribution of six vowel types in the F1/F2 plane shown as a centroid and an equal-probability contour for each vowel type. Data taken from a single male talker of Australian English, continuous speech.

The conditional probability of a token with parameter vector $\boldsymbol{v}$ of type t_i using the multidimensional Gaussian distribution is

$$P(\boldsymbol{v}|t_i) = (2\pi)^{-m/2}|\boldsymbol{W}_i|^{-1/2}e^{-(\boldsymbol{v}-\boldsymbol{\mu}_i)'\boldsymbol{W}_i^{-1}(\boldsymbol{v}-\boldsymbol{\mu}_i)/2}, \tag{9.11}$$

where $|\boldsymbol{W}_i|$ is the determinant of the covariance matrix $\boldsymbol{W}_i$ for the type t_i, $\boldsymbol{\mu}_i$ is the vector of means for each parameter for this type, and m is the number of dimensions.[1] Using this equation, the conditional probability of any unknown token with a parameter vector $\boldsymbol{v}$ can be calculated for a type with mean vector $\boldsymbol{\mu}_i$ and covariance matrix $\boldsymbol{W}_i$.

With two or more parameters it is not possible to visualise the probability density directly as it was with one parameter. In the case of two parameters, we can show the class centroids and a contour of equal probability around each centroid. Such a plot is known as an *ellipse plot* because the equi-probability contour is an ellipse. Figure 9.6 shows data from a number of six vowels for the parameters F1 and F2 taken from a database of Australian English continuous speech produced by a male talker. The contour is usually drawn at about 2.45 times the standard deviation from the mean; this value includes about 95% of the data points in the class on average.

For more than two parameters it is difficult to visualise the associated probability distribution. However, techniques exist for reducing the dimensionality

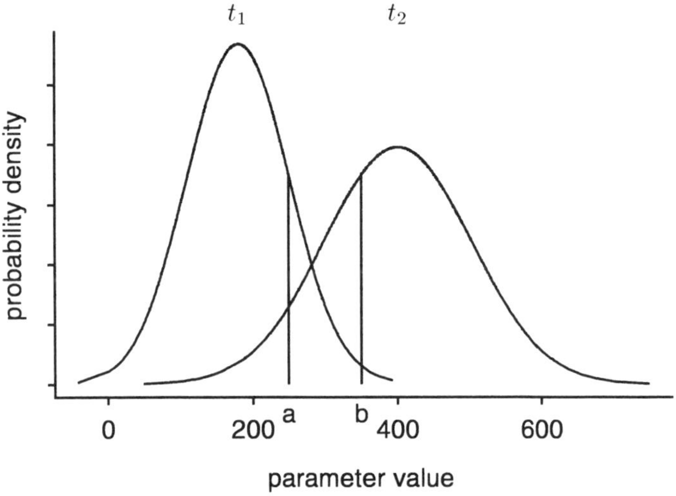

Figure 9.7: The conditional probability density for two types on a single parameter. Point a would be assigned to type t_1 and point b to type t_2.

of a data set that can allow informative plots to be generated. These techniques will be discussed in a later section.

9.3 Discriminant functions and classification

We are now in a position to consider the task of assigning an unknown token to one of a number of types defined by mean vectors and covariance matrices. The basic problem is to find a *decision rule* that compares the parameter vector of the unknown token to the Gaussian model associated with each type and decides which type the token should be assigned to.

As an example, consider a problem where the object is to classify an unknown token as one of two types based on a single parameter. Figure 9.7 shows the conditional probability density for each class along this hypothetical variable. It can be shown that in order to make as few misclassifications as possible on average, an unknown token with parameter value v should be assigned to the class t_i for which $P(t_i|\boldsymbol{v})$ is largest: that is, it should be assigned to the class that the observation is *most likely* to belong to. This is perhaps obvious but it is useful to know that, mathematically, this decision rule can be shown to give rise to the least number of errors on average (Duda & Hart, 1973).

In Section 9.2.1, we saw that the posterior probability was not directly observable but could be calculated from the conditional probability and the prior probability for the type (Equation 9.7). In this example there are two types, t_1 and t_2 and so the posterior probability that an unknown token belongs to type

t_1 given that it has parameter value v is

$$P(t_1|v) = \frac{P(v|t_1)P(t_1)}{P(v)}$$

The decision rule we are using assigns the unknown token to class t_1 if $P(t_1|v) > P(t_2|v)$, which we can expand to

$$\frac{P(v|t_1)P(t_1)}{P(v)} > \frac{P(v|t_2)P(t_2)}{P(v)}$$

The term $P(v)$ is common to both sides of the inequality and so can be cancelled out leaving the rule to assign v to type t_1 if

$$P(v|t_1)P(t_1) > P(v|t_2)P(t_2) \tag{9.12}$$

and to t_2 otherwise. In many speech experiments that use artificial populations such as samples from databases or elicited utterances, the relative frequency of each type is assumed to be equal so as not to bias an experiment towards any one token. In such a case, the prior probabilities ($P(t_i)$) of each type can be assumed to be equal and so can be cancelled from the inequality above.

This decision rule can be visualised by looking at the plot of the conditional probability density for the two types (for example, Figure 9.7): the rule corresponds to choosing the type whose probability curve is highest at any given value of the parameter. Thus, in the figure, point a would be assigned to class t_1, while point b would be assigned to class t_2. The point at which the two curves cross corresponds to $P(v|t_1) = P(v|t_2)$ and is known as the *decision boundary*. In one dimension this boundary is a single point, but with more dimensions it becomes a line or surface in space. It is often informative to look at the decision boundaries between classes since they divide up the feature space and can show the relative shapes and sizes of each class in the space.

To simplify the discussion of decision rules, we define the *discriminant function*, $g_i(v)$, for type t_i to be the quantity that measures the goodness of fit of a parameter value v to the type. In the above case this corresponds to the posterior probability for the type, and so we write

$$g_i(v) = P(v|t_i)P(t_i),$$

and the decision rule above reduces to choosing type t_1 if $g_1(v) > g_2(v)$ and type t_2 otherwise. Since only the relative magnitudes of the discriminant functions are important we can perform any operation on the function that preserves these and still guarantee the optimal Bayes's decision procedure. In particular we can use the log of the above function,

$$g_i(v) = \log(P(v|t_i)) + \log(P(t_i)), \tag{9.13}$$

which turns out to be a useful simplification when using the Gaussian curve to model the conditional probability density.

If we model the type conditional probability distributions with Gaussian curves, and assume that the prior probabilities for each type are equal, the discriminant function of Equation 9.13 can be rewritten as

$$
\begin{aligned}
g_i(v) &= \log\left(\frac{1}{\sigma_i\sqrt{2\pi}}e^{-\frac{(v-\mu_i)^2}{2\sigma_i^2}}\right) \\
&= -\log(\sigma_i) - \frac{(v-\mu_i)^2}{2\sigma_i^2} \qquad (9.14)
\end{aligned}
$$

where the constant factor of $-\log(\sqrt{2\pi})$ has been removed from the final form of the function (note that since we are using the log probability, the expression for the discriminant function is much simpler). The decision rule can now use this discriminant function to decide which type an unknown token should be assigned to given the means and variances of the parameter for each type.

As an numerical example, we can calculate the discriminant functions for two types with means $\mu_1 = 10$ and $\mu_2 = 20$ and variances $\sigma_1 = 5$ and $\sigma_2 = 10$ for the parameter value $v = 15$. By Equation 9.14,

$$
\begin{aligned}
g_1(15) &= -\log(5) - \frac{(15-10)^2}{2\times 5^2} \\
&= -2.1 \\
g_2(15) &= -\log(10) - \frac{(15-20)^2}{2\times 10^2} \\
&= -2.4,
\end{aligned}
$$

and so in this case the unknown token is assigned to type t_1 since $g_1(15) > g_2(15)$.

Note that the decision boundary between types t_1 and t_2 in the this case corresponds to

$$g_1(v) = g_2(v).$$

By substituting for $g_i(v)$ from Equation 9.14 we could obtain an expression for the value of the parameter that lies on the decision boundary. The decision rule could then be implemented by checking whether v was above or below this boundary value.

The multivariate case

The decision rule derived above in terms of conditional probability is also valid when the token is characterised by a parameter *vector* rather than a single parameter. As discussed above, the decision boundary in the multivariate case is a line (for two parameters) or a surface (for more than two) that divides the parameter space between the two types. The discriminant function in the multivariate case must of course use the *multivariate* Gaussian of Equation 9.11. The simplified discriminant function of Equation 9.13, assuming all type prior probabilities are equal, is the logarithm of the multivariate Gaussian distribution,

$$g_i(v) = -\log|\boldsymbol{W_i}| - (\boldsymbol{v}-\boldsymbol{\mu_i})'\boldsymbol{W_i}^{-1}(\boldsymbol{v}-\boldsymbol{\mu_i}),$$

where constant terms have been eliminated from the function. The second term in this function is known as the squared *Mahalanobis distance* between the parameter vector $\boldsymbol{v}$ and the class mean $\boldsymbol{\mu}_i$ (Duda & Hart, 1973; Devijver & Kittler, 1986):

$$D_i(v)^2 = (\boldsymbol{v} - \boldsymbol{\mu}_i)'\boldsymbol{W}_i^{-1}(\boldsymbol{v} - \boldsymbol{\mu}_i).$$

More than two types

When there are more than two types to which an unknown token could be assigned, the above decision rule needs to be extended to choose the type that gives the largest value of the discriminant function. Thus a token with parameter vector $\boldsymbol{v}$ will be assigned to type t_i if

$$g_i(\boldsymbol{v}) > g_j(\boldsymbol{v}) \quad \text{for all } j \neq i, \tag{9.15}$$

where $g_i(\boldsymbol{v})$ has the same form as in the two type case above.

The decision boundary in this case becomes very complicated because of the interaction between the probability densities of multiple types. One way of visualising the way that the discriminant function divides up the space (at least for a two parameter example) is to classify all points on a grid in a defined region of the space. The resulting labels can then be plotted giving a picture of the shape of the decision boundaries within the space. Figure 9.8 gives an example.

9.3.1 Variations on the Bayesian rule

The rule of Equation 9.15 guarantees that an unknown token is assigned to the correct type with the minimum number of errors on average: that is, no other rule could make fewer errors as long as the assumption of Gaussian probability distributions holds. This is certainly the optimal decision rule to use; however, because of the computational expense of calculating the Bayesian discriminant function, a number of simplifications are often made in the name of speed of computation.

Mahalanobis distance

If the assumption is made that all covariance matrices $\boldsymbol{W}_i$ are equal — that is, that tokens of all types have the same distribution in the parameter space: for example, if the ellipses on an ellipse plot of the data have the same shape, size, and orientation — then the decision rule can be reduced to just a comparison of the squared Mahalanobis distances between the unknown and the class centroid,

$$g_i(\boldsymbol{v}) = D_i(\boldsymbol{v})^2 = (\boldsymbol{v} - \boldsymbol{\mu}_i)'\boldsymbol{W}^{-1}(\boldsymbol{v} - \boldsymbol{\mu}_i),$$

where $\boldsymbol{W}$ is the common covariance matrix.

Euclidean distance

A further assumption can be made that all parameters are statistically independent (all covariances are zero) and have the same variance σ^2. In this case,

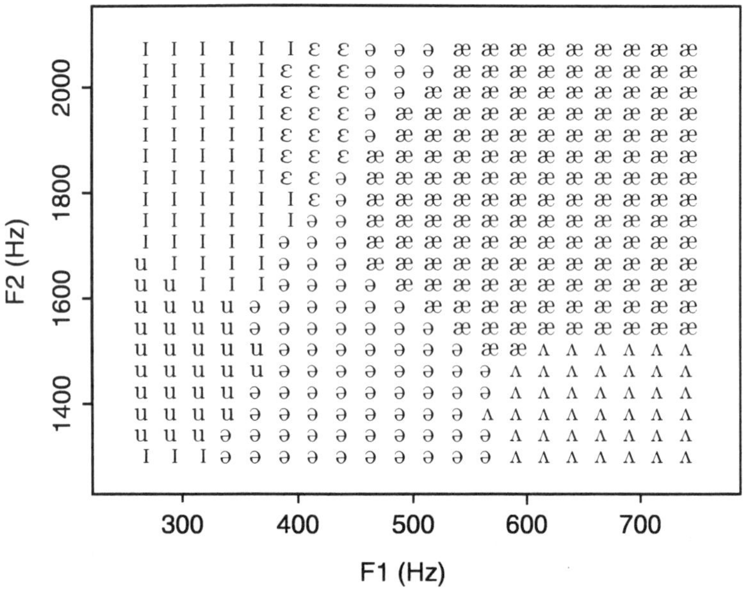

Figure 9.8: A plot showing the decision boundaries in the space defined by the first two formants for nine vowel types. Data taken from a single male talker of Australian English, continuous speech.

the common covariance matrix $\boldsymbol{W}$ is just the identity matrix $\boldsymbol{I}$ multiplied by the constant σ^2 (since all off-diagonal elements are zero and all diagonal elements are equal to σ^2). The decision rule now reduces to finding the type with the smallest *Euclidean distance* (Equation 9.5) to the unknown parameter vector. Note that, although these assumptions are highly unlikely to be valid, the Euclidean distance measure is often used in speech recognition applications because of its simplicity and speed (Sakoe & Chiba, 1977; Rabiner, Levinson, Rosenberg, & Wilpon, 1979).

9.4 Classification experiments

Having covered the technical details of classification, we can now begin to look at applying classification as an experimental method in speech research. Many questions in speech research relate to the efficacy of a set of parameters for classification of a small set of phonetic, prosodic or other linguistic segment types: examples might include the use of filter bank amplitudes in classifying vowel tokens or of spectral slope for classifying fricatives. The question reduces to how effectively the space defined by the given parameter set can be divided up such that tokens are correctly assigned to one of the segment types. This issue is discussed below.

9.4.1 Training and testing

Before any classification can take place using the decision rules based on Gaussian probability densities, the means and covariance matrices for each type must be established. This is often called the *training* phase of the experiment since the Gaussian model is being trained on the labelled speech data to find the means and covariance matrices for each type. Having done this, the model can be used to classify speech tokens; this is called the *testing* phase of the experiment. In an experiment, the types of the tokens being classified are known since the result of the classification must be checked to see if the model is performing well. Two classes of test can be distinguished: in *closed tests*, the same speech tokens are used in both the training and testing phases whereas in *open tests* the two sets of tokens are different. The merits of both types of test will be discussed later.

Training a Gaussian model consists of estimating the means (by averaging the parameter vectors of all tokens) and covariance matrices (from Equation 9.10) for each type. These values then constitute a *model* of the data and are used to classify further tokens in the testing phase. The set of tokens used in the testing phase is known as the *testing set*. In the testing phase, each token in the testing set is classified according to the decision rule being used; the result is a list of labels for each token. This list can then be compared with the known correct labels for the testing set to generate the results of the classification experiment. These results are often summarised in a *confusion matrix* that shows both the number of correct classifications for each type and the types that were confused with each other.

As an example, [s] and [z] tokens from a database of continuous speech from a single male speaker of Australian English will be classified using the parameters *duration* and *RMS energy*. The training set consists of 429 tokens of [s] and 249 token of [z]; Figure 9.9 shows that the distributions of these tokens overlap considerably, and so we should not expect perfect classification scores in this example. The type means are estimated as follows:

Class	Duration	RMS
[s]	104	661.2
[z]	69.9	664.2

and the class covariance matrices are

$$\sigma_{[s]} = \begin{bmatrix} 1589.6869 & 298.4831 \\ 298.4831 & 52715.0959 \end{bmatrix} \qquad \sigma_{[z]} = \begin{bmatrix} 1164.3082 & -1133.8741 \\ -1133.8741 & 81491.8799 \end{bmatrix}$$

For simplicity, the same tokens are used to test the model (a closed test). To test the model, the discriminant function is evaluated for each token in the test set, and the token is given the label corresponding to the type with the largest value for the discriminant function (Equation 9.15). The first test token has the parameter vector [87.27, 732.70]; the discriminant functions for each type

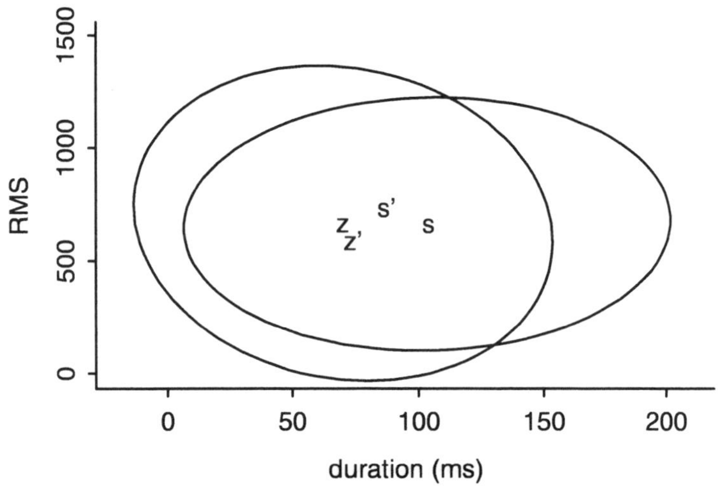

Figure 9.9: The distribution of [s] and [z] on parameters *duration* and *RMS energy* used in the classification example. The centroids of the types are marked with s and z, the points plotted as s' and z' correspond to two tokens used in the example in the text.

evaluate to

$$g_{[s]}([87.27, 732.70]) = -18.53$$
$$g_{[z]}([87.27, 732.70]) = -18.70$$

and hence the decision rule would assign this token to type [s], which is the correct type for this token. A second test token has the parameter vector [74.0, 587.032]; the discriminant functions for this token evaluate to $g_{[s]} = -18.90$ and $g_{[z]} = -18.44$: in this case the decision rule correctly assigns the token to the type [z]. The positions of both of these tokens are shown in Figure 9.9. When this procedure is repeated for each token in the set, the result is a list of labels for the tokens, which we can compare with the known labels to assess the performance of the model. The usual way to summarise this performance is as a confusion matrix:

	[s]	[z]
[s]	249	180
[z]	37	212

which shows that 249 [s] tokens were correctly classified as [s] and 212 [z] tokens were also correctly classified. The off-diagonal elements show that, while only 37 [z] tokens were misclassified as [s], 180 [s] tokens were misclassified as [z]. A total classification figure can also be given by dividing the number of correctly classified tokens by the total number of tokens (0.68 or 68% in this case). Figure 9.9 shows that there is a very large overlap between the two types that may explain the asymmetry in the results: in cases like this it is common for the type with the broader distribution (ellipse) to dominate the parameter space and hence the results.

9.4.2 Open and closed tests

It is important to realise that in this kind of experiment we are modelling a *sample* and not the population as a whole. Hence a *closed test*, where the training data is also used to test a model, will be valid only to the extent that the training sample is representative of the population as a whole. An *open test* uses a second random sample from the population to evaluate the model and hence shows how well the model generalises beyond the set of tokens it was trained on.

To perform an open test on the example data from [z] and [s], the available tokens must be divided between a training set and a testing set, and the model must be retrained on the training set tokens alone. Splitting the tokens roughly into two parts gives a training set with 217 [s] and 122 [z] tokens and a testing set with 212 [s] and 127 [z] tokens. Repeating the closed test we get the following confusion matrix:

	[s]	[z]
[s]	120	97
[z]	19	103

or 65% correct, while performing an open test using the separate testing set of tokens gives the result

	[s]	[z]
[s]	133	79
[z]	20	107

or 70% correct. In this case the results are, unusually, better in the open test than the closed test. This might be because there is, accidentally, more variability in the training data than the test data, but in any case it indicates that the model generalises well outside of the set of tokens it was trained on.

9.4.3 Statistical tests

The results from a closed or open test of any model must be interpreted with respect to the original goals of the experiment. The questions that are typically asked are: does this set of parameters separate (classify) these types well, or which of two competing sets of parameters separate these types better? In both cases the question could be answered by quoting the percentage scores for one or both sets of parameters; however, the experimenter will normally want some measure of the significance of these figures or the differences between them.

Performance relative to chance

The simplest measure for evaluating the performance of a model is to assess the extent to which the classification performance is better than chance: that is, how much better does the model perform than one which just guesses? For a model trained on two label types, as in the above example, chance performance is 50%; this means that by guessing the identity of every token you could expect to get half right (for example, by making the same response each time). More generally, for a model trained on T label types, the chance classification performance is $100/T\%$. The performance can be normalised for the number of types to give the following performance estimate (Klecka, 1980):

$$\tau = \frac{n_c - \sum_{i=1}^{T} p_i n_i}{N - \sum_{i=1}^{T} p_i n_i}$$

where n_c is the number of correctly labelled tokens, T is the number of label types, n_i is the number of tokens of type i, and p_i is the prior probability of type i. If we assume, as we have for most of this chapter, that the types are equally likely, then $p_i = 1/T$ for all i. The τ statistic varies between one for perfect classification performance and zero for chance performance. Negative scores are also possible, and they indicate worse than chance performance — a very bad result indeed!

In the example above, the open test scored 70% correct but the corresponding τ statistic evaluates to

$$\tau = \frac{240 - (212/2 + 127/2)}{339 - (212/2 + 127/2)} = 0.41$$

assuming that the types are equally likely. This reflects the fact that in this case chance is at 50% and so the score of 70% is just less than halfway between chance and the perfect score.

Performance relative to another classifier

The problem of evaluating the performance of one classifier relative to another on the same sets of training and testing tokens is more complicated and has no definitive solution. The usual statistical procedure is to adopt the null hypothesis that the two sets of results are equal and try to reject this hypothesis with some statistical test.

Consider a hypothetical experiment to compare two different sets of parameters for the identification of stop consonants; the first parameter set, voice onset time and burst duration (denoted by VOTB), results in 75 out of 110 correct classifications; the second, the first and second formant locus frequencies (LOC), gives 85 out of 110 (in each case the same training and testing data has been used). The difference between the two sets of results could have arisen for one of two reasons: either LOC is better than VOTB for identifying stop consonants, or the testing sample was such that there happened to be 10 cases that fitted the model better when LOC was used. In the second case we would expect a different outcome if a different test set were chosen. In order to decide between these two cases, we ask how likely it was that the observed difference occurred by chance: if the probability is small (below some threshold, usually 0.05), then we can conclude that the difference is a result of an underlying difference between the two parameter sets. This is done by adopting a *null hypothesis* that the two parameter sets score equally well and finding the probability of the observed pattern of scores if this were the case.

There are a number of standard methods for finding the probability of a given set of scores in an experiment in order to evaluate the null hypothesis. Each incorporates a number of assumptions about the design of the experiment. The main concern in choosing between them is to ensure that their assumptions about the *sources of variance* in the data are justified given the experimental design. We cannot give an exhaustive review of these methods here, but we will give two examples that relate to the kinds of classification experiments described in this chapter; many texts give more detailed coverage of these issues (e.g., Bethea, Duran, & Boullion, 1995; Mosteller & Rourke, 1973; Hays, 1963).

The data in the example cited above could be presented as a 2 × 2 contingency table as follows:

	Correct	Incorrect	Total
VOTB	75	35	110
LOC	85	25	110
Total	160	60	220

A test that is often used with such a table is the χ^2 (or chi-squared) test of independence that tests the null hypothesis that the cells of the table are independent of one another: in this case the null hypothesis is that the number of correct and incorrect scores is independent of the parameter set used. If there is independence, then the observed difference between the parameter sets must have arisen by chance. The procedure used is to calculate a value for χ^2 and then determine the probability of this value in the appropriate χ^2 distribution.[2] In this case, $\chi^2 = 2.29$, which has a probability of 0.32 according to the χ^2 distribution for two degrees of freedom.[3] Since this probability is larger than the threshold value of 0.05 the null hypothesis must be accepted and the conclusion drawn that the difference between the VOTB and LOC is not significant.

One problem with the use of the χ^2 test is that it embodies an assumption that the data in each cell come from a uniform distribution (in fact a binomial distribution). This means in this example that for every stop token, the probability of a correct answer should be the same. In many cases that we are interested in, this assumption might not hold. For example, if the original speech data contains stop tokens produced by a number of speakers, then stops from some speakers might be closer to the class norm than others; this leads to a nonuniformity in the probability distribution, which can make the χ^2 test invalid. In this case, a more appropriate statistical test is one that retains the results from the different speakers and compares the two parameter sets based on their performance across speakers.

An appropriate way to evaluate this data is to use the t-test, which tests whether the means of two distributions of scores are significantly different from one another. If we retabulate the results of our experiment as follows,

Speaker	VOTB	LOC	Total
1	19	18	25
2	13	20	25
3	22	24	30
4	21	23	30
Total	75	85	110

then we now have two samples from populations of speakers tested using VOTB and LOC parameters. We can use a *paired* t-test (since our data involves pairs of trials: the same stop tokens are classified by two different methods) to determine whether the means of these populations differ significantly. Here we adopt the null hypothesis that the two parameter sets give rise to the same results and hence that the *means* of the two result sets are equal. The t-test looks at the mean of the differences between VOTB and LOC for each speaker; if the two sets of results are the same, then this mean should be close to zero. The test finds the probability that the observed mean of differences occurred assuming the null hypothesis; if this value is less than the criterion value, we will reject the null hypothesis and conclude that LOC is better than VOTB. Using the data above we find that the probability of the observed mean of differences is $p = 0.23$ — that is, that the observed difference could occur 23 times out of 100 by chance if the two parameter sets were the same. This probability is larger than the criterion value, and so we conclude that the two parameter sets are equally good.

It is difficult to prescribe a statistical test for every classification experiment that might arise in the course of a research program. The reader should be aware, however, that different statistical tests are appropriate in different situations and that care should be taken to ensure that the experimental design does not violate the assumptions of the statistical test. These tests are important since, as we have seen in the simple example above, an apparent difference between test cases can be due to chance. Theoretical conclusions should not be drawn without some statistical support.

9.5 Classifying signals in time

The technique of Gaussian classification described so far has been used to classify speech tokens based on a parameter vector taken at a *single* time point in the token. The technique does not take into account the time-varying nature of speech; for example, in Chapter 4 it was shown that place of articulation in stop consonants is cued by the shape of the formants transitions into the following vowel (Section 4.2.2). We will briefly mention here some techniques for encoding time-based information in a Gaussian classifier.

Perhaps the simplest way to encode the changing spectral shape of a speech token is to sample it at a number of points in time. So, for example, the changing spectral shape of a diphthongal vowel could be encoded by taking three separate parameter vectors (for example, three sets of 10 cepstral coefficients, at 25%, 50%, and 75% of the duration of the vowel). These three parameter vectors are then concatenated into one larger parameter vector with 30 elements from which a Gaussian model can be trained and tested as described above. In this way the parameter vector contains some information about the time varying characteristics of the speech token (Huang, 1992; Harrington & Cassidy, 1994). The position of the individual parameter vectors within the token can be varied depending on how the cues are presumed to be distributed in the signal; for example, stop consonants might be sampled on and just after the burst, whereas vowels might be sampled at the onset, target, and offset. Since this method tends to result in very large parameter vectors, the data reduction techniques of the next section become very relevant. This approach has been used in some speech recognition systems (Poritz & Richter, 1986; Bahl, Brown, Souze, & Mercer, 1989).

This technique records only snapshots through the speech token, and so it could be said that it does not preserve the dynamic nature of the signal. Another way to encode the time varying nature of the signal is to look at the rate of change of a parameter at some point in the signal. A simple way of doing this is to include the difference between the parameters for two consecutive time points; this then encodes whether the parameter is increasing, decreasing, or relatively flat at the time the parameter vector is taken. Differencing can be applied to any kind of parameter. The time between the differenced parameter vectors would normally be small, in which case the result is a crude estimate of the gradient of the parameter vector; the larger the time difference the more approximate the estimate of gradient. The differenced parameters are usually combined with the original parameters to give a larger parameter vector that encodes both the absolute value and the rate of change of the original parameters.

Differencing is a crude method of finding the gradient of a parameter that varies in time. A more precise method is to fit a curve to the parameter and to then differentiate the curve by analytic methods. The problem with this approach is that it is likely to be difficult to fit a curve to the entire parameter track over the length of the speech token; the usual solution is to take small groups of points (for example, four points at a time) and fit a curve to these points. Rabiner and Juang (1993) describe this technique in more detail.

The entire trajectory of a parameter in time can be modeled by fitting a curve to the parameter and using the coefficients of the curve as input to the classifier rather than the parameter itself. The simplest example would be to model a parameter by its mean value though time by averaging over the length of the token: here the "curve" is a horizontal line. Better models might result from using different families of curves to approximate the trajectory of the parameter in time; for example, Legendre polynomials (Pols & van Son, 1993) or the discrete cosine transform (Zahorian & Jagharghi, 1993). These curves are characterised by a small number of coefficients that can be used as input to a classifier. A brief review of these methods is given by Milner (1996).

It is important to consider the reasons for using techniques such as these in a classification experiment when selecting from the range of methods which are available. If a specific hypothesis is being tested, then the chosen model must clearly fit the hypothesis. If the hypothesis is merely that the variation of the parameter over time might be important, then the experimenter must carefully consider what it is about this variation that might be relevant and select an encoding method that captures just that feature. Some relevant work is discussed in more detail in Chapter 4, Section 4.1.6.

9.6 Data reduction

It was mentioned earlier, in the context of visualising a probability distribution, that methods exist for reducing the dimensionality of a set of data. These techniques are not only useful as an aid to visualisation of the data but can also increase classification performance for a data set by concentrating the useful information in a parameter set down to the first few parameters. Two classes of data reduction will be discussed here: *principal components analysis* and *discriminant analysis*.

9.6.1 Principal components analysis

Principal components analysis, or PCA, is a data-reduction method that finds an alternative set of parameters for a set of tokens such that most of the variability in the data is compressed down to the first few parameters. The transformed dimensions in PCA are called *principal components*, and the new dimensions are guaranteed to be orthogonal (at right angles to each other) and uncorrelated.

The reason that such a transformation can be beneficial to classification performance lies in an analysis of the sources of variability in speech data. If a given sample of speech tokens is parameterised in an appropriate way, the majority of the variability in the parameter vectors should be due to differences in label type. The kind of variability that speech scientists are more normally concerned with is variability *within* a label type, and, for the most part this should be smaller in magnitude than the variability *between* label types. Even within a label type, variability can be seen to come from a number of different sources such as coarticulation effects, speaker differences, and even experimental effects due perhaps to differing recording conditions.

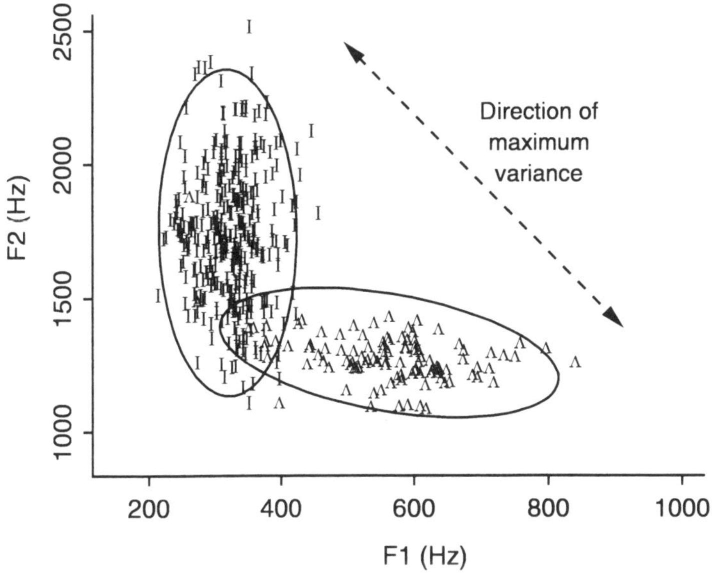

Figure 9.10: Tokens from two vowel types plotted on the F1/F2 plane.

An important property of PCA is that since the transformed dimensions are orthogonal and uncorrelated, any *independent* sources of variability in the data will tend to be resolved into distinct principal components. The largest sources of variability, which are likely to be caused by the differences *between* label types, will make up the lower order principal components, while the smaller sources, such as those due to experimental methods, will make up the higher-order ones. By discarding these higher-order principal components we can reduce the number of parameters in the data set while retaining the salient differences between label types. This is the essence of data reduction using PCA.

As an example, consider the data shown in Figure 9.10, which shows two vowel types on two parameters F1 and F2 (the data was taken from continuous speech of a single male talker of Australian English). The plot shows that the two types are well separated on the two parameters but less so on either parameter alone. We can observe however that most of the variability in the data lies on the diagonal line shown and that if this were the x-axis of the plot, the data would be very well separated along that dimension. What is needed is a *rotation* of the data such that the direction of most variation lies along the x-axis; this new transformed parameter separates the data much more effectively than either F1 or F2 alone and almost as well as both parameters in combination. PCA performs just such a rotation giving a new set of transformed parameters. The effect can be seen in Figure 9.11, which is a result of a PCA transformation of the data of Figure 9.10. In this figure, the first principal component, $PC1$ on the x-axis, contains most of the variability in the data and separates the data

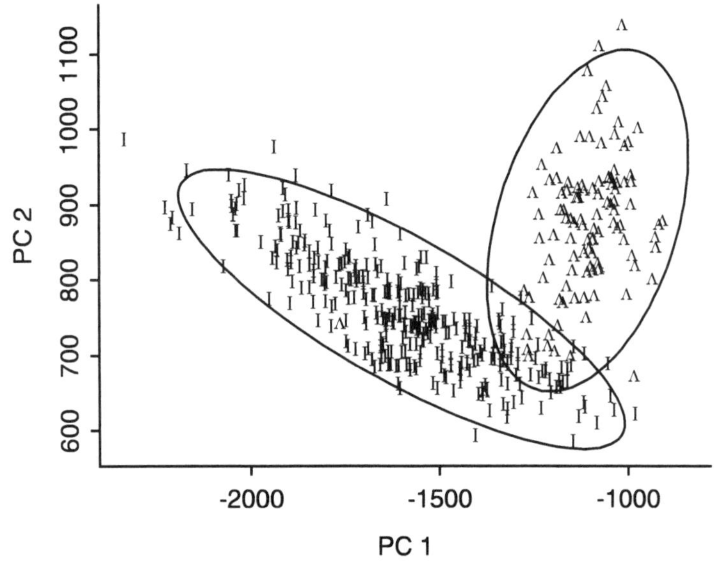

Figure 9.11: The same tokens as shown in Figure 9.10 plotted on the transformed dimensions resulting from a principal components analysis.

quite effectively on its own. The second principal component, $PC2$, contains less variability and adds very little to the separation of the two types in this example.

Although this two parameter example allows easy visualisation of the transformation, PCA is more useful when it is applied to many more parameters. The results are analogous — a rotation of the n-space defined by the n parameters and a concentration of the variability in the lower numbered parameters — but harder to visualise because of the increased dimensionality. One benefit of concentrating the variability into the lower-order principal components is that the first two principal components can be plotted to give a rough picture of the distribution of data and the separation of token types. Figure 9.12 (on page 268) shows the first two principal components of spectral data from a set of vowels where the original parameters were the energies in 20 Bark scaled spectral bands.

The mathematics of PCA

It is not necessary for our purposes to go into detail about the mathematical properties of principal components; there are many sources for such details in the literature of multivariate statistics (Jackson, 1991; Dillon & Goldstein, 1984). This section provides an overview of this material.

The starting point for PCA is the covariance matrix of the data (as introduced in Section 9.2.2), which encodes the variance and covariance of the data, and is used in PCA to find the optimal rotation of the parameter space. PCA finds the *eigenvectors* or *characteristic vectors* and the *eigenvalues* or *characteristic roots* of the data covariance matrix. These have the property that

$$\boldsymbol{U}'\boldsymbol{S}\boldsymbol{U} = \boldsymbol{L},$$

where $\boldsymbol{S}$ is the data covariance matrix, the columns of the square matrix $\boldsymbol{U}$ are the eigenvectors $\boldsymbol{u}_1, \boldsymbol{u}_2 \ldots \boldsymbol{u}_n$, and the diagonal elements of the square matrix $\boldsymbol{L}$ are the eigenvalues $l[1], l[2] \ldots l[n]$. Here n is the number of parameters in the original data. Hence, there is one eigenvector and one eigenvalue for every parameter in the original data, and the eigenvectors also have one element per original parameter.

As an example we will look at the eigenvectors and eigenvalues derived from the data in Figure 9.10. In this case the covariance matrix was

$$\boldsymbol{S} = \begin{bmatrix} 14492.28 & -20760.14 \\ -20760.14 & 14492.28 \end{bmatrix}$$

The eigenvectors of this matrix are the columns of $\boldsymbol{U}$:

$$\boldsymbol{U} = \begin{bmatrix} 0.26 & 0.96 \\ -0.96 & 0.26 \end{bmatrix}$$

and the eigenvalues are

$$\boldsymbol{l} = [302.84448, 94.39923].$$

Having found the eigenvectors and eigenvalues, the principal components are found by matrix multiplying the eigenvector matrix ($\boldsymbol{U}$ above) with the parameter vector for a token. This transformation can be described using a symbolic example. A token with parameter values $[\mathrm{F1}, \mathrm{F2}] = [x, y]$ is transformed as

$$[x, y] \begin{bmatrix} 0.26 & 0.96 \\ -0.96 & 0.26 \end{bmatrix} = [0.26x - 0.96y, 0.96x + 0.26y]$$

that is, the principal components are linear combinations of the two original parameters: 0.26 times the first parameter plus -0.96 times the second gives the first principal component. Thus the eigenvectors (the columns of $\boldsymbol{U}$) give an idea of the "importance" of each of the original parameters in accounting for the variance in the data. In this case, the first principal component uses nearly all of the F2 parameter (the multiplier is negated but it is the magnitude which is important in this case) but less than a third of the F1 parameter. We will say more about this kind of interpretation of the eigenvectors in a later section.

To give a numerical example, a token with $[\mathrm{F1}, \mathrm{F2}] = [495.113, 914.84]$ would be transformed as

$$[495.113, 914.84] \begin{bmatrix} 0.26 & 0.96 \\ -0.96 & 0.26 \end{bmatrix} = [-754.93, 715.64].$$

The eigenvectors, then, define the way in which the original parameters are combined to make the principal components. We can visualise the transformation as a rotation of the space defined by the parameters. It is a well-known fact of geometry that rotation can be achieved by matrix multiplying a set of coordinates with an appropriate rotation matrix; in PCA this matrix is the eigenvector matrix $\boldsymbol{U}$, which defines just the right rotation of the space to make the first transformed dimension lie in the direction of greatest variance for the data.

The eigenvalues correspond to the variances of the principal components of the original data. Since the purpose of performing the principal components analysis is to encode the variance in the data in the first few principal components, the variances, and hence the eigenvalues, of these are larger than those of higher order principal components. The proportion of the total variance accounted for by any principal component can be found by dividing the corresponding eigenvalue by the sum of all the eigenvalues. So, in the example above, the first principal component accounts for 0.76 or 76% of the variance in the data with the remaining 23% accounted for by the second component.

Standardised data

Since the goal of PCA is to encode the variance in the data in a small number of dimensions, problems may arise if one parameter has a much greater variance than any other. This can happen if different units are used for each original parameter: for example, the variance of duration measured in milliseconds would be numerically much greater than if it were measured in seconds. In this case, the parameter with the larger variance dominates the first principal component, but since this variance is not due to the variation across label types, the discrimination power of the first principal component is unlikely to be good. For example, if the original parameters are F1 and duration, and the variance of the duration is 10 times that of F1, then the first principal component will be made up almost exclusively of the duration.

The solution to this problem is to standardise all data prior to performing PCA by subtracting the mean from each dimension and dividing by the standard deviation. This ensures that each parameter has a standard deviation of 1 so that no single parameter will dominate the analysis. Once this standardisation has been performed, the remaining variance in the data should be due to the distribution of types in the parameter space; the principal components analysis will then rotate the space in order to compress the variance into the first few principal components.

The problem with standardisation of data is that the transformed dimensions are not in the same units as the original data. If the original data had been formant frequencies, then the transformed dimensions would also have been in units of frequency. The transformed dimensions of standardised data have no units; this is not usually a problem if the goal is to classify tokens on the transformed dimensions.

Using PCA in classification experiments

In the previous discussion of classification, it was suggested that adding more parameters in an experiment generally improves the classification score obtained, as long as the parameter is useful in distinguishing the types under study. In general this is true, but as the number of parameters increases, the usefulness of each additional parameter is lessened (this is sometimes called the 'curse of dimensionality', Duda & Hart, 1973). The reason is essentially that the accuracy, or goodness of fit, of a Gaussian model depends on having a large number of tokens on which to base the estimates of the mean and covariance matrix; if we increase the number of dimensions (parameters) without increasing the number of tokens, the model is less accurate. Hence classification performance can *fall* when a new parameter is added, not because the parameter has some confounding effect, but because we are no longer able to build an accurate Gaussian model. This is why principal components analysis is useful in classification experiments: it allows the number of dimensions to be reduced while retaining all, or most, of the important variance in the data.

As an example, we will use some data from 10 vowel types on 17 critical (Bark) bands spanning the frequency range 200–5250 Hz and vowel duration in milliseconds, 18 parameters in all. The data are taken from 10 different speakers of Australian English and are split between a training set (four speakers and 648 tokens) and a testing set (six speakers and 780 tokens). The principal components analysis on the standardised data gives the following eigenvalues:

$$\begin{aligned} l = \ & [5.68, 5.34, 1.48, 1.05, 0.87, 0.58, 0.53, 0.39, 0.33, \\ & 0.27, 0.25, 0.21, 0.18, 0.16, 0.13, 0.11, 0.10, 0.06]. \end{aligned}$$

As a point of reference in the following discussion, a classification experiment using the raw data (that is, all 18 untransformed parameters) yields a performance of 74% in an open test with the confusion matrix shown in Table 9.3. The data are shown in Figure 9.12 on the first two principal components. It is interesting to observe here that these two principal components seem to encode the phonetic dimensions of vowel height and backness; this is, in fact, a common observation on PCA transformed data (Pols et al., 1973).

How many principal components?

Given that we have decided to use fewer than the full set of principal components, the question arises as to how many should be kept and how many discarded. Unfortunately there is no clear answer to this question. Statistics can provide a few guidelines that are imprecise at best (Jackson, 1991; Dillon & Goldstein, 1984). One simple method of making this decision is to look at the amount of variance accounted for by successive numbers of principal components (by looking at the eigenvalues, see Section 9.6.1). The aim of this method is to retain enough principal components to account for, say, 95% of the variance, on the assumption that that the remaining 5% is likely to be due to random processes (since 5% is the standard *significance level* used in statistics). By

	æ	ɛ	ɪ	ɒ	ʊ	a	ɛə	i	ɔ	u
æ	99.2	0	0	0.8	0	0	0	0	0	0
ɛ	3	68.2	22.7	0	0	0	3	0	3	0
ɪ	0	7.9	84.9	0	0	0	4.8	2.4	0	0
ɒ	6.1	0	0	84.8	0	6.1	0	0	3	0
ʊ	0	1.5	10.6	6.1	60.6	0	3	0	18.2	0
a	24.2	0	0	0	0	75.8	0	0	0	0
ɛə	9.1	10.6	0	0	0	1.5	57.6	0	7.6	13.6
i	0	3	25.8	0	0	0	15.2	39.4	0	16.7
ɔ	0	0	0	15.2	0	1.5	0	0	83.3	0
u	0	1.5	25.8	0	0	0	16.7	3	0	53

Table 9.3: Confusion matrix for 10 vowels classified on 17 Bark scaled spectral parameters and duration. Scores are percentages.

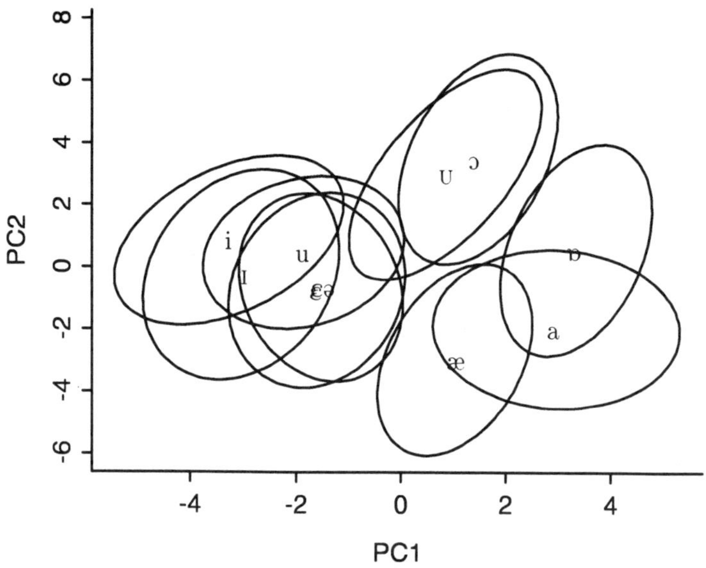

Figure 9.12: A set of vowel tokens plotted on the first two principal components of Bark scaled spectral data. Ellipses show the variance of the tokens of each type. Each ellipse is labelled at the centroid for the type and includes around 95% of the tokens for that type.

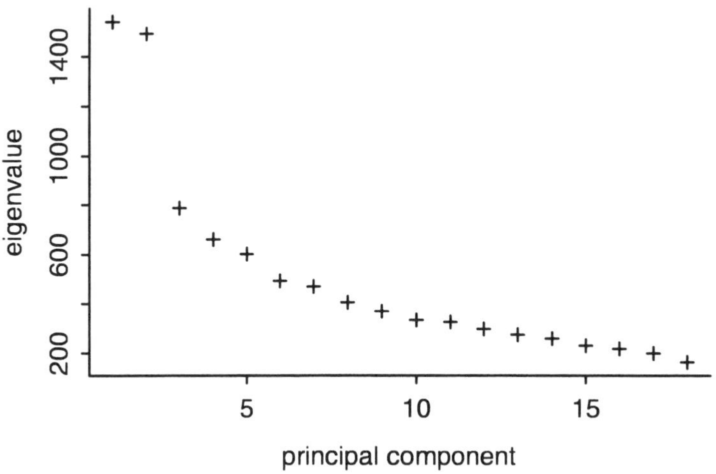

Figure 9.13: A SCREE plot of the eigenvalues for the vowel data in the example. The elbow in the curve shows the number of principal components to take.

this method we would retain 16 of the 18 principal components in our example. Another method suggests that we retain only those principal components whose eigenvalues are above one (this assumes that the data has been standardised so that each original parameter has a variance of one), since all others account for less variance than one of the original parameters; in our example, this leaves four principal components. Finally, we can use what is called a SCREE[4] plot of the eigenvalues to look for a suitable cutoff point. A SCREE plot shows the eigenvalues plotted against their index number. Figure 9.13 shows the plot for the present example. Typically there is a kink or elbow in the curve, as there is at index number 3 in the figure; by this method, all principal components up to and including the first one after the elbow are retained (Jackson, 1991, p. 45). In the example, this suggests retaining 4 principal components. A common problem is that there is often no elbow or more than one elbow, and so this method can be inconclusive. In the case of two elbows, it has been suggested that the two groups of PCs so defined (up to the first elbow and between the first and second elbows) are due to separate sources of variability. For example, the first group might account for the variability between types under study, while the second group might account for instrumental and/or other identifiable testing and measurement variability (Jackson, 1991, p. 46). However, any such inference could only be suggestive at best and should not be the basis for any theoretical statements.

Fortunately, within a classification experiment, there is another principled way to decide on how many principal components to retain: we retain the number that give the maximum classification performance. If we run a classification experiment using PCA transformed data, we can train and test a model on any number of principal components. After training and testing on one, two, three, and so on, components, we can plot the overall classification accuracy vs. the number of components used. In doing this, we often observe a peak in the accuracy: fewer components do not contain sufficient information and more components make the model less accurate for the reasons outlined above. This peak number then gives the number of components to use. Figure 9.14 shows an example of such a curve for the current example.

This method raises issues to do with the independence of the sets of data used to estimate both the model and the number of parameters used in testing. As we said before, an *open* test provides the best measure of the generality of a model by testing on unseen data. In this situation, the *training* of the model must include selecting how many principal components to use; this step then should not be carried out using the *testing* set but on the training set or perhaps on a third *halting* set. The motivation for using a third set of data is that we would like to choose the number of components that is best suited to classifying unseen data; since we can't use the testing data without biasing the experiment, a third set of data must be formed.

The results shown in Figure 9.14 are obtained by using the example training data to evaluate a series of models using increasing numbers of principal components. We can see here that a rapid increase in performance is achieved by adding the first 6 principal components when the score reaches 97.1%. The scores continue to increase slowly as more components are added, reaching a maximum of 99.69% at the 14th component. Since the difference between the scores for six and 14 components is small, we might choose to use 6 components here to give a simpler model.

Hence in this case we have two competing recommendations for how many principal components to retain: the SCREE plot and examination of the eigenvalues suggest that four components are appropriate, whereas the above evaluation suggests six. There is no sure way of selecting between these two alternatives but since the latter estimate is based on empirical measurement, we will choose it in this case (it happens to give a slightly better result when the model is tested, but using this information in a research context would inappropriate for the reasons outlined above).

With six principal components retained, the final classification score is 83.1%, an improvement of 9.1% over the untransformed case. The confusion matrix is shown in Table 9.4.

9.6.2 Discriminant analysis

Principal components analysis rotates the parameter space such that the first principal component lies along the direction of maximum variance. However, in many cases, this also corresponds to the direction that best separates the types under study because most of the variance is due to differences between types.

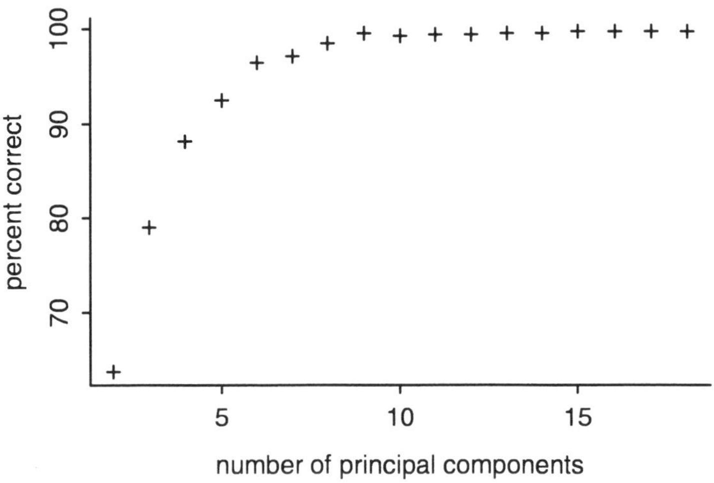

Figure 9.14: The performance curve for the example vowel data shows the overall performance of a Gaussian model trained on 2, 3, ... 13 principal components. This evaluation was carried out by performing a series of closed tests on successive Gaussian models.

	æ	ɛ	ɪ	ɒ	ʊ	a	ɛə	i	ɔ	u
æ	96.8	0	0	0	0	3.2	0	0	0	0
ɛ	0	69.7	0	0	7.6	9.1	4.5	0	9.1	0
ɪ	0	15.1	80.2	0	0	0	0	4	0.8	0
ɒ	7.6	0	0	77.3	0	15.2	0	0	0	0
ʊ	1.5	1.5	0	7.6	77.3	4.5	0	0	7.6	0
a	7.6	0	0	0	0	92.4	0	0	0	0
ɛə	1.5	4.5	0	0	0	3	75.8	0	13.6	1.5
i	0	0	4.5	0	0	0	16.7	72.7	0	6.1
ɔ	1.5	0	0	10.6	0	0	0	0	87.9	0
u	0	1.5	0	0	0	0	16.7	3	0	78.8

Table 9.4: Confusion matrix for 10 vowels on the first six principal components of 17 Bark scaled spectral bands and duration. Scores are percentages.

There is, though, another technique that is relevant to the separation of types more directly: *discriminant analysis*. The goal of discriminant analysis is to find a linear combination of the original parameters that provides the optimal separation between groups of points defined by the different types. This method presupposes that the label type of each point is known; it analyses the differences between and within types to find a parameter set that will keep members of a type together while making the different types distinct.

This technique is based on the principles of Bayesian probability and the maximum likelihood rule discussed earlier in this chapter. It uses analytic methods to find a new set of dimensions that retain all of the discrimination power of the original parameters and therefore is very well suited to classification applications. The number of transformed dimensions available from a discriminant analysis is one less than the number of types in the original data. Unfortunately, some assumptions about the distributions of the data are necessary: first, that the data covariance matrices are equal for every label type; and second, that the data are normally distributed. This second condition is fundamental to the classification procedures discussed above and so presents no additional constraint. However the assumption of equal covariance matrices would seem to limit the usefulness of this technique. If the assumption of equal covariance matrices is grossly invalid, the resulting transformation is unlikely to improve classification performance compared with a PCA treatment. However, if improved results are observed, then the assumption would seem to be vindicated (that is, the covariance matrices are similar enough that a useful transformation resulted from applying this technique).

When a discriminant analysis is performed on the Bark scaled vowel data used in the earlier PCA example, eight transformed dimensions are derived since there are nine vowel types in the sample. Figure 9.15 shows the first two transformed dimensions resulting from this analysis. It is instructive to compare this plot with the analogous plot of the first two principal components (Figure 9.12). It is clear that the discriminant analysis has found a slightly better separation of the vowel types in these first two dimensions. However, the striking feature is the similarity of the two plots: the first two transformed dimensions encode vowel height and backness once again.

Mathematics

Discriminant analysis again finds a set of eigenvectors and eigenvalues, but this time from (the inverse of) the product of the within-group sum-of-squares matrix $\boldsymbol{W}$ and the between-group sum-of-squares matrix $\boldsymbol{B}$. The within-group sum-of-squares matrix is a measure of the variability within the label types and is defined as

$$\boldsymbol{W} = \boldsymbol{W}_1 + \boldsymbol{W}_2 + \ldots + \boldsymbol{W}_N, \tag{9.16}$$

where $\boldsymbol{W}_i$ is the within-group sum-of-squares matrix for label type i:

$$\boldsymbol{W}_i = \sum_{j=1}^{n_i} (\boldsymbol{x}_j - \mu_i), (\boldsymbol{x}_j - \mu_i)'$$

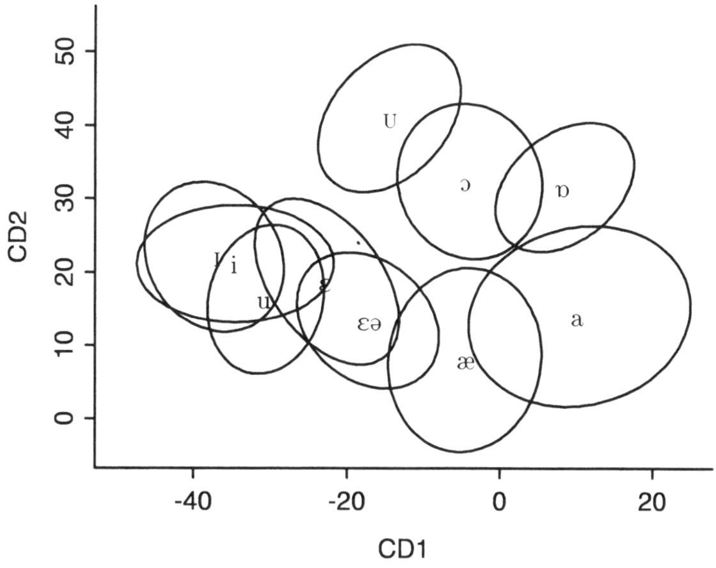

Figure 9.15: A set of vowel tokens plotted on the first two discriminant dimensions derived from speaker normalised Bark scaled spectral data.

where n_i is the number of tokens of type t_i, $\boldsymbol{x}_j$ is the parameter vector of the jth token of type t_i, and μ_i is the mean vector for group i.

The between-group sum-of-squares matrix is defined as

$$\boldsymbol{B} = \boldsymbol{T} - \boldsymbol{W}, \tag{9.17}$$

where $\boldsymbol{T}$ is the total mean corrected sum-of-squares of cross-products for all of the tokens:

$$\boldsymbol{T} = \sum_{i=1}^{N} \sum_{j=1}^{n_i} (\boldsymbol{x}_j - \mu)(\boldsymbol{x}_j - \mu)',$$

where N is the number of label types or groups in the data and μ is the grand mean vector for all types.

In discriminant analysis the eigenvectors and eigenvalues of the square matrix,

$$\boldsymbol{W}^{-1}\boldsymbol{B},$$

are found. However, as discussed earlier, only the first $N - 1$ eigenvectors and eigenvalues are retained where N is the number of types in the training data.

Discriminant analysis is a rotation of the axes, as is PCA, but this time without the constraint that the new axes be orthogonal, although they remain uncorrelated.

How many discriminant dimensions?

Since one of the aims of discriminant analysis, in common with principal components analysis, is to reduce the dimensionality of the data, we again face the question of how many of the transformed dimensions to use in any classification experiment. The question is much easier to answer in the case of discriminant analysis for two reasons: first, there are often only a small number of dimensions since this number is limited by the number of types in the data; second, there exists a statistical test that can be used to estimate the effectiveness of successive numbers of dimensions.

The effectiveness of dimensions can be evaluated by determining the *residual discrimination* in the data *prior* to extracting $1, 2, \ldots, N-1$ transformed dimensions. The motivation for this test is that we keep extracting dimensions while there are still residual differences between the groups. When the groups become statistically indistinguishable (that is, not significantly different) in the space formed by the remaining dimensions, no additional discriminant dimensions can distinguish the groups further. The residual discrimination in the data can be estimated by a statistic called *Wilk's lambda* (Klecka, 1980; Dillon & Goldstein, 1984), which is derived from the eigenvalues of the discriminant analysis. For k retained dimensions,

$$\Lambda_k = \prod_{i=k+1}^{N-1} \frac{1}{1+l_i} \tag{9.18}$$

where N is the number of different label types or groups in the data, and l_i is the ith eigenvalue (the $\prod$ is analogous to the summation symbol $\sum$ and denotes the product of a number of terms). The value of Λ_k for any given k lies between zero and one. A value of zero implies that the centroids of the different types are very different; as Λ_k increases towards one, so the types become more difficult to distinguish.

The statistical significance of the difference between types can be determined by converting Λ_k into an approximation of a χ^2 distribution. The χ^2 version of Wilk's lambda is

$$V_k = -\left(n - \frac{p+N}{2} - 1\right) \log(\Lambda_k) \tag{9.19}$$

where p is the original number of parameters in the data and n is the number of samples. V_k has $(p-k)(N-k-1)$ degrees of freedom. If the probability of V_k under the χ^2 distribution is less than the criterion value (usually 0.05), then we can conclude that some differences between the types remain and that at least k dimensions should be taken.

To make the decision as to how many dimensions to retain, V_k is evaluated for $k = 0, 1, 2 \ldots, N-1$ dimensions in turn. The significance level for each value

k	Λ_k	V_k	Degrees of Freedom	Significance Level
0	0.0002	5237.5	153	< 0.01
1	0.0043	3447.8	128	< 0.01
2	0.0260	2309.5	105	< 0.01
3	0.0926	1505.7	84	< 0.01
4	0.2464	886.6	65	< 0.01
5	0.5077	429.0	48	< 0.01
6	0.7259	202.7	33	< 0.01
7	0.8895	74.0	20	< 0.01
8	0.9493	32.9	9	< 0.01

Table 9.5: Statistical measures on the eigenvalues of the Bark scaled vowel data.

of V_k is determined, and those dimensions that exceed the desired significance level are retained.

The relevant statistics have been calculated for the vowel data used earlier and are tabulated in Table 9.5. We can see from this table that Λ_k has a very small value for $k = 0$, which means that the raw data, before any discriminant dimensions have been extracted, has large differences between the types. The value of $V_0 = 5238.5$ (with 153 degrees of freedom) has a probability of less than 0.01; hence the raw data has *significant* differences between groups and it will be useful to extract the first discriminant dimension. After extracting the first discriminant dimension ($k = 1$), the residual discrimination in the data is less (the value of $\Lambda_1 = 0.0043$ is larger than for $k = 0$ but still very small), which is to be expected since the first discriminant dimension should account for a large proportion of the type differences. However, the remaining differences between groups are still significant, and so we are justified in taking a second discriminant dimension. In fact, in this example, all discriminant dimensions contribute to significant differences between groups and so according to these statistics, we should take all nine discriminant dimensions from this data.

We can now perform a classification experiment using the first nine discriminant dimensions to train and test a Gaussian model. The overall performance on this transformed data is 81.4%, which is an improvement compared with the corresponding classification score from the original data. The confusion matrix is shown in Table 9.6 and shows the same patterns of confusion as were seen in the untransformed data (Table 9.3). In this example, the best overall classification performance came from the first five PCA dimensions at 88.3%. However, discriminant analysis has provided performance that is almost as good (remembering that we haven't tested the statistical significance of any of these differences) with only three transformed dimensions.

	æ	ɛ	ɪ	ɒ	ʊ	a	ɛə	i	ɔ	u
æ	96	0.8	0	2.4	0	0	0.8	0	0	0
ɛ	6.1	71.2	9.1	0	0	0	10.6	0	0	3
ɪ	0	7.9	83.3	0	0	0	4	4	0.8	0
ɒ	0	0	0	98.5	0	1.5	0	0	0	0
ʊ	0	0	0	12.1	81.8	0	1.5	0	4.5	0
a	4.5	0	0	0	0	95.5	0	0	0	0
ɛə	16.7	7.6	0	0	0	1.5	66.7	0	1.5	6.1
i	0	1.5	21.2	0	0	0	16.7	53	0	7.6
ɔ	0	0	0	22.7	0	0	0	0	77.3	0
u	0	0	6.1	0	0	0	13.6	4.5	0	75.8

Table 9.6: Confusion matrix for 10 vowels classified on nine discriminant dimensions derived from Bark scaled spectral data and duration. Scores are percentages.

Notes

1. This equation introduces some matrix algebra terms that have not been used so far in this book. A brief summary of these operations is included here.

A *vector* is an ordered list of numbers. In this book, vectors are denoted in bold type as in $\boldsymbol{x}$; the ith element of vector $\boldsymbol{x}$ is written as $x[i]$. A vector can equivalently be treated as a *row vector*

$$\boldsymbol{x} = [x[1], x[3], x[3]],$$

or as a *column vector*,

$$\begin{bmatrix} x[1] \\ x[3] \\ x[3] \end{bmatrix}$$

The difference is important mainly when vectors are multiplied by matrices.

A *matrix* is a rectangular array of numbers arranged in rows and columns and is denoted in uppercase bold $\boldsymbol{X}$ in this book. An $n \times n$ matrix has n rows and n columns: the following is a 3×2 matrix:

$$\boldsymbol{X} = \begin{bmatrix} x_{11} & x_{12} \\ x_{21} & x_{22} \\ x_{31} & x_{32} \end{bmatrix}$$

The element in the second row and the first column is denoted x_{21}. A *square matrix* is a matrix with the same number of rows and columns.

The *determinant* of a square matrix, $|\boldsymbol{X}|$, can be thought of as a measure of the "size" or magnitude of the matrix. The method for working out the determinant is complicated because finding the determinant of a large matrix involves finding the determinants of many submatrices. In the simple case of a 2×2 matrix, the determinant is

$$\begin{vmatrix} x_{11} & x_{12} \\ x_{21} & x_{22} \end{vmatrix} = x_{11}x_{22} + x_{12}x_{21}$$

The reader is referred to any linear algebra text for further details.

The *diagonal* elements of a square matrix are those with equal row and column indexes, x_{11}, x_{22}, and so on. All other elements are said to be *off-diagonal*. In a *diagonal matrix*, all off-diagonal elements are zero.

Two matrices can be multiplied if the number of columns of the first is equal to the number of rows of the second: for example a 3×4 matrix can be multiplied by a 4×5 matrix but not by a 5×4 matrix. The result is a matrix with the same number of rows as the first matrix and the same number of columns as the second. If we multiply two matrices $\boldsymbol{A}$ and $\boldsymbol{B}$ to give $\boldsymbol{C}$, then element c_{ij} of the resulting matrix is given by

$$c_{ij} = a_{i1}b_{1j} + a_{i2}b_{2j} + \ldots + a_{in}b_{nj}.$$

So for example if $\boldsymbol{A}$ is a 2×3 matrix and $\boldsymbol{B}$ is a 3×2 matrix the result is

$$\begin{bmatrix} a_{11}b_{11} + a_{12}b_{21} & a_{11}b_{12} + a_{12}b_{22} \\ a_{21}b_{11} + a_{22}b_{21} & a_{21}b_{12} + a_{22}b_{22} \end{bmatrix}$$

We often multiply a vector by a matrix to get another vector. In this case we are regarding the vector of length n as a $1 \times n$ matrix (a row vector), which, when multiplied by a $n \times m$ matrix, yields a $1 \times m$ matrix or a vector of length m. In most cases in this book, the matrix is symmetric, and so the length of the result vector is the same as that of the initial one. For this example, the first element of the result vector is the sum of first element of the initial vector multiplied by the first column of the matrix.

The *identity* matrix $\boldsymbol{I}$ is a diagonal matrix where all the diagonal elements are 1. It has the property that if a matrix is multiplied by $\boldsymbol{I}$ it remains unchanged. So

$$\boldsymbol{AI} = \boldsymbol{A}.$$

The *inverse* of a matrix is denoted by $\boldsymbol{A}^{-1}$ and has the property that

$$\boldsymbol{A}^{-1}\boldsymbol{A} = \boldsymbol{I},$$

so multiplying by the inverse of a matrix is like dividing by the matrix.

The *transpose* of a matrix, written as $\boldsymbol{A}^T$ is obtained by swapping the columns and rows of a matrix, so the transpose of a 4×2 matrix is a 2×4 matrix. A vector is often transposed before being multiplied by a matrix — this assumes then that the vector is a *column vector* (for example, $n \times 1$) and so needs to be transposed into a *row vector* ($1 \times n$) before multiplying. As we said above a vector can be equivalently written as a row or column vector.

2. The χ^2 distribution is a probability distribution (or more accurately a family of distributions since the shape of the probability density varies with the *degree of freedom* of the χ^2 variable) similar to the normal distribution introduced earlier. The distribution shows the probability of each value of χ^2 for a given degree of freedom. χ^2 can be calculated from a number of independent variables; see Mosteller and Rourke (1973) or any statistics text for more details.

3. There are two degrees of freedom here because there are two cells in the table that we are free to choose values for. Having chosen these values, the other two cells are then predetermined by subtracting from the total number of tokens in the test set.

4. The SCREE plot is named after slopes of loose rock called Scree found in England. The rock tends to fall down the steep face of the hill and collect at the bottom.

References

Abbs, M. S., & Minifie, F. D. (1969). Effects of acoustic cues in fricatives on perceptual confusions in preschool children. *Journal of the Acoustical Society of America, 70*, 1535–1542.

Abercrombie, D. (1964). Syllable quantity and enclitics in English. In D. Abercrombie, D. Fry, N. MacCarthy, & J. Trim (Eds.), *In Honour of Daniel Jones* (pp. 255–309). London: Longmans.

Abercrombie, D. (1967). *Elements of General Phonetics.* Edinburgh: Edinburgh University Press.

Ainsworth, W. A. (1975). Intrinsic and extrinsic factors in vowel judgements. In G. Fant & M. Tatham (Eds.), *Auditory Analysis and Perception of Speech* (pp. 103–113). London: Academic Press.

Ainsworth, W. A. (1988). *Speech Recognition by Machine.* London: Peregrinus.

Allen, J., Hunnicutt, S., & Klatt, D. (1987). *From Text to Speech: the MITalk System.* Cambridge: Cambridge University Press.

Ananthaphadmanabha, T. V. (1984). Acoustic analysis of voice source dynamics. *Speech Transmission Laboratory, Quarterly Progress Status Report, 2–3*, 1–24.

Andruski, J. E., & Nearey, T. M. (1992). On the sufficiency of compound target specification of isolated vowels in /bVb/ syllables. *Journal of the Acoustical Society of America, 91*, 390–410.

Assmann, P. F., Nearey, T. M., & Hogan, J. T. (1982). Vowel identification: orthographic, perceptual, and acoustic aspects. *Journal of the Acoustical Society of America, 71*, 975–989.

Atal, B., & Hanauer, S. (1971). Speech analysis and synthesis by linear prediction of the speech wave. *Journal of the Acoustical Society of America, 50*, 637–655.

Atal, B. S. (1985). Linear predictive coding of speech. In F. Fallside & W. A. Woods (Eds.), *Computer Speech Processing* (pp. 81–124). Englewood Cliffs NJ: Prentice-Hall.

Aull, A., & Zue, V. (1985). Lexical stress determination and its application to large vocabulary speech recognition. In *Institute of Electrical and Electronic Engineers, International Conference on Acoustics Speech and Signal Processing* (pp. 1549–1552).

Badin, P. (1989). Acoustics of voiceless fricatives: production theory and data. *Speech Transmission Laboratory, Quarterly Progress Status Report, 3*, 33–55.

Badin, P. (1991). Fricative consonants: acoustic and X-ray measurements. *Journal of Phonetics, 19*, 397–408.

Badin, P., & Fant, G. (1984). Notes on vocal tract computation. *Speech Transmission Laboratory, Quarterly Progress Status Report, 2–3*, 53–108.

Badin, P., & Fant, G. (1987). Fricative production modelling: aerodynamic and acoustic data. In *Proceedings of EUROSPEECH 1987* (Vol. 3, pp. 23–26). Edinburgh.

Bahl, L. R., Brown, P. F., Souze, P. V. de, & Mercer, R. L. (1989). Speech Recognition with continuous-parameter hidden Markov models. *Computer Speech and Language*, *2*(3/4).

Bailly, G., Benoît, C., & Sawallis, T. R. (Eds.). (1992). *Talking Machines — Theories, Models and Designs.* Amsterdam: North Holland.

Barry, W. J. (1983). On the perception of juncture in English. In M. P. R. van den Broecke & A. Cohen (Eds.), *Proceedings of the Tenth International Congress of Phonetic Sciences* (pp. 526–536). Dordrecht: Foris.

Beckman, M. E. (1986). *Stress and Non-Stress Accent.* Dordrecht: Foris.

Beckman, M. E. (1996). The parsing of prosody. *Language and Cognitive Processes*, *11*, 17–67.

Beckman, M. E., & Ayers, G. M. (1994). *Guidelines for ToBI Labeling version 2.0.* `tobi@ling.ohio-state.edu` and `http://ling.ohio-state.edu/Phonetics/Phonetics.html`.

Beckman, M. E., & Edwards, J. R. (1994). Articulatory evidence for differentiating stress categories. In P. A. Keating (Ed.), *Between the Grammar and the Physics of Speech: Papers in Laboratory Phonology III* (pp. 7–33). Cambridge: Cambridge University Press.

Beckman, M. E., Edwards, J. R., & Fletcher, J. (1991). Prosodic structure and tempo in a sonority model of articulatory dynamics. In G. Docherty & D. R. Ladd (Eds.), *Papers in Laboratory Phonology II: Segment, Gesture, and Tone* (pp. 68–86). Cambridge: Cambridge University Press.

Beckman, M. E., Edwards, J. R., & Fletcher, J. (1992). Prosodic structure and tempo in a sonority model of articulatory dynamics. In G. Docherty & D. R. Ladd (Eds.), *Papers in Laboratory Phonology II: Gesture, Segment and Prosody* (pp. 68–86). Cambridge: Cambridge University Press.

Beckman, M. E., Jung, T.-P., Lee, S., Jong, K. de, Krishnamurthy, A., Ahalt, S., Cohen, K. B., & Collins, M. (1995). Variability in the production of quantal vowels revisited. *Journal of the Acoustical Society of America*, *97*, 471–490.

Beckman, M. E., & Pierrehumbert, J. (1986). Intonational structure in Japanese and English. *Phonology Yearbook*, *3*, 255–309.

Behrens, S. J., & Blumstein, S. E. (1988). On the role of the amplitude of the fricative noise in the perception of place of articulation in voiceless fricative consonants. *Journal of the Acoustical Society of America*, *84*, 861–867.

Bell, A. G. (1879). Vowel theories. *American Journal of Otolaryngology*, *1*, 163–180.

Benguerel, A.-P., & McFadden, T. U. (1989). The effect of coarticulation on the role of transitions in vowel perception. *Phonetica*, *46*, 80–96.

Berdan, R. (1978). Multidimensional analysis of vowel variation. In D. Sankoff (Ed.), *Language Variation: Models and Methods* (pp. 149–160). New York: Academic Press.

Bethea, R. M., Duran, B. S., & Boullion, T. L. (1995). *Statistical Methods for Engineers and Scientists* (3rd ed.). New York: Marcel Dekker.

Bladon, R. A. W. (1983). Two-formant models of vowel perception: shortcomings and enhancements. *Speech Communication, 2*, 305–313.

Bladon, R. A. W. (1985). Diphthongs: a case study of dynamic auditory processing. *Speech Communication, 4*, 145–154.

Bladon, R. A. W., & Al-Bamerni, A. (1976). Coarticulation resistance in English /l/. *Journal of Phonetics, 4*, 137–150.

Bladon, R. A. W., & Fant, G. (1978). A two-formant model and the cardinal vowels. *Speech Transmission Laboratory, Quarterly Progress Status Report, 1*, 1–8.

Bladon, R. A. W., Henton, C. G., & Pickering, J. B. (1984). Towards an auditory theory of speaker normalisation. *Language and Communication, 4*, 59–69.

Blumstein, S., & Stevens, K. (1979). Acoustic invariance in speech production: evidence from measurements of the spectral characteristics of stop consonants. *Journal of the Acoustical Society of America, 66*, 1001–1017.

Blumstein, S., & Stevens, K. (1980). Perceptual invariance and onset spectra for stop consonants in different vowel environments. *Journal of the Acoustical Society of America, 67*, 648–662.

Blumstein, S. E., Isaacs, E., & Mertus, J. (1982). The role of gross spectral shape as a perceptual cue to place of articulation in initial stop consonants. *Journal of the Acoustical Society of America, 72*, 43–50.

Boë, L.-J., Schwartz, J.-L., & Vallée, N. (1994). The prediction of vowel systems: Perceptual contrast and stability. In E. Keller (Ed.), *Fundamentals of Speech Synthesis and Speech Recognition* (pp. 185–214). Chichester: Wiley.

Bolinger, D. (1958). A theory of pitch accent in English. *Word, 14*, 109–149.

Bolinger, D. (1965). Pitch accent and sentence rhythm. In I. Abe & T. Kanekiya (Eds.), *Forms of English: Accent, Morpheme, Order* (pp. 139–180). Cambridge, MA: Harvard University Press.

Bolinger, D. (1972). Accent is predictable (If you're a mind-reader). *Language, 48*, 633–644.

Bolinger, D. (1975). *Aspects of Language.* New York: Harcourt Brace Jovanovich.

Bond, Z. S. (1982). Experiments with synthetic diphthongs. *Journal of Phonetics, 10*, 259–264.

Borden, G., Harris, K. S., & Raphael, L. J. (1994). *Speech Science Primer: Physiology, Acoustics, and Perception of Speech* (3rd ed.). Baltimore: Williams & Wilkins.

Boucher, V. (1988). A parameter of syllabification for VstopV and relative-timing invariance. *Journal of Phonetics, 16*, 299–326.

Broad, D., & Fertig, R. H. (1970). Formant-frequency trajectories in selected CVC-syllable nuclei. *Journal of the Acoustical Society of America, 47*, 1572–1582.

Broad, D. J., & Wakita, H. (1977). Piecewise-planar representation of vowel formant frequencies. *Journal of the Acoustical Society of America, 62*, 1467–1473.

Bruce, G. (1977). *Swedish Word Accents in Sentence Perspective.* Lund, Sweden: University of Lund (Travaux de l'Institut de Linguistique de Lund).

Butterfield, S., & Cutler, A. (1988). Segmentation errors by human listeners: evidence for a prosodic segmentation strategy. In *Proceedings of SPEECH '88 (Seventh Conference of the Federation of Acoustic Societies of Europe)* (Vol. 3, pp. 827–833). Edinburgh.

Campbell, W. N. (1990). Evidence for a syllable based model of speech timing. In *Proceedings of the International Conference on Spoken Language Processing* (pp. 9–12). Kobe, Japan.

Carlson, R., Fant, G., & Granström, B. (1975). Two-formant models, pitch, and vowel perception. In G. Fant & M. A. A. Tatham (Eds.), *Auditory Analysis and Perception of Speech* (pp. 55–82). London: Academic Press.

Carlson, R., & Granström, B. (1975). A phonetically oriented programming language for rule description of speech. In G. Fant (Ed.), *Speech Communication* (Vol. 2, pp. 245–253). Uppsala, Sweden: Almqvist & Wiksell.

Carlson, R., & Granström, B. (1976). A text-to-speech systems based entirely on rules. In *Institute of Electrical and Electronic Engineers, International Conference on Acoustics Speech and Signal Processing* (Vol. 76, pp. 686–688).

Carlson, R., Granström, B., & Hunnicutt, S. (1982). Bliss communication with speech or text output. In *Institute of Electrical and Electronic Engineers, International Conference on Acoustics Speech and Signal Processing* (pp. 747–750).

Carlson, R., Granström, B., & Klatt, D. H. (1979). Some notes of the perception of temporal patterns in speech. In *9th International Congress of Phonetic Sciences.* Copenhagen, Denmark.

Carroll, J. D., & Chang, J. J. (1970). Analysis of individual differences in multidimensional scaling via an N-way generalization of 'Eckhart-Young' decomposition. *Psychometrika, 31*, 33–42.

Cassidy, S., & Harrington, J. (1995). The place of articulation distinction in voiced stops: evidence from burst spectra and formant transitions. *Phonetica, 52*, 263–284.

Catford, J. C. (1977). *Fundamental Problems in Phonetics.* Edinburgh: Edinburgh University Press.

Chafe, W. (1974). Language and consciousness. *Language, 50*, 111–133.

Chandra, S., & Lin, W. C. (1974). Experimental comparisons between stationary and non-stationary formulations of linear prediction. *Institute of Electrical and Electronic Engineers, Transactions on Acoustics Speech and Signal Processing, 22*, 403–415.

Chandra, S., & Lin, W. C. (1977). Linear prediction with a variable analysis frame size. *Institute of Electrical and Electronic Engineers, Transactions on Acoustics Speech and Signal Processing, 25*, 322–330.

Charpentier, & Stella. (1986). Diphone synthesis using an overlap–add technique for speech waveform concatentation. In *Institute of Electrical and Electronic Engineers, International Conference on Acoustics Speech and Signal Processing* (pp. 2015–2018).

Chasaide, A. N., & Gobl, C. (1997). Voice source variation. In W. Hardcastle & J. Laver (Eds.), *The Handbook of Phonetic Sciences* (pp. 427–461). Oxford: Blackwell.

Chatfield, C. (1989). *The Analysis of Time Series : an Introduction* (4th ed.). London: Chapman and Hall.

Chiba, T., & Kajiyama, M. (1941). *The Vowel: its Nature and Structure.* Tokyo: Tokyo Publishing.

Childers, D. G., & Ahn, C. (1995). Modeling the glottal volume-velocity waveform for three voice types. *Journal of the Acoustical Society of America, 97*, 505–519.

Childers, D. G., & Hu, H. T. (1994). Speech synthesis by glottal excited linear prediction. *Journal of the Acoustical Society of America, 96*, 2026–2036.

Chistovich, L. (1985). Central auditory processing of peripheral vowel spectra. *Journal of the Acoustical Society of America, 77*, 789–805.

Chistovich, L., & Lublinskaya, V. (1979). The center of gravity effect in vowel spectra and critical distance between the formants: psychoacoustical study of vowel-like stimuli. *Hearing Research, 1*, 185–195.

Chomsky, N., & Halle, M. (1968). *The Sound Pattern of English.* New York: Harper and Row.

Christie, W. M. (1977). Some multiple cues for juncture in English. *General Linguistics, 17*, 213–222.

Christophe, A., Dupoux, E., Bertoncini, J., & Mehler, J. (1994). Do infants perceive word boundaries? An empirical study of the bootstrapping of lexical acquisition. *Journal of the Acoustical Society of America, 95*, 1570–1580.

Church, K. W. (1983). *Phrase-Structure Parsing: a Method of Taking Advantage of Allophonic Constraints.* Bloomington, Indiana: Indiana University Linguistics Club.

Church, K. W. (1987). Phonological parsing and lexical retrieval. In U. H. Frauenfelder & L. K. Tyler (Eds.), *Spoken Word Recognition* (pp. 53–70). Cambridge, MA: MIT Press.

Clark, J., & Yallop, C. (1995). *An Introduction to Phonetics and Phonology* (2nd ed.). Oxford: Basil Blackwell.

Cole, R. A., & Cooper, W. E. (1975). Perception of voicing in English affricates and fricatives. *Journal of the Acoustical Society of America, 58*, 1280–1287.

Cole, R. A., & Scott, B. (1974). The phantom is the phoneme: Invariant cues for stop consonants. *Perception and Psychophysics, 15*, 101–107.

Collier, R., & 't Hart, J. (1975). The role of intonation in speech perception. In A. Cohen & S. G. Nooteboom (Eds.), *Structure and Process in Speech Perception* (pp. 107–122). New York: Springer-Verlag.

Cooper, F. S., Liberman, A. M., & Borst, J. M. (1951). The interconversion of audible and visible patterns as a basis for research in the perception of speech. *Proceedings National Academy Sciences (US), 37*, 318–325.

Cooper, W. E., & Paccia-Cooper, J. (1980). *Syntax and Speech.* Cambridge MA: Harvard University Press.

Cooper, W. E., & Sorensen, J. M. (1981). *Fundamental Frequency in Sentence Production.* New York: Springer-Verlag.

Crystal, T. H., & House, A. S. (1988). The duration of American-English vowels: an overview. *Journal of Phonetics, 16*, 263–284.

Crystal, T. H., & House, A. S. (1990). Articulation rate and the duration of syllables and stress groups in connected speech. *Journal of the Acoustical Society of America, 88*, 101–112.

Curtis, J. F. (1942). *An experimental study of the wave-composition of nasal voice quality.* Unpublished doctoral dissertation, University of Iowa.

Cutler, A., & Butterfield, S. (1990). Durational cues to word boundaries in clear speech. *Speech Communication, 9*, 485–495.

Cutler, A., & Butterfield, S. (1991). Word boundary cues in clear speech: a supplementary report. *Speech Communication, 10*, 335–353.

Cutler, A., & Carter, D. (1987). The predominance of strong initial syllables in the English vocabulary. *Computer Speech and Language, 2*, 133–142.

Cutler, A., & Foss, D. J. (1979). On the role of sentence stress in sentence processing. *Language and Speech, 10*, 1–10.

Cutler, A., & Norris, D. (1988). The role of strong syllables in segmentation for lexical access. *Journal of Experimental Psychology: Human Perception and Performance, 14*, 113–121.

Cutting, D., & Rosner, B. (1974). Categories and boundaries in speech and music. *Perception and Psychophysics, 16*, 564–570.

Cutting, D., & Rosner, B. (1976). Discrimination functions predicted from categories in speech and music. *Perception and Psychophysics, 20*, 87–88.

Dalston, R. M. (1975). Acoustic characteristics of English /w,r,l/ spoken correctly by young children and adults. *Journal of the Acoustical Society of America, 57*, 462–469.

Davis, S. B., & Mermelstein, P. (1980). Comparison of parametric representations of monosyllabic word recognition in continuously spoken sentences. *Institute of Electrical and Electronic Engineers, Transactions on Acoustics Speech and Signal Processing, 28*(4), 357–366.

de Jong, K. (1995). The supraglottal articulation of prominence in English: linguistic stress as localized hyperarticulation. *Journal of the Acoustical Society of America, 97*, 491–504.

Dechovitz, D. (1977). Information conveyed by vowels: a confirmation. *Haskins Laboratories Status Reports on Speech Research, 51/52*, 213–219.

Delattre, P. (1951). The physiological interpretation of sound spectrograms. *PMLA, 66*, 864–875.

Delattre, P. (1954). Les attributs acoustiques de la nasalité vocalique et consonantique. *Studia Linguistica, 8*, 103–109.

Delattre, P. (1969). An acoustic and articulatory study of vowel reduction in four languages. *Int. Rev. Appl. Linguist. Lang. Teach., 8*, 295–325.

Delattre, P., Liberman, A. M., & Cooper, F. S. (1955a). Acoustic loci and transitional cues for consonants. *Journal of the Acoustical Society of America, 27*, 769–773.

Delattre, P., Liberman, A. M., & Cooper, F. S. (1955b). Formant transitions and loci as acoustic correlates of place of articulation in American fricatives. *Studia Linguistica, 5*, 104–121.

Delattre, P., Liberman, A. M., Cooper, F. S., & Gerstman, F. J. (1952). An experimental study of the acoustic determinants of vowel color: observations on one- and two-formant vowels synthesised from spectrographic patterns. *Word, 8*, 195–210.

Denes, P. (1955). Effect of duration on the perception of voicing. *Journal of the Acoustical Society of America, 27*, 761–764.

Derr, M. A., & Massaro, D. W. (1980). The contribution of vowel duration, F0 contour, and frication duration as cues to the /juz/-/jus/ distinction. *Perception and Psychophysics, 27*, 51–59.

Devijver, P. A., & Kittler, J. (1986). *Pattern Recognition* (Vol. 30). Berlin: Springer-Verlag.

Di Benedetto, M.-G. (1989). Vowel representation: some observations on temporal and spectral properties of the first formant frequency. *Journal of the Acoustical Society of America, 86*, 55–66.

Diehl, R. L., McCusker, S. B., & Chapman, L. S. (1981). Perceiving vowels in isolation and in consonantal context. *Journal of the Acoustical Society of America, 69*, 239–248.

Dillon, W. R., & Goldstein, M. (1984). *Multivariate Analysis: Methods and Applications*. New York: Wiley.

Disner, S. F. (1980). Evaluation of vowel normalisation procedures. *Journal of the Acoustical Society of America, 67*, 253–261.

Disner, S. F. (1983). Vowel quality: the relation between universal and language-specific factors. *UCLA Working Papers in Phonetics. University of California, Los Angeles, 58.*

Dixon, N., & Maxey, H. (1968). Terminal analog synthesis of continuous speech using the diphone method of segment assembly. *Institute of Electrical and Electronic Engineers, Transactions on Audio and Electroacoustics, AU-16*, 40–50.

Dorman, M. F., Raphael, L. J., & Isenberg, D. (1980). Acoustic cues for a fricative-affricate contrast in word-final position. *Journal of Phonetics, 8*, 397–405.

Dorman, M. F., Studdert-Kennedy, M., & Raphael, L. J. (1977). Stop-consonant recognition: release bursts and formant transitions as functionally equivalent, context-dependent cues. *Perception and Psychophysics, 22*, 109–122.

Duda, R. O., & Hart, P. E. (1973). *Pattern Classification and Scene Analysis*. New York: Wiley.

Dudley, H. (1939). The Vocoder. *Bell Labs. Record, 17*, 122–126.

Dudley, H., Risez, R., & Watkins, S. (1939). A synthetic speaker. *Journal of Franklin Institute, 227*, 739–764.

Dudley, H., & Tarnóczy, T. (1950). The speaking machine of Wolfgang Kempelen. *Journal of the Acoustical Society of America, 22*, 151–166.

Dunn, H. (1950). The calculation of vowel resonances in an electrical vocal tract. *Journal of the Acoustical Society of America, 22*, 740–753.

Edwards, J., Beckman, M. E., & Fletcher, J. (1991). The articulatory kinematics of final lengthening. *Journal of the Acoustical Society of America, 89*, 369–382.

Edwards, T. (1981). Multiple feature analysis of intervocalic English plosives. *Journal of the Acoustical Society of America, 69*, 535–547.

Engstrand, O. (1988). Articulatory correlates of stress and speaking rate in Swedish VCV utterances. *Journal of the Acoustical Society of America, 83*, 1863–1875.

Espy-Wilson, C. Y. (1992). Acoustic measures for linguistic features distinguishing the semivowels in American English. *Journal of the Acoustical Society of America, 92*, 736–757.

Espy-Wilson, C. Y. (1994). A feature-based semivowel recognition system. *Journal of the Acoustical Society of America, 96*, 65–72.

Essner, C. (1947). Recherche sur la structure des voyelles orales. *Archives Néerlandaises de Phonétique Expérimentale, 20*, 40–77.

Fant, G. (1953). Speech communication research. *Royal Swedish Academy of Engineering Sciences, 24*, 734–742.

Fant, G. (1959). Acoustic analysis and synthesis of speech with applications to Swedish. *Ericsson Technics, 15*, 1–106.

Fant, G. (1960). *The Acoustic Theory of Speech Production.* The Hague: Mouton.

Fant, G. (1966). A note on vocal tract size factors and non-uniform F-pattern scalings. *Speech Transmission Laboratory, Quarterly Progress Status Report, 4*, 22–30.

Fant, G. (1968). Analysis and synthesis of speech processes. In B. Malmberg (Ed.), *Manual of Phonetics* (pp. 173–277). Amsterdam: North-Holland.

Fant, G. (1972). Vocal tract wall effects, losses, and resonance bandwidths. *Speech Transmission Laboratory, Quarterly Progress Status Report, 2-3*, 28–52.

Fant, G. (1973). *Speech Sounds and Features.* Cambridge, MA: MIT Press.

Fant, G. (1975). Non-uniform vowel normalization. *Speech Transmission Laboratory, Quarterly Progress Status Report, 2-3*, 28–52.

Fant, G. (1980). The relations between area functions and the acoustic signal. *Phonetica, 37*, 55–86.

Fant, G. (1985). The vocal tract in your pocket calculator. *Speech Transmission Laboratory, Quarterly Progress Status Report, 2-3*, 1–20.

Fant, G. (1993). Some problems in voice source analysis. *Speech Communication, 13*, 7–22.

Fant, G. (1995). The LF-model revisited. Transformations and frequency domain analysis. *Speech Transmission Laboratory, Quarterly Progress Status Report, 2-3*, 119–156.

Fant, G., & Kruckenberg, A. (1995). The voice source in prosody. In *Proceedings of the Thirteenth International Conference of Phonetic Sciences* (pp. 622–625).

Fant, G., Liljencrants, J., & Lin, Q. G. (1985). A four-parameter model of glottal flow. *Speech Transmission Laboratory, Quarterly Progress Status Report, 4*, 1–13.

Fant, G., & Lin, Q. G. (1988). Frequency domain interpretation and derivation of glottal flow parameters. *Speech Transmission Laboratory, Quarterly Progress Status Report, 2-3*, 1–21.

Farnsworth, D. W. (1940). High-speed motion pictures of human vocal cords. *Bell Laboratories Record, 18*, 203–208.

Fear, B. D., Cutler, A., & Butterfield, S. (1995). The strong/weak syllable distinction in English. *Journal of the Acoustical Society of America, 97*, 1893–1904.

Fischer-Jørgensen, E. (1954). Acoustic analysis of stop consonants. *Miscellanea Phonetica, 2*, 42–59.

Flanagan, J. L. (1957). Note on the design of terminal analog speech synthesisers. *Journal of the Acoustical Society of America, 29*, 306–310.

Flanagan, J. L. (1958). Some properties of the glottal sound source. *Journal of Speech and Hearing Research, 1*, 99–116.

Flanagan, J. L. (1972). *Speech Synthesis, Analysis, and Perception.* New York: Springer-Verlag.

Flanagan, J. L., & Rabiner, L. R. (1973). *Speech Synthesis.* Stroudsburg, Pennsylvania: Dowden, Hutchinson and Ross.

Flege, J., & Hillenbrand, J. (1986). Differential use of temporal cues to the /s/-/z/ contrast by native and non-native speakers of English. *Journal of the Acoustical Society of America, 79*, 508–517.

Forrest, K., Weismer, G., Milenkovic, P., & Dougall, R. N. (1988). Statistical analysis of word-initial voiceless obstruents: preliminary data. *Journal of the Acoustical Society of America, 84*, 115–123.

Fourakis, M. (1991). Tempo, stress, and vowel reduction in American English. *Journal of the Acoustical Society of America, 90*, 1816–1827.

Fourier, J. (1822). *Théorie Analytique de la Chaleur.* Paris.

Fowler, C. A. (1983). Converging sources of evidence on spoken and perceived rhythms in speech: cyclic productions of vowels in monosyllabic stress feet. *Journal of Experimental Psychology: General, 112*, 386–412.

Fowler, C. A. (1984). Segmentation of coarticulated speech in perception. *Perception and Psychophysics, 36*, 359–368.

Fowler, C. A. (1986). An event approach to the study of speech perception from a direct-realist perspective. *Journal of Phonetics, 14*, 3–28.

Fowler, C. A. (1987). Perceivers as realists, talkers too: commentary on papers by Strange, Diehl et al., and Rakerd and Verbrugge. *Journal of Memory and Language, 26*, 574–587.

Fowler, C. A., & Housum, J. (1987). Talkers' signaling of "new" and "old" words in speech and listeners' perception and use of the distinction. *Journal of Memory and Language, 26*, 489–504.

Fox, R. (1983). Perceptual structure of monophthongs and diphthongs in English. *Language and Speech, 26*, 21–49.

Fox, R. (1985). Multidimensional scaling and perceptual features: evidence of stimulus processing or memory prototypes? *Journal of Phonetics, 13*, 205–217.

Fox, R. (1989). Dynamic information in the identification and discrimination of vowels. *Phonetica, 46*, 97–116.

Fry, D. (1979). *The Physics of Speech.* Cambridge: Cambridge University Press.

Fujimura, O. (1962). Analysis of nasal consonants. *Journal of the Acoustical Society of America, 34*, 1865–1875.

Fujimura, O., & Lindqvist, J. (1971). Sweep-tone measurements of vocal-tract characteristics. *Journal of the Acoustical Society of America, 49*, 541–548.

Fujimura, O., & Lovins, J. (1978). Syllables as concatenative phonetic elements. In A. Bell & J. Hooper (Eds.), *Syllables and Segments* (pp. 107–120). New York: North Holland.

Fujisaki, H., & Kunisaki, O. (1976). Analysis, recognition, and perception of voiceless fricative consonants in Japanese. *Annual Bulletin, RILP, 10*, 145–156.

Gabioud, B. (1994). Articulatory models in speech synthesis. In E. Keller (Ed.), *Fundamentals of Speech Synthesis and Speech Recognition: Basic Concepts, State-of-the-Art and Future Challenges* (pp. 215–230). Chichester: Wiley.

Garcia, E. (1966). The identification and discrimination of synthetic nasals. *Haskins Laboratories Status Reports on Speech Research, 7/8*, 3.1–3.16.

Garcia, E. (1967). Discrimination of three-formant nasal-vowel syllables. *Haskins Laboratories Status Reports on Speech Research, 12*, 143–153.

Gårding, E. (1967). *Internal Juncture in Swedish. (Travaux de l'Institut de Phonétique de Lund, VI)*. Lund: Berlingska Boktryckeriet.

Gay, T. (1968). Effect of speaking rate on diphthong formant movement. *Journal of the Acoustical Society of America, 44*, 1570–1573.

Gay, T. (1970). A perceptual study of American English diphthongs. *Language and Speech, 13*, 65–88.

Gay, T. (1978). Effect of speaking rate on vowel formant movements. *Journal of the Acoustical Society of America, 63*, 223–230.

Gay, T. (1981). Mechanisms in the control of speech rate. *Phonetica, 38*, 148–158.

Gerstman, L. H. (1957). *Cues for distinguishing among fricatives, affricates, and stop consonants*. Unpublished doctoral dissertation, New York University.

Gerstman, L. H. (1968). Classification of self-normalized vowels. *Institute of Electrical and Electronic Engineers, Transactions on Audio and Electroacoustics, 16*, 78–80.

Giles, S. B., & Moll, K. L. (1975). Cinefluorographic selected allophones of /l/. *Phonetica, 31*, 206–227.

Gimson, A. C., & Cruttenden, A. (1994). *Gimson's Pronunciation of English. (Revised by Alan Cruttenden)* (5th ed.). London: Edward Arnold.

Gold, B., & Rabiner, L. (1968). Analysis of digital and analog formant synthesisers. *Institute of Electrical and Electronic Engineers, Transactions on Audio and Electroacoustics, AU-16*, 81–94.

Goldsmith, J. (1990). *Autosegmental and Metrical Phonology*. Oxford: Basil Blackwell.

Gottfried, M., Miller, J. D., & Meyer, D. J. (1993). Three approaches to the classification of American English diphthongs. *Journal of Phonetics, 21*, 205–229.

Gottfried, T. L., & Strange, W. (1980). Identification of coarticulated vowels. *Journal of the Acoustical Society of America, 68*, 1626–1635.

Greenberg, S. (1988). *Representation of speech in the auditory periphery.* [Special issue] *Journal of Phonetics, 16, 1.*

Grosjean, F., & Gee, J. P. (1987). Prosodic structure and spoken word recognition. *Cognition, 25*, 157–188.

Gurlekian, J. A. (1981). Recognition of the Spanish fricatives /s/ and /f/. *Journal of the Acoustical Society of America, 70*, 1624–1627.

Haggard, M. (1978). The devoicing of voiced fricatives. *Journal of Phonetics, 6*, 613–617.

Haggard, M., Summerfield, Q., & Roberts, M. (1981). Psychoacoustical and cultural determinants of phoneme boundaries: evidence from trading F0 cues in the voiced-voiceless distinction. *Journal of Phonetics, 9*, 49–62.

Halle, M., Hughes, W., & Radley, J. (1957). Acoustic properties of stop consonants. *Journal of the Acoustical Society of America, 29*, 107–116.

Halliday, M. A. K. (1967). *Intonation and Grammar in British English.* The Hague: Mouton.

Halliday, M. A. K. (1980). *A Course in Spoken English: Intonation.* Oxford: Oxford University Press.

Hamming, R. W. (1989). *Digital Filters.* New Jersey: Prentice-Hall.

Hardcastle, W. J., & Marchal, A. (Eds.). (1990). *Speech Production and Speech Modelling.* Dordrecht: Kluwer.

Harrington, J. (1994). The contribution of the murmur and vowel to the place of articulation distinction in nasal consonants. *Journal of the Acoustical Society of America, 96*, 19–32.

Harrington, J., & Cassidy, S. (1994). Dynamic and target theories of vowel classification: evidence from monophthongs and diphthongs in Australian English. *Language and Speech, 37*, 357–373.

Harrington, J., Cassidy, S., Fletcher, J., & McVeigh, A. (1993). The mu+ system for corpus based speech research. *Computer Speech and Language, 7*, 305–331.

Harrington, J., Fletcher, J., & Beckman, M. E. (in press). Manner and place conflicts in the articulation of accent in Australian English. In M. Broe (Ed.), *Papers in Laboratory Phonology 5.* Cambridge University Press: Cambridge.

Harris, K. S. (1958). Cues for the discrimination of American English fricatives in spoken syllables. *Language and Speech, 1*, 1–7.

Harris, K. S. (1978). Vowel duration change and its underlying physiological mechanisms. *Language and Speech, 21*, 354–361.

Harris, M. S., & Umeda, N. (1974). Effect of speaking mode on temporal factors in speech: vowel duration. *Journal of the Acoustical Society of America, 56*, 1016–1018.

Harshman, R. (1970). Foundations of the PARAFAC procedure: models and conditions for an "explanatory" multi-modal factor analysis. *UCLA Working Papers in Phonetics. University of California, Los Angeles, 16.*

Harshman, R., Ladefoged, P., & Goldstein, L. (1977). Factor analysis of tongue shapes. *Journal of the Acoustical Society of America, 62*, 693–707.

Hattori, S., Yamamoto, K., & Fujimura, O. (1958). Nasalisation of vowels in relation to nasals. *Journal of the Acoustical Society of America, 30*, 267–274.

Hayes, B. (1984). The phonology of English rhythm. *Linguistic Inquiry, 15*, 33–74.

Hays, W. L. (1963). *Statistics for Psychologists.* New York: Holt, Rinehard and Winston.

Hedrick, M. S., & Ohde, R. N. (1994). Effect of relative amplitude of frication on perception of place of articulation. *Journal of the Acoustical Society of America, 94*, 2005–2026.

Heinz, J. M., & Stevens, K. N. (1961). On the properties of voiceless fricative consonants. *Journal of the Acoustical Society of America, 33*, 589–596.

Helmholtz, H. L. F. (1863). *Die Lehre von den Tonempfindungen als physiologische Grundlage für die Theorie der Musik.* Braunschweig: Vieweg und Sohn.

Henton, C. G. (1983). Changes in the vowels of received pronunciation. *Language and Speech, 11*, 353–371.

Hillendbrand, J., Getty, L., Clark, M., & Wheeler, K. (1995). Acoustic characteristics of American English vowels. *Journal of the Acoustical Society of America, 97*, 3099–3111.

Hirano, M., Kakita, Y., Kawasaki, H., Gould, W. J., & Lambiase, A. (1981). Data from high-speed motion picture studies. In K. N. Stevens & M. Hirano (Eds.), *Vocal Fold Physiology* (pp. 85–93). Tokyo: University of Tokyo Press.

Hoard, J. E. (1966). Juncture and syllable structure in English. *Phonetica, 15*, 96–109.

Hoard, J. E. (1971). Aspiration, tenseness, and syllabification in English. *Language, 47*, 133–140.

Hogan, J. T., & Rozsypal, A. J. (1980). Evaluation of vowel duration as a cue for the voicing distinction in the following word-final consonant. *Journal of the Acoustical Society of America, 67*, 1764–1771.

Holbrook, A., & Fairbanks, G. (1962). Diphthong formants and their movements. *Journal of Speech and Hearing Research, 5*, 38–58.

Holmes, J. (1986). Normalization in vowel perception. In J. Perkell & D. Klatt (Eds.), *Invariance and Variability in Speech Processes* (pp. 346–357). New York: Academic Press.

Holmes, J., Mattingley, I., & Shearme, J. (1964). Speech synthesis by rule. *Language and Speech, 7*, 127–143.

Holmes, J. N. (1973). The influence of the glottal waveform on the naturalness of speech from a parallel formant synthesiser. *Institute of Electrical and Electronic Engineers, Transactions on Audio and Electroacoustics, AU-21*, 298–305.

Holmes, J. N. (1980). The JSRU channel vocoder. *Proceedings of the Institute of Electrical Engineers, 127*, 53–60.

Holmes, J. N. (1983). Formant synthesisers: cascade or parallel? *Speech Communication, 2*, 251–273.

Holmes, J. N. (1985). A parallel formant synthesiser for machine voice output. In F. Fallside & W. A. Woods (Eds.), *Computer Speech Processing* (pp. 163–188). Englewood Cliffs, NJ: Prentice-Hall.

House, A. S. (1957). Analog studies of nasal consonants. *Journal of Speech and Hearing Disorders, 22*, 190–204.

House, A. S. (1961). On vowel duration in English. *Journal of the Acoustical Society of America, 33*, 1174–1178.

House, A. S., & Fairbanks, G. (1953). The influence of consonantal environment on the secondary acoustical characteristics of vowels. *Journal of the Acoustical Society of America, 25*, 105–113.

House, A. S., & Stevens, K. N. (1956). Analog studies of the nasalisation of vowels. *Journal of Speech and Hearing Disorders, 21*, 218–231.

Howell, P., & Rosen, S. (1983). Production and perception of rise time in the voiceless affricate/fricative distinction. *Journal of the Acoustical Society of America, 73*, 976–984.

Huang, C. B. (1986). The effect of formant trajectory and spectral shape on the tense/lax distinction in American vowels. In *Institute of Electrical and Electronic Engineers, International Conference on Acoustics Speech and Signal Processing* (pp. 893–896). Tokyo, Japan.

Huang, C. B. (1992). Modeling human vowel identificatin using aspects of format trajectory and context. In Y. Tohkura, E. Vatikiotis-Bateson, & Y. Sagisaka (Eds.), *Speech Perception, Production and Linguistic Structure* (pp. 43–61). Amsterdam: IOS Press.

Hughes, G. W., & Halle, M. (1956). Spectral properties of fricative consonants. *Journal of the Acoustical Society of America, 28*, 303–310.

Hunnicutt, S. (1985). Intelligibility versus redundancy - conditions of dependency. *Language and Speech, 28*, 47–56.

Hunnicutt, S. (1987). Acoustic correlates of redundancy and intelligibility. *Speech Transmission Laboratory, Quarterly Progress Status Report, 2-3*, 7–14.

Huss, V. (1978). English word stress in postnuclear position. *Phonetica, 35*, 86–105.

Huttenlocher, D. O., & Zue, V. (1984). A model of lexical access based on partial phonetic information. In *Institute of Electrical and Electronic Engineers, International Conference on Acoustics Speech and Signal Processing* (pp. 26.4.1–26.4.4).

Isard, S., & Millar, D. (1986). Diphone synthesis techniques. In *Proceedings of the Institute of Electronic Engineers Speech Input/Output Conference* (pp. 77–82).

Ishizaka, K., Matsudaira, M., & Kaneko, T. (1976). Input acoustic-impedance measurement of the subglottal system. *Journal of the Acoustical Society of America, 60*, 190–197.

Itakura, F., & Saito, S. (1970). A statistical method for estimation of speech spectral density and formant frequencies. *Electronics and Communication, Japan, 53-A*, 36–43.

Itakura, F., & Saito, S. (1973a). Analysis synthesis method based on the maximum likelihood method. In J. L. Flanagan & L. R. Rabiner (Eds.), *Speech Synthesis* (pp. 287–292). Stroudsberg, Pennsylvania: Dowden, Hutchinson and Ross.

Itakura, F., & Saito, S. (1973b). On the optimum quantisation of feature parameters in the PARCOR speech synthesiser. In J. L. Flanagan & L. R. Rabiner (Eds.), *Speech Synthesis* (pp. 301–304). Stroudsberg, Pennsylvania: Dowden, Hutchinson and Ross.

Jackson, J. E. (1991). *A User's Guide to Principal Components.* New York: Wiley.

Jakobson, R., Fant, G., & Halle, M. (1952). *Preliminaries to Speech Analysis (MIT Acoustics Laboratory Technical Report, 13).* Cambridge, MA: MIT Press.

Jassem, V. (1965). The formants of fricative consonants. *Language and Speech, 8*, 1–16.

Jenkins, J. J., Strange, W., & Miranda, S. (1994). Vowel identification in mixed-speaker silent-center syllables. *Journal of the Acoustical Society of America, 95*, 1030–1043.

Jha, S. K. (1985). Acoustic analysis of Maithilli diphthongs. *Journal of Phonetics, 13*, 107–115.

Johnson, K. (1990). Contrast and normalisation in vowel perception. *Journal of Phonetics, 18*, 229–254.

Johnson, T. L., & Strange, W. (1982). Perceptual constancy of vowels in rapid speech. *Journal of the Acoustical Society of America, 72*, 1761–1770.

Jones, D. (1917). *An English Pronouncing Dictionary* (1st ed.). London: Dent's Modern Language Series.

Jones, D. (1956). The hyphen as a phonetic sign. *Zeitschrift für Phonetik Sprachwissenschaft und Kommunikationsforschung, 9*, 99–107.

Jongman, A. (1989). Duration of frication noise required for identification of English fricatives. *Journal of the Acoustical Society of America, 85*, 1718–1725.

Joos, M. (1948). Acoustic Phonetics. *Language, 24*, 1–136.

Just, M. A., Suslick, R. L., Michaels, S., & Shockey, L. (1978). Acoustic cues and psychological processes in the perception of natural stop consonants. *Perception and Psychophysics, 24*, 327–336.

Kahn, D. (1976). *Syllable-based generalisations in English phonology.* Unpublished doctoral dissertation, MIT, Cambridge, MA.

Kaplan, H. M. (1960). *Anatomy and Physiology of Speech.* New York: McGraw-Hill.

Karlsson, I. (1990). Voice source dynamics for female speakers. In *Proceedings of the International Conference on Spoken Language Processing* (pp. 69–72).

Karlsson, I. (1991). Female voices in speech synthesis. *Journal of Phonetics, 19*, 111–120.

Karlsson, I. (1992). Modelling voice variations in female speech synthesis. *Speech Communication, 11*, 491–495.

Karlsson, I., & Neovius, L. (1993). Speech synthesis experiments with the GLOVE synthesiser. In *Proceedings of Eurospeech 93, 4th European Conference on Speech Communication and Technology* (pp. 925–928).

Kelso, J. A. S., Vatikiotis-Bateson, E., Saltzman, E., & Kay, B. (1985). A qualitative dynamic analysis of reiterant speech production: phase portraits, kinematics, and dynamic modeling. *Journal of the Acoustical Society of America, 77*, 266–280.

Kent, R. D., & Read, C. (1992). *The Acoustic Analysis of Speech.* San Diego, California: Singular.

Kewley-Port, D. (1982). Measurements of formant transitions in naturally produced stop consonant-vowel syllables. *Journal of the Acoustical Society of America, 72*, 379–389.

Kewley-Port, D. (1983). Time-varying features as correlates of place of articulation in stop consonants. *Journal of the Acoustical Society of America, 73*, 322–335.

Kewley-Port, D., & Atal, B. S. (1989). Perceptual differences between vowels located in a limited phonetic space. *Journal of the Acoustical Society of America, 85*, 1726–1740.

Kewley-Port, D., Pisoni, D., & Studdert-Kennedy, M. (1983). Perception of static and dynamic acoustic cues to place of articulation in initial stop consonants. *Journal of the Acoustical Society of America, 73*, 1779–1793.

Kingdon, R. (1958). *The Groundwork of English Intonation.* London: Longmans.

Klatt, D. (1980). Software for a cascade/parallel formant synthesiser. *Journal of the Acoustical Society of America, 67*, 971–995.

Klatt, D. H. (1973). Interaction between two factors that influence vowel duration. *Journal of the Acoustical Society of America, 54*, 1102–1104.

Klatt, D. H. (1975). Vowel length is syntactically determined in a connected discourse. *Journal of Phonetics, 3*, 129–140.

Klatt, D. H. (1982). The Klattalk text-to-speech system. In *Institute of Electrical and Electronic Engineers, International Conference on Acoustics Speech and Signal Processing* (pp. 1589–1592).

Klatt, D. H. (1987). Review of text-to-speech conversion for English. *Journal of the Acoustical Society of America, 82*, 737–793.

Klatt, D. H., & Klatt, L. C. (1990). Analysis, synthesis, and perception of voice quality variations among female and male talkers. *Journal of the Acoustical Society of America, 87*, 820–857.

Klecka, W. R. (1980). *Discriminant Analysis.* Beverly Hills: Sage Publications.

Klein, W., Plomp, R., & Pols, L. C. W. (1970). Vowel spectra, vowel spaces, and vowel identification. *Journal of the Acoustical Society of America, 48*, 999–1009.

Koenig, W., Dunn, H. K., & Lacy, L. Y. (1946). The sound spectrograph. *Journal of the Acoustical Society of America, 18*, 19–49.

Kohler, K. J. (1985). F0 in the perception of lenis and fortis plosives. *Journal of the Acoustical Society of America, 78*, 21–32.

Koopmans-van Beinum, F. J., & van Bergem, D. R. (1989). The role of given and new in the production and perception of vowel contrasts in read text and in spontaneous speech. In *European Conference on Speech Communication and Technology* (Vol. 2, pp. 113–116).

Krull, D. (1988). Acoustic properties as predictors of perceptual responses: a study of Swedish voiced stops. *Phonetic Experimental Research at the Institute of Linguistics, University of Stockholm (PERILUS), 7*, 66–70.

Krull, D. (1989). Second formant locus patterns and consonant-vowel coarticulation in spontaneous speech. *Phonetic Experimental Research at the Institute of Linguistics, University of Stockholm (PERILUS), 10*, 87–108.

Kruskal, J. B. (1964). Multidimensional scaling by optimizing goodness of fit to a non-metric hypothesis. *Psychometrika, 29*, 115–129.

Kuehn, D. P., & Moll, K. L. (1976). A cineradiographic study of VC and CV articulatory velocities. *Journal of Phonetics, 4*, 303–320.

Kurowski, K., & Blumstein, S. (1984). Perceptual integration of the murmur and formant transitions for place of articulation in nasal consonants. *Journal of the Acoustical Society of America, 76*, 383–390.

Kurowski, K., & Blumstein, S. (1987). Acoustic properties for place of articulation in nasal consonants. *Journal of the Acoustical Society of America, 81*, 1917–1927.

Lacerda, F. (1982). Acoustic perceptual study of the Portuguese voiceless fricatives. *Journal of Phonetics, 10*, 11–22.

Ladd, D. (1986). Intonational phrasing: the case for recursive prosodic structure. *Journal of the Acoustical Society of America, 3*, 307–370.

Ladd, D. (1988). Declination "reset" and the hierarchical organization of utterances. *Journal of the Acoustical Society of America, 84*, 530–544.

Ladd, D., & Campbell, N. (1991). Theories of prosodic structure: evidence from syllable duration. In *Proceedings of the 12th International Conference of Phonetic Sciences, Aix-en-Provence.*

Ladd, D. R. (1996). *Intonational Phonology.* Cambridge: Cambridge University Press.

Ladefoged, P. (1962). *Elements of Acoustic Phonetics.* Edinburgh: Oliver & Boyd.

Ladefoged, P. (1967). *Three Areas of Experimental Phonetics.* Oxford: Oxford University Press.

Ladefoged, P. (1971). *Preliminaries to Linguistic Phonetics.* Chicago: University of Chicago Press.

Ladefoged, P. (1985). The phonetic basis for computer speech processing. In F. Fallside & W. A. Woods (Eds.), *Computer Speech Processing* (pp. 3–27). Englewood Cliffs, NJ: Prentice-Hall.

Ladefoged, P. (1993). *A Course in Phonetics* (3rd ed.). London: Harcourt Brace Jovanovich.

Ladefoged, P., & Bladon, A. (1982). Attempts by human speakers to reproduce Fant's nomograms. *Journal of the Acoustical Society of America, 1*, 185–197.

Ladefoged, P., & Broadbent, D. (1957). Information conveyed by vowels. *Journal of the Acoustical Society of America, 29*, 98–104.

Ladefoged, P., Harshman, R., Goldstein, L., & Rice, L. (1978). Generating vocal tract shapes from formant frequencies. *Journal of the Acoustical Society of America, 64*, 1027–1035.

Lahiri, A., Gewirth, L., & Blumstein, S. (1984). A reconsideration of acoustic invariance for place of articulation in diffuse stop consonants: evidence from a cross-language study. *Journal of the Acoustical Society of America, 76*, 391–404.

LaRiviere, C., Winitz, H., & Herriman, E. (1975). The distribution of perceptual cues in English prevocalic fricatives. *Journal of Speech and Hearing Research, 18*, 613–622.

Larkey, L., Wald, J., & Strange, W. (1978). Perception of synthetic nasal consonants in initial and final syllable position. *Perception and Psychophysics, 23*, 299–311.

Lass, N. J. (Ed.). (1996). *Principles of Experimental Phonetics.* St. Louis, Missouri: Mosby.

Laver, J. (1980). *The Phonetic Description of Voice Quality.* Cambridge: Cambridge University Press.

Laver, J. (1994). *Principles of Phonetics.* Cambridge: Cambridge University Press.

Lawrence, W. (1953). The synthesis of speech from signals which have a low information rate. In W. Jackson (Ed.), *Communication Theory.* Butterworths.

Lehiste, I. (1960). An acoustic-phonetic study of internal open juncture. *Phonetica, 5 (Supplement).*

Lehiste, I. (1964). *Acoustical Characteristics of Selected English Consonants.* Indiana University: Bloomington.

Lehiste, I. (Ed.). (1967). *Readings in Acoustic Phonetics.* Cambridge, MA: MIT Press.

Lehiste, I. (1970). *Suprasegmentals.* Cambridge, MA: MIT Press.

Lehiste, I. (1973). Phonetic disambiguation of syntactic ambiguity. *Glossa, 7,* 107–122.

Lehiste, I. (1977). Isochrony reconsidered. *Journal of Phonetics, 5,* 253–263.

Lehiste, I. (1979). Perception of sentence and paragraph boundaries. In B. Lindblom & S. E. G. Öhman (Eds.), *Frontiers of Speech Communication Research* (pp. 191–201). New York: Academic Press.

Lehiste, I., & Peterson, G. (1961a). Some basic considerations in the analysis of intonation. *Journal of the Acoustical Society of America, 33,* 419–425.

Lehiste, I., & Peterson, G. (1961b). Transitions, glides and diphthongs. *Journal of the Acoustical Society of America, 33,* 268–277.

Lehiste, I., & Peterson, G. E. (1959). Vowel amplitude and phonemic stress in American English. *Journal of the Acoustical Society of America, 31,* 428–435.

Liberman, A. M., Delattre, P., & Cooper, F. S. (1952). The role of selected stimulus-variables in the perception of the unvoiced stop consonants. *American Journal of Psychology, 65,* 497–516.

Liberman, A. M., Delattre, P. C., & Cooper, F. S. (1958). The role of selected stimulus variables in the perception of voiced and voiceless stops in initial position. *Language and Speech, 1,* 153–167.

Liberman, A. M., Delattre, P. C., Cooper, F. S., & Gerstman, L. J. (1954). The role of consonant-vowel transitions in the perception of the stop and nasal consonants. *Psychological Monographs: General and Applied, 68,* 1–13.

Liberman, M., Harris, K. S., Kinney, J. A., & Lane, H. (1961). The discrimination of relative onset times of the components of certain speech and nonspeech patterns. *Journal of Experimental Psychology, 61,* 379–388.

Liberman, M., & Prince, A. (1977). On stress and linguistic rhythm. *Linguistic Inquiry, 8,* 249–336.

Lieberman, P. (1963). Some effects of semantic and grammatical context on the production and perception of speech. *Language and Speech, 6,* 172–187.

Lieberman, P. (1968). Vocal cord motion in man. In A. Bouhuys (Ed.), *Sound Production in Man* (pp. 28–38). New York: Annals of the New York Academy of Science.

Liljencrants, J., & Lindblom, B. (1972). Numerical simulations of vowel quality systems: The role of perceptual contrast. *Language, 48,* 839–862.

Lin, Q. (1992). Vocal tract computation: how to make it robuster and faster. *Speech Transmission Laboratory, Quarterly Progress Status Report, 4,* 29–42.

Lindau-Webb, M. (1985). Hausa vowels and diphthongs. *UCLA Working Papers in Phonetics. University of California, Los Angeles, 60,* 40–54.

Lindblom, B. (1963). Spectrographic study of vowel reduction. *Journal of the Acoustical Society of America, 35,* 1773–1781.

Lindblom, B. (1964). Articulatory activity in vowels. *Speech Transmission Laboratory, Quarterly Progress Status Report, 2*, 1–5.

Lindblom, B., & Rapp, K. (1973). *Some temporal regularities of spoken Swedish* (Vol. 21; Tech. Rep.). Institute of Linguistics, University of Stockholm.

Lindblom, B., & Studdert-Kennedy, M. (1967). On the role of formant transitions in vowel recognition. *Journal of the Acoustical Society of America, 42*, 830–843.

Lindblom, B., & Sundberg, J. (1971). Acoustical consequences of lip, tongue, jaw, and larynx movement. *Journal of the Acoustical Society of America, 50*, 1166–1179.

Linggard, R. (1985). *Electronic Synthesis of Speech.* Cambridge: Cambridge University Press.

Lisker, L. (1957). Closure duration and the intervocalic voiced-voiceless distinction in English. *Language, 33*, 42–49.

Lisker, L. (1975). Is it VOT or a first formant detector? *Journal of the Acoustical Society of America, 57*, 1547–1551.

Lisker, L. (1978). Rapid vs. rabid: a catalogue of acoustic features that may cue the distinction. *Haskins Laboratories Status Reports on Speech Research, 54*, 127–132.

Lisker, L. (1984). On reconciling monophthongal vowel percepts and continuously varying F-patterns. *Haskins Laboratories Status Reports on Speech Research, 79/80*, 167–174.

Lisker, L., & Abramson, A. S. (1964). A cross-language study of voicing in initial stops. *Word, 20*, 384–422.

Lisker, L., & Abramson, A. S. (1967). Some effects of context on voice onset time in English stops. *Language and Speech, 10*, 1–28.

Lloyd, R. J. (1890). *Some Researches into the Nature of Vowel-Sound.* Liverpool, U.K.: Turner and Dunnett.

Lobanov, B. M. (1971). Classification of Russian vowels spoken by different speakers. *Journal of the Acoustical Society of America, 49*, 606–608.

Local, J. (1994). Phonological structure, parametric phonetic interpretation and natural-sounding synthesis. In E. Keller (Ed.), *Fundamentals of Speech Synthesis and Speech Recognition: Basic Concepts, State-of-the-Art and Future Challenges* (pp. 253–270). Chichester: Wiley.

Löfqvist, A. (1975). Intrinsic and extrinsic F0 variations in Swedish tonal accents. *Phonetica, 31*, 228–247.

Loizou, P., Dorman, M., & Spanias, A. (1995). Automatic recognition of syllable-final nasals preceded by /ɛ/. *Journal of the Acoustical Society of America, 97*, 1925–1928.

Lyberg, B. (1977). Some observations on the timing of Swedish utterances. *Journal of Phonetics, 5*, 49–59.

Macchi, M. J. (1980). Identification of vowels spoken in isolation versus vowels spoken in consonantal context. *Journal of the Acoustical Society of America, 68*, 1636–1642.

Mack, M., & Blumstein, S. E. (1983). Further evidence of acoustic invariance in speech production: the stop-glide contrast. *Journal of the Acoustical Society of America, 73*, 1739–1750.

Maeda, S. (1990). Compensatory articulation during speech: evidence from the analysis and synthesis of vocal-tract shapes using an articulatory model. In W. J. Hardcastle & A. Marchal (Eds.), *Speech Production and Speech Modelling* (pp. 131–149). Dordrecht: Kluwer.

Makhoul, J. (1973). Spectral analysis of speech by linear prediction. *Institute of Electrical and Electronic Engineers, Transactions on Audio and Electroacoustics, 21*, 140–148.

Makhoul, J. (1975). Linear prediction - a tutorial review. In *Proceedings Institute of Electrical and Electronic Engineers* (Vol. 63, pp. 561–580).

Makhoul, J. (1977). Stable and efficient lattice methods for linear prediction. *Institute of Electrical and Electronic Engineers, Transactions on Acoustics Speech and Signal Processing, 25*, 423–428.

Makhoul, J., & Wolf, J. (1972). *Linear Prediction and the Spectral Analysis of Speech* (BBN Report No. 2304). Cambridge, MA: Bolt Beranek and Newman Inc.

Malécot, A. (1956). Acoustic cues or nasal consonants: an experimental study involving a tape-splicing technique. *Language and Speech, 32*, 274–284.

Malme, C. (1959). Detectability of small irregularities in a broadband noise spectrum. *Research Lab. of Electronics, Quarterly Progress Report, MIT, 52*, 139–141.

Mann, V. A., & Repp, B. H. (1980). Influence of vocalic context on perception of the [s]-[ʃ] distinction. *Perception and Psychophysics, 28*, 213–228.

Manrique, A., & Massone, M. (1981). Acousic analysis and perception of Spanish fricative consonants. *Journal of the Acoustical Society of America, 69*, 1145–1153.

Markel, J. (1972a). Digital inverse filtering: a new tool for formant trajectory estimation. *Institute of Electrical and Electronic Engineers, Transactions on Audio and Electroacoustics, AU-20*, 129–137.

Markel, J. (1972b). The SIFT algorithm for fundamental frequency estimation. *Institute of Electrical and Electronic Engineers, Transactions on Audio and Electroacoustics, 20*, 367–377.

Markel, J., & Gray, A. H. (1976). *Linear Prediction of Speech.* Berlin: Springer-Verlag.

Marslen-Wilson, W. (1987). Functional parallelism in spoken-word recognition. In U. H. Frauenfelder & L. K. Tyler (Eds.), *Spoken Word Recognition* (pp. 71–102). Cambridge, MA: MIT Press.

Marslen-Wilson, W. D. (1986). Aspects of human speech understanding. In F. Fallside & W. W. Woods (Eds.), *Computer Speech Processing* (pp. 383–404). Englewood Cliffs, New Jersey: Prentice-Hall.

Marslen-Wilson, W. D., & Welsh, A. (1978). Processing interactions during word recognition in continuous speech. *Cognitive Psychology, 10*, 29–63.

Massaro, D. W., & Cohen, M. M. (1976). The contribution of fundamental frequency and voice onset time to the /zi/-/si/ distinction. *Journal of the Acoustical Society of America, 60*, 704–717.

McClelland, J. L., & Elman, J. L. (1986). The TRACE model of speech perception. *Cognitive Psychology, 18*, 1–86.

Mermelstein, P. (1973). Articulatory model for the study of speech production. *Journal of the Acoustical Society of America, 53*, 1070–1082.

Meyer-Eppler, W. (1953). Zum Erzeugungsmechanismus der Gerauschläute. *Zeitschrift für Phonetik, 7*, 196–212.

Millar, J. D. (1989). Auditory-perceptual interpretation of the vowel. *Journal of the Acoustical Society of America, 85*, 2114–2134.

Millar, R. L. (1959). Nature of the vocal cord wave. *Journal of the Acoustical Society of America, 31*, 667–677.

Miller, J. L. (1981). Effects of speaking rate on segmental distinctions. In P. D. Eimas & J. L. Miller (Eds.), *Perspectives on the Study of Speech* (pp. 39–74). Hillsdale, NJ: Lawrence Erlbaum.

Miller, J. L. (1987). Rate-dependent processing in speech perception. In A. Ellis (Ed.), *Progress in the Psychology of Language vol. 3* (pp. 119–157). Hillsdale, NJ: Lawrence Erlbaum.

Milner, B. (1996). Inclusion of temporal information into features for speech recognition. In *Proceedings of the International Conference on Spoken Language Processing* (pp. 256–259).

Mohr, B. (1971). Intrinsic variations in the speech signal. *Phonetica, 23*, 65–93.

Monsen, R., & Engebretson, A. (1977). Study of variation in the male and female glottal wave. *Journal of the Acoustical Society of America, 62*, 981–993.

Moon, S.-J., & Lindblom, B. (1994). Interaction between duration, context, and speaking style in English stressed vowels. *Journal of the Acoustical Society of America, 96*, 40–55.

Moore, B., & Glasberg, B. (1983). Suggested formulae for calculating auditory-filter bandwidths and excitation patterns. *Journal of the Acoustical Society of America, 74*, 750–753.

Moore, B., & Glasberg, B. (1990). Derivation of auditory filter shapes from notched-noise data. *Hearing Research, 47*, 103–138.

Moore, B. J. C. (1989). *An Introduction to the Psychology of Hearing.* London: Academic Press.

Moore, P., & van Leden, H. (1958). Dynamic variation of the vibratory pattern in normal larynx. *Folia Phoniatrica, 10*, 205–238.

Mosteller, F., & Rourke, R. E. K. (1973). *Sturdy Statistics.* Reading, MA: Addison-Wesley.

Mullenix, J. M., Pisoni, D. B., & Martin, C. S. (1989). Some effects of talker variability on spoken word recognition. *Journal of the Acoustical Society of America, 85*, 365–378.

Nâbélek, A. K., & Dagenais, P. A. (1986). Vowel errors in noise and in reverberation by hearing-impaired listeners. *Journal of the Acoustical Society of America, 80*, 741–748.

Nakata, K. (1959). Synthesis and perception of nasal consonants. *Journal of the Acoustical Society of America, 31*, 661–666.

Nakatani, L. H., & Dukes, K. D. (1977). Locus of segmental cues for word juncture. *Journal of the Acoustical Society of America, 62*, 714–719.

Nakatani, L. H., & Schaffer, J. A. (1978). Hearing 'words' without words: prosodic cues for word perception. *Journal of the Acoustical Society of America, 63*, 234–245.

Nearey, T. M. (1977). *Phonetic feature systems for vowels.* Unpublished doctoral dissertation, University of Connecticut, Storrs, CT.

Nearey, T. M. (1989). Static, dynamic, and relational properties in vowel perception. *Journal of the Acoustical Society of America, 85*, 2088–2113.

Nearey, T. M., & Assmann, P. F. (1986). Modeling the role of inherent spectral change in vowel identification. *Journal of the Acoustical Society of America, 80*, 1297–1308.

Nespor, M., & Vogel, I. (1986). *Prosodic Phonology.* Dordrecht: Foris.

Nittrouer, S. (1995). Children learn separate aspects of speech production at different rates: evidence from spectral moments. *Journal of the Acoustical Society of America, 97*, 520–530.

Nolan, F. J. (1983). *The Phonetic Bases of Speaker Recognition.* Cambridge: Cambridge University Press.

Nooteboom, S. G., & Kruyt, J. G. (1987). Accents, focus distribution, and the perceived distribution of given and new information: an experiment. *Journal of the Acoustical Society of America, 82*, 1512–1524.

Nord, L. (1976). Perceptual experiments with nasals. *Speech Transmission Laboratory, Quarterly Progress Status Report, 2/3*, 5–8.

Nossair, Z. B., & Zahorian, S. A. (1991). Dynamic spectral shape features as acoustic correlates for initial stop consonants. *Journal of the Acoustical Society of America, 89*, 2978–2991.

Nyquist, H. (1928). Certain topics in telegraph transmission theory. *Transactions of the American Institute of Electrical Engineers, 47*, 617–644.

Ochiai, Y., Fukimura, T., & Nakatani, K. (1957). Timbre studies of nasalics, part II. *Memoirs of the Faculty of Engineering, Nagoya University, 9*, 160–173.

O'Connor, J., Gerstman, L., Liberman, A. M., Delattre, P., & Cooper, F. S. (1957). Acoustic cues for the perception of initial /w,j,r,l/ in English. *Word, 13*, 24–43.

Ohde, R. (1988). Revisiting stop-consonant perception for two-formant stimuli. *Journal of the Acoustical Society of America, 84*, 1551–1555.

Ohde, R., & Sharf, D. J. (1977). Order effect of acoustic segments of VC and CV syllables on stop and vowel identification. *Journal of Speech and Hearing Research, 20*, 543–554.

Ohde, R., & Stevens, K. N. (1983). Effect of burst amplitude on the perception of stop consonant place of articulation. *Journal of the Acoustical Society of America, 74*, 706–714.

Ohde, R. N. (1994). The development of the perception of cues for the [m] - [n] distinction in CV syllables. *Journal of the Acoustical Society of America, 96*, 675–686.

Öhman, S. E. G. (1966). Coarticulation in VCV utterances: spectrographic measurements. *Journal of the Acoustical Society of America, 39*, 151–168.

Olive, J. P., Greenwood, A., & Coleman, J. (1993). *Acoustics of American English Speech: A Dynamic Approach.* New York: Springer Verlag.

O'Shaughnessy, D. (1981). A study of French vowel and consonant durations. *Journal of Phonetics, 9*, 385–406.

O'Shaughnessy, D. (1987). *Speech Communication: Human and Machine.* Reading, MA: Addison-Wesley.

O'Shaughnessy, D., & Allen, J. (1983). Linguistic modality effects on fundamental frequency in speech. *Journal of the Acoustical Society of America, 74*, 1155–1171.

O'Shaughnessy, D., Barbeau, L., Bernardi, D., & Archambault, D. (1988). Diphone speech synthesis. *Speech Communication, 7*, 55–65.

Owens, F. J. (1993). *Signal Processing of Speech.* New York: McGraw-Hill.

Paget, R. (1923). The production of artificial vowel sounds. *Proceedings of the Royal Society A., 102*, 752–765.

Palethorpe, S. (1992). Speech intelligibility in communicative difficulty. In J. Pittam (Ed.), *Proceedings of the 4th International Conference on Speech Science and Technology (Brisbane)* (pp. 420–424).

Paliwal, K., Lindsay, D., & Ainsworth, W. (1983). A study of two-formant models for vowel identification. *Speech Communication, 2*, 295–303.

Parker, E. M., & Diehl, R. L. (1984). Identifying vowels in CVC syllables: effects of inserting silence and noise. *Perception and Psychophysics, 36*, 369–380.

Parker, F. (1974). The coarticulation of vowels and stop consonants. *Journal of Phonetics, 2*, 211–221.

Parsons, T. W. (1987). *Voice and Speech Processing.* New York: McGraw Hill.

Peeters, W. J. M., & Barry, W. J. (1989). Diphthong dynamics: production and perception in Southern British English. In *European Conference on Speech Communication and Technology* (Vol. 1, pp. 55–58). Paris.

Perkell, J., & Klatt, D. (Eds.). (1986). *Invariance and Variability in Speech Processes.* New York: Academic Press.

Perkell, J., & Nelson, W. (1985). Variability in the production of the vowels /i/ and /a/. *Journal of the Acoustical Society of America, 77*, 1889–1895.

Peterson, G., & Barney, H. (1952). Control methods used in a study of vowels. *Journal of the Acoustical Society of America, 24*, 175–184.

Peterson, G. E. (1952). The information-bearing elements of speech. *Journal of the Acoustical Society of America, 24*, 629–637.

Peterson, G. E. (1961). Parameters of vowel quality. *Journal of Speech and Hearing Research, 4*, 10–29.

Peterson, G. E., & Lehiste, I. (1960). Duration of syllable nuclei in English. *Journal of the Acoustical Society of America, 32*, 693–703.

Pickett, J. M. (1980). *The Sounds of Speech Communication.* Baltimore: University Park Press.

Pierrehumbert, J. (1980). *The phonology and phonetics of English intonation.* Unpublished doctoral dissertation, Massachusetts Institute of Technology, Cambridge MA.

Pierrehumbert, J., & Beckman, M. E. (1988). *Japanese Tone Structure.* Cambridge, MA: MIT Press.

Pierrehumbert, J., & Hirschberg, J. (1990). The meaning of intonation contours in the interpretation of discourse. In P. R. Cohen, J. Morgan, & M. E. Pollack (Eds.), *Intentions in Communication.* Cambridge, MA: MIT Press.

Pijper, J. de, & Sanderman, A. (1994). On the perceptual strength of prosodic boundaries and its relation to suprasegmental cues. *Journal of the Acoustical Society of America, 96*, 2037–2047.

Pike, K. L. (1945). *The Intonation of American English.* Ann Arbor: Michigan.

Pitrelli, J., Beckman, M., & Hirschberg, J. (1994). Evaluation of prosodic transcription labeling reliability in the ToBI framework. In *Proceedings of the Third International Conference on Spoken Language Processing* (pp. 123–126).

Plomp, R. (1975). Auditory analysis and timbre perception. In G. Fant & M. A. A. Tatham (Eds.), *Auditory Analysis and Perception of Speech* (pp. 7–22). London: Academic Press.

Polka, L., & Strange, W. (1985). Perceptual equivalence of acoustic cues that differentiates /r/ and /l/. *Journal of the Acoustical Society of America, 78*, 1187–1197.

Pols, L., Tromp, H., & Plomp, R. (1973). Frequency analysis of Dutch vowels from 50 male speakers. *Journal of the Acoustical Society of America, 53*, 1093–1101.

Pols, L., & van Son, R. (1993). Acoustics and perception of dynamic vowel segments. *Speech Communication, 13*, 135–147.

Pols, L. C. W. (1977). *Spectral analysis and identification of Dutch vowels.* Unpublished doctoral dissertation, University of Amsterdam, the Netherlands.

Poritz, A. B., & Richter, G. (1986). On hidden Markov models in isolated word recognition. In *Institute of Electrical and Electronic Engineers, International Conference on Acoustics Speech and Signal Processing* (pp. 705–708). Tokyo, Japan.

Port, R. F. (1976). *The influence of speaking tempo on the duration of stressed vowel and median stop in English trochee words.* Unpublished doctoral dissertation, University of Connecticut.

Port, R. F. (1979). The influence of tempo on stop closure duration as a cue for voicing and place. *Journal of Phonetics, 7*, 45–56.

Potter, R., Kopp, G., & Green, H. (1947). *Visible Speech.* New York: Dover. (Reprinted from Bell Labs publication.)

Potter, R. K., & Steinberg, J. C. (1950). Toward the specification of speech. *Journal of the Acoustical Society of America, 22*, 807–819.

Price, P. J., Ostendorf, M., Shattuck-Hufnagel, S., & Fong, C. (1991). The use of prosody in syntactic disambiguation. *Journal of the Acoustical Society of America, 90*, 2956–2970.

Pulgrum. (1970). *Syllable, Word, Nexus, Cursus.* the Hague: Mouton.

Qi, Y., & Fox, R. A. (1992). Analysis of nasal consonants using perceptual linear prediction. *Journal of the Acoustical Society of America, 91*, 1718–1726.

Quené, H. (1985). Word boundary perception in fluent speech: a listening experiment. *Progress report, Institute of Phonetics, University of Amsterdam, 10*, 69–85.

Quené, H. (1992). Durational cues for word segmentation in Dutch. *Journal of Phonetics, 20*, 331–350.

Quené, H. (1993). Segment durations and accent as cues to word segmentation in Dutch. *Journal of the Acoustical Society of America, 94*, 2027–2035.

Rabiner, L., & Juang, B. (1993). *Fundamentals of Speech Recognition.* New Jersey: Prentice-Hall.

Rabiner, L. R. (1968). Digital-formant synthesiser for speech synthesis studies. *Journal of the Acoustical Society of America, 43*, 822–828.

Rabiner, L. R., Levinson, S. E., Rosenberg, A. E., & Wilpon, J. G. (1979). Speaker Independent Recognition of Isolated Words using Clustering Techniques. *Institute of Electrical and Electronic Engineers, Transactions on Acoustics Speech and Signal Processing, 27*(4), 336–349.

Rabiner, L. R., & Schafer, R. W. (1978). *Digital Processing of Speech Signals.* Englewood Cliffs: NJ.: Prentice-Hall.

Rakerd, B., & Verbrugge, R. R. (1985). Linguistic and acoustic correlates of the perceptual structure found in an individual differences scaling study of vowels. *Journal of the Acoustical Society of America, 71*, 296–301.

Rakerd, B., & Verbrugge, R. R. (1987). Evidence that the dynamic information for vowels is talker independent in form. *Journal of Memory and Language, 26*, 558–563.

Rakerd, B., Verbrugge, R. R., & Shankweiler, D. P. (1984). Monitoring for vowels in isolation and in a consonantal context. *Journal of the Acoustical Society of America, 76*, 27–31.

Recasens, D. (1983). Place cues for nasal consonants with special reference to Catalan. *Journal of the Acoustical Society of America, 73*, 1346–1353.

Ren, H. (1986). On the acoustic structure of diphthongal syllables. *UCLA Working Papers in Phonetics. University of California, Los Angeles, 85.*

Repp, B. H. (1979). Relative amplitude of aspiration noise as a voicing cue for syllable-initial stop consonants. *Language and Speech, 27*, 173–189.

Repp, B. H. (1986). Perception of the [m]-[n] distinction in CV syllables. *Journal of the Acoustical Society of America, 79*, 1987–1999.

Repp, B. H. (1987). On the possible role of auditory short-term adaptation in perception of the prevocalic [m]-[n] contrast. *Journal of the Acoustical Society of America, 82*, 1525–1538.

Repp, B. H., Liberman, A. M., Eccardt, T., & Pesetsky, M. (1978). Perceptual integration of acoustic cues for stop, fricative and affricate manner. *Journal of Experimental Psychology: Human Perception and Performance, 4*, 621–637.

Repp, B. H., & Lin, H.-B. (1989). Acoustic properties and perception of stop consonant release transients. *Journal of the Acoustical Society of America, 85*, 379–396.

Repp, B. H., & Svastikula, K. (1988). Perception of the [m]-[n] distinction in VC syllables. *Journal of the Acoustical Society of America, 83*, 237–247.

Revoile, S., Pickett, J. M., & Holden, L. D. (1982). Acoustic cues to final stop voicing for impaired and normal-hearing listeners. *Journal of the Acoustical Society of America, 72*, 1145–1154.

Rosen, G. (1956). Dynamic analog speech synthesiser. *Journal of the Acoustical Society of America, 30*, 201–209.

Rosen, S., & Howell, P. (1992). *Signals and Systems for Speech and Hearing.* London: Academic Press.

Rosenberg, A. E. (1971). Effect of glottal pulse shape on the quality of natural vowels. *Journal of the Acoustical Society of America, 49*, 583–590.

Rothenberg, M. (1973). A new inverse filtering technique for deriving the glottal air flow during voicing. *Journal of the Acoustical Society of America, 53*, 1632–1645.

Rothenberg, M. (1977). Measurement of air flow in speech. *Journal of Speech and Hearing Research, 20*, 155–176.

Rothenberg, M. (1981). Some relations between glottal air flow and vocal fold contact area. *ASHA Reports, 11*, 88–96.

Rubin, P., Baer, T., & Mermelstein, P. (1981). An articulatory synthesiser for perceptual research. *Journal of the Acoustical Society of America, 70*, 321–328.

Sakoe, H., & Chiba, S. (1977). Dynamic Programiming Algorithm Optimisation of Spoken Word Recognition. *Institute of Electrical and Electronic Engineers, Transactions on Acoustics Speech and Signal Processing, 26*(1), 43–49.

Saltzman, E., & Munhall, K. (1989). A dynamical approach to gestural patterning in speech production. *Ecological Psychology, 1*, 333–382.

Sanderman, A., & Collier, R. (1996). Prosodic rules for the implementation of phrase boundaries in synthetic speech. *Journal of the Acoustical Society of America, 100*, 3390–3397.

Sawashima, M., & Hirose, H. (1968). New laryngoscopic technique by use of fiber optics. *Journal of the Acoustical Society of America, 43*, 168–169.

Scharf, B. (1970). Critical bands. In J. V. Tobias (Ed.), *Foundations of Modern Auditory Theory* (pp. 157–200). New York: Academic Press.

Schouten, M. E. H. (1992). *The Auditory Processing of Speech.* Berlin: de Gruyter.

Schouten, M. E. H., & Pols, L. C. W. (1979a). Vowel segments in consonantal contexts: a spectral study of coarticulation - part I. *Journal of Phonetics, 7*, 1–23.

Schouten, M. E. H., & Pols, L. C. W. (1979b). CV and VC transitions: a spectral study of coarticulation - part II. *Journal of Phonetics, 7*, 205–224.

Schroeder, M. R., Atal, B. S., & Hall, J. L. (1979). Auditory analysis and timbre perception. In B. Lindblom & S. E. G. Öhman (Eds.), *Frontiers of Speech Communication Research* (pp. 217–229). London: Academic Press.

Scott, D. (1982). Duration as a cue to the perception of a phrase boundary. *Journal of the Acoustical Society of America, 71*, 996–1007.

Seitz, P., McCormick, M., Watson, I. M. C., & Bladon, R. A. W. (1990). Relational spectral features for place of articulation in nasal consonants. *Journal of the Acoustical Society of America, 87*, 351–358.

Selkirk, E. O. (1982). The syllable. In H. van der Hulst & N. Smith (Eds.), *The Structure of Phonological Representations (part 2).* Dordrecht: Foris.

Selkirk, E. O. (1984). *Phonology and Syntax: the Relation Between Sound and Structure.* Cambridge, MA: MIT Press.

Shadle, C. (1997). The aerodynamics of speech. In W. Hardcastle & J. Laver (Eds.), *The Handbook of Phonetic Sciences* (pp. 33–64). Oxford: Blackwell.

Shadle, C. H. (1990). Articulatory-acoustic relationships in fricative consonants. In W. Hardcastle & A. Marchal (Eds.), *Speech Production and Speech Modelling*. Dordrecht: Kluwer.

Shadle, C. H. (1991). The effect of source geometry on source mechanisms in fricative consonants. *Journal of Phonetics*, *19*, 409–424.

Shattuck-Hufnagel, S., Ostendorf, M., & Ross, K. (1994). Stress shift and early pitch accent placement in lexical items in American English. *Journal of Phonetics*, *22*, 357–388.

Shepard, R. N. (1972). Psychological representation of speech sounds. In E. D. David & D. P. Denes (Eds.), *Human Communication: a Unified View* (pp. 67–113). New York: McGraw-Hill.

Silverman, K. (1984). F0 perturbations as a function of voicing of prevocalic and postvocalic stops and fricatives, and of syllable stress. In *Proceedings of the Institute of Acoustics (Autumn Conference, Windermere)* (Vol. 6, pp. 445–452).

Slis, I., & Cohen, A. (1969). On the complex regulating the voiced-voiceless distinction. *Language and Speech*, *12*, 80–102.

Sluijter, A., & Heuven, V. van. (1996). Spectral balance as an acoustic correlate of linguistic stress. *Journal of the Acoustical Society of America*, *100*, 2471–2485.

Smith, S. (1951). Vocalisation and added nasal resonance. *Folia Phoniatrica*, *3*, 165–169.

Soli, S. (1981). Second formants in fricatives: acoustic consequences of fricative-vowel coarticulation. *Journal of the Acoustical Society of America*, *70*, 976–984.

Soli, S. (1982). Structure and duration of vowels together specify fricative voicing. *Journal of the Acoustical Society of America*, *72*, 366–378.

Sondhi, M., & Resnick, J. (1983). The inverse problem for the vocal tract: numerical methods, acoustical experiments and speech synthesis. *Journal of the Acoustical Society of America*, *73*, 985–1002.

Sondhi, M. M. (1979). Estimation of vocal tract areas: the need for acoustical measurements. *Institute of Electrical and Electronic Engineers, Transactions on Acoustics Speech and Signal Processing*, *27*, 268–273.

Sonesson, B. (1959). A method of studying the vibratory movements of the vocal cords: A preliminary report. *Journal of Laryngology*, *73*, 732–737.

Sonesson, B. (1960). On the anatomy and vibratory pattern of the human vocal folds. *Acta Otolaryngologica*, *156*, 1–80.

Sonoda, Y. (1987). Effect of speaking rate on articulatory dynamics and motor event. *Journal of Phonetics*, *15*, 145–156.

Steiglitz, K. (1996). *A Digital Signal Processing Primer*. Menlo Park, California: Addison-Wesley.

Stevens, K. (1997). Articulatory-acoustic-auditory relationships. In W. Hardcastle & J. Laver (Eds.), *The Handbook of Phonetic Sciences* (pp. 462–506). Oxford: Blackwell.

Stevens, K. N. (1972a). Airflow and turbulent noise for fricative and stop consonants: static considerations. *Journal of the Acoustical Society of America*, *50*, 1182–1192.

Stevens, K. N. (1972b). The quantal nature of speech: Evidence from articulatory-acoustic data. In E. E. Davis & P. B. Denes (Eds.), *Human Communication: A Unified View* (pp. 51–66). New York: McGraw-Hill.

Stevens, K. N. (1985). Evidence for the role of acoustic boundaries in the perception of speech sounds. In V. Fromkin (Ed.), *Phonetic Linguistics: Essays in Honor of Peter Ladefoged* (pp. 243–254). New York: Academic Press.

Stevens, K. N. (1989). On the quantal nature of speech. *Journal of Phonetics, 17*, 3–45.

Stevens, K. N., Blumstein, S. E., Glicksman, L., Burton, M., & Kurowski, K. (1992). Acoustic and perceptual characteristics of voicing in fricatives and fricative clusters. *Journal of the Acoustical Society of America, 91*, 2979–3000.

Stevens, K. N., & House, A. S. (1955). Development of a quantitative description of vowel articulation. *Journal of the Acoustical Society of America, 27*, 484–493.

Stevens, K. N., & House, A. S. (1956). Studies of formant transitions using a vocal tract analog. *Journal of the Acoustical Society of America, 28*, 578–585.

Stevens, K. N., & House, A. S. (1963a). An acoustical theory of vowel production and some of its implications. *Journal of Speech and Hearing Research, 4*, 303–320.

Stevens, K. N., & House, A. S. (1963b). Perturbation of vowel articulations by consonantal context: an acoustical study. *Journal of Speech and Hearing Research, 6*, 111–127.

Stevens, K. N., House, A. S., & Paul, A. P. (1966). Acoustical description of syllabic nuclei: an interpretation in terms of a dynamic model of articulation. *Journal of the Acoustical Society of America, 40*, 123–132.

Stevens, K. N., Kasowski, S., & Fant, G. (1953). An electrical analog of the vocal tract. *Journal of the Acoustical Society of America, 25*, 734–742.

Stevens, K. N., & Klatt, D. H. (1974). Role of formant transitions in the voiced-voiceless distinction for stops. *Journal of the Acoustical Society of America, 55*, 653–659.

Stewart, J. Q. (1922). An electrical analogue of the vocal organs. *Nature, 110*, 311–312.

Strange, W. (1987). Information for vowels in formant transitions. *Journal of Memory and Language, 26*, 550–557.

Strange, W. (1989a). Dynamic specification of coarticulated vowels spoken in sentence context. *Journal of the Acoustical Society of America, 85*, 2135–2153.

Strange, W. (1989b). Evolving theories of vowel perception. *Journal of the Acoustical Society of America, 85*, 2081–2087.

Strange, W., Jenkins, J. J., & Johnson, T. L. (1983). Dynamic specification of coarticulated vowels. *Journal of the Acoustical Society of America, 74*, 695–705.

Strange, W., Verbrugge, R. R., Shankweiler, D. P., & Edman, T. R. (1976). Consonant environment specifies vowel identity. *Journal of the Acoustical Society of America, 60*, 213–224.

Strevens, P. (1960). Spectra of fricative noise in human speech. *Language and Speech, 3*, 32–49.

Summerfield, Q., & Haggard, M. (1975). Vocal tract normalization as demonstrated by reaction time. In G. Fant & M. A. A. Tatham (Eds.), *Auditory Analysis and Perception of Speech* (pp. 115–141). London: Academic Press.

Summerfield, Q., & Haggard, M. (1977). On the dissociation of spectral and temporal cues to the voicing distinction in initial stop consonants. *Journal of the Acoustical Society of America, 62*, 435–448.

Summers, W. (1987). Effects of stress and final-consonant voicing on vowel production: articulatory and acoustic analyses. *Journal of the Acoustical Society of America, 82*, 847–863.

Suomi, K. (1985). The vowel-dependence of gross spectral cues to place of articulation of stop consonants in CV syllables. *Journal of Phonetics, 13*, 267–285.

Suomi, K. (1987). On spectral coarticulation in stop-vowel-stop syllables: implications for automatic speech recognition. *Journal of Phonetics, 15*, 85–100.

Sussman, H. (1990). Acoustic correlates of the front/back vowel distinction: a comparison of transition onset versus 'steady-state'. *Journal of the Acoustical Society of America, 88*, 87–96.

Sussman, H., Fruchter, D., & Cable, A. (1995). Locus equations derived from compensatory articulation. *Journal of the Acoustical Society of America, 97*, 3112–3124.

Sussman, H. M., Hoemeke, K. A., & Ahmed, F. S. (1993). A cross-linguistic investigation of locus equations as a phonetic descriptor of place of articulation. *Journal of the Acoustical Society of America, 94*, 1256–1268.

Sussman, H. M., McCaffrey, H. A., & Matthews, S. A. (1991). An investigation of locus equations as a source of relational invariance for stop place categorization. *Journal of the Acoustical Society of America, 90*, 1309–1325.

Syrdal, A., & Gopal, H. S. (1986). A perceptual model of vowel recognition based on the auditory representation of American English vowels. *Journal of the Acoustical Society of America, 79*, 1086–1100.

Syrdal, A. K. (1985). Aspects of a model of the auditory representation of American English vowels. *Speech Communication, 4*, 121–135.

Tarnóczy, T. (1948). Resonance data concerning nasals, laterals, and trills. *Word, 4*, 71–77.

Taylor, H. C. (1933). The fundamental pitch of English vowels. *Journal of Experimental Psychology, 16*, 565–582.

Terbeek, D. (1977). Cross-language multidimensional scaling study of vowel perception. *UCLA Working Papers in Phonetics. University of California, Los Angeles, 37.*

Thorsen, N. (1985). Intonation and text in standard Danish. *Journal of the Acoustical Society of America, 77*, 1205–1216.

Thorsen, N. (1986). Sentence intonation in textual context - supplementary data. *Journal of the Acoustical Society of America, 80*, 1041–1047.

Tiffany, W. R. (1959). Non-random sources of variation in vowel quality. *Journal of Speech and Hearing Research, 2*, 305–317.

Traunmüller, H. (1981). Perceptual dimensions of openness in vowels. *Journal of the Acoustical Society of America, 69*, 1465–1475.

Traunmüller, H., & Laçerda, F. (1987). Perceptual relativity in identification of two-formant vowels. *Speech Communication, 6*, 143–157.

Umeda, N. (1975). Vowel duration in American English. *Journal of the Acoustical Society of America, 58*, 434–445.

Umeda, N. (1981). Influence of segmental factors on fundamental frequency in fluent speech. *Journal of the Acoustical Society of America, 70*, 350–355.

van Bergem, D. R. (1993). Acoustic vowel reduction as a function of sentence accent, word stress, and word class. *Speech Communication, 12*, 1–23.

van Bergem, D. R. (1994). A model of coarticulatory effects on the schwa. *Speech Communication, 14*, 143–162.

van Bergem, D. R., Pols, L. C. W., & Koopmans-van Beinum, F. J. (1988). Perceptual normalisation of the vowels of a man and a child in various contexts. *Speech Communication, 7*, 1–20.

van den Berg, J. (1958). Myoelastic-aerodynamic theory of voice production. *Journal of Speech and Hearing Research, 1*, 227–244.

van den Berg, J., Zantema, J., & Doornebal, P. (1957). On the air resistance and the Bernoulli effect of the human larynx. *Journal of the Acoustical Society of America, 29*, 626–631.

van Heuven, V., & Pols, L. C. W. (1993). *Analysis and Synthesis of Speech - Strategic Research Towards High Quality Text-to-Speech Generation.* Berlin: Mouton de Gruyter.

van Son, R. J. J. H., & Pols, L. C. W. (1990). Formant frequencies of Dutch vowels in a text, read at normal and fast rate. *Journal of the Acoustical Society of America, 88*, 1683–1693.

van Son, R. J. J. H., & Pols, L. C. W. (1992). Formant movements of Dutch vowels in a text, read at normal and fast rate. *Journal of the Acoustical Society of America, 92*, 121–127.

Vanderslice, R., & Ladefoged, P. (1972). Binary suprasegmental features and transformational word-accentuation rules. *Language, 48*, 819–838.

Verbrugge, R. R., & Rakerd, B. (1986). Evidence of talker-independent information for vowels. *Language and Speech, 29*, 39–55.

Verbrugge, R. R., & Shankweiler, D. (1977). Prosodic information for vowel identity. *Haskins Laboratories Status Reports on Speech Research, 51/52*, 27–35.

Verbrugge, R. R., Strange, W., Shankweiler, D. P., & Edman, T. R. (1976). What information enables a listener to map a talker's vowel space? *Journal of the Acoustical Society of America, 60*, 198–212.

von Kempelen, W. (1791). *Mechanismus der menschlichen Sprache nebst der Beschreibung seiner sprechenden Maschine.* Vienna.

Wakita, H. (1972). *Estimation of the vocal tract shape by optimal inverse filtering and acoustic/articulatory conversion methods* (Tech. Rep. No. 9). Santa Barbara, California: Speech Commmunication Research Laboratory.

Wakita, H. (1973). Direct estimation of the vocal tract shape by inverse filtering of the acoustic speech waveform. *Institute of Electrical and Electronic Engineers, Transactions on Audio and Electroacoustics, 21*, 417–427.

Wakita, H. (1979). Estimation of vocal tract shapes from acoustical analysis of the speech wave: the state of the art. *Institute of Electrical and Electronic Engineers, Transactions on Acoustics Speech and Signal Processing, 27*, 281–285.

Wakita, H., & Fant, G. (1978). Toward a better vocal tract model. *Speech Transmission Laboratory, Quarterly Progress Status Report, 1*, 9–29.

Wakita, H., & Gray, A. (1974). Some theoretical considerations for linear prediction of speech and applications. In G. Fant (Ed.), *Speech Communication, Proceedings of the Speech Communication Seminar, Stockhom* (pp. 45–50). New York: Wiley.

Wakita, H., & Gray, A. (1975). Numerical determination of the lip impendance and vocal tract area functions. *Institute of Electrical and Electronic Engineers, Transactions on Acoustics Speech and Signal Processing, 23*, 574–580.

Walley, A., & Carrell, T. (1983). Onset spectra and formant transitions in the adult's and child's perception of place of articulation in stop consonants. *Journal of the Acoustical Society of America, 73*, 1011–1022.

Walsh, T., Parker, F., & Miller, C. J. (1987). The contribution of rate of F1 decline to the perception of [± voice]. *Journal of Phonetics, 15*, 101–103.

Wardrip-Fruin, C. (1982). On the status of temporal cues to phonetic categories: preceding vowel duration as a cue to voicing in final stop consonants. *Journal of the Acoustical Society of America, 71*, 187–195.

Weibel, E. S. (1955). Vowel synthesis by means of resonant circuits. *Journal of the Acoustical Society of America, 22*, 858–865.

Weigelt, L. F., Sadoff, S. J., & Miller, J. D. (1993). Plosive/fricative distinction: the voiceless case. *Journal of the Acoustical Society of America, 87*, 2729–2737.

Weinstein, C., McCandless, S., Mondshein, L., & Zue, V. (1975). A system for acoustic-phonetic analysis of continuous speech. *Institute of Electrical and Electronic Engineers, Transactions on Acoustics Speech and Signal Processing, 23*, 54–67.

Wells, J. C. (1982). *Accents of English: Beyond the British Isles.* Cambridge: Cambridge University Press.

Wiener, N. (1947). *Extrapolation, Interpolation and Smoothing of Stationary Time Series.* Cambridge, MA: MIT Press.

Wightman, S., Shattuck-Hufnagel, S., Ostendorf, M., & Price, P. (1992). Segmental durations in the vicinity of prosodic phrase boundaries. *Journal of the Acoustical Society of America, 91*, 1707–1717.

Willis, R. (1829). On the vowel sounds and on reed organ pipes. *Trans. Camb. Phil. Soc., 3.*

Winitz, H., Scheib, M. E., & Reeds, J. A. (1972). Identification of stops and vowels for the burst portion of /p, t, k/ isolated from conversational speech. *Journal of the Acoustical Society of America, 51*, 1309–1317.

Wish, M., & Carroll, D. (1982). Multidimensional scaling and its applications. In P. R. Krishnaiah & L. N. Kanal (Eds.), *Handbook of Statistics* (pp. 317–345). Amsterdam: North-Holland Publishing Company.

Witten, I. H. (1982). *Principles of Computer Speech.* London: Academic Press.

Wolf, C. G. (1978). Voicing cues in English final stops. *Journal of Phonetics, 6*, 299–309.

Wood, S. (1979). A radiographic analysis of constriction locations for vowels. *Journal of Phonetics, 7*, 25–43.

Yeni-Komshian, G., & Soli, S. (1981). Recognition of vowels from information in fricatives: perceptual evidence of fricative-vowel coarticulation. *Journal of the Acoustical Society of America, 70*, 966–975.

Zahorian, S. A., & Jagharghi, A. J. (1993). Spectral-shape features versus formants as acoustic correlates for vowels. *Journal of the Acoustical Society of America, 94*, 1966–1982.

Zee, E. (1981). Effect of vowel quality on perception of post-vocalic nasal consonants in noise. *Journal of Phonetics, 9*, 35–48.

Zemlin, W. R. (1981). *Speech and Hearing Science* (2nd ed.). Englewood Cliffs New Jersey: Prentice-Hall.

Zue, V. W. (1976). *Acoustic Characteristics of Stop Consonants: a Controlled Study.* Bloomington, Indiana: Indiana University Linguistics Club.

Zwicker, E. (1961). Subdivision of the audible frequency range into critical bands (Frequenzgruppen). *Journal of the Acoustical Society of America, 33*, 248.

Zwicker, E., & Feldtkeller, R. (1967). *Das Ohr als Nachrichtenempfänger.* Stuttgart: Hirzel Verlag.

Zwicker, E., & Terhardt, E. (1980). Analytical expressions for critical-band rate and critical bandwidth as a function of frequency. *Journal of the Acoustical Society of America, 68*, 1523–1525.

INDEX